Contributions to the ecology of halophytes

Tasks for vegetation science 2

Series Editor

HELMUT LIETH

University of Osnabrück, F.R.G.

DR W. JUNK PUBLISHERS   THE HAGUE / BOSTON / LONDON 1982

Contributions to the ecology of halophytes

edited by

DAVID N. SEN and KISHAN S. RAJPUROHIT

DR W. JUNK PUBLISHERS THE HAGUE / BOSTON / LONDON 1982

Distributors:

for the United States and Canada

Kluwer Boston, Inc.
190 Old Derby Street
Hingham, MA 02043
USA

for all other countries

Kluwer Academic Publishers Group
Distribution Center
P.O. Box 322
3300 AH Dordrecht
The Netherlands

Library of Congress Cataloging in Publication Data   CIP

Main entry under title:

Contributions to the ecology of halophytes.

    (Tasks for vegetation science ; 2)
    Includes bibliographies and index.
    1. Halophytes. 2. Botany--Ecology. I. Sen,
David N., 1934-   II. Rajpurohit, Kishan S.
III. Series.
QK922.C66   581.5'265   81-12381
                            AACR2

ISBN 90-6193-942-9   (this volume)
ISBN 90-6193-897-X   (series)

PRINTED IN THE NETHERLANDS

Preface

The ecology of halophytes has a wide scope of interest, appealing to people of many disciplines. It covers widely different fields such as climatology, soil science, phytogeography, adaptive biology and agriculture. Ecologists study these specialized plants in relation to estuarine ecosystems, biology of dominant genera, germination ecology, water relations, salt secretion, and senescence.

The present volume is divided into three parts and attempts to elucidate new aspects of the problems faced by this special group of plants. It tries to give the reader an overall view of saline environments and the ecology of plants found therein.

In the first chapter of part one Zahran presents the halophytic vegetation of Egypt, which includes the inland and the littoral (Red Sea and Mediterranean Sea) salt marshes. The plants he describes have been classified as succulents, excretives and cumulatives, according to their adaptability to saline soils and according to their different life-forms.

The second chapter throws light on the estuarine ecosystem of India. The estuaries are described by Joshi, and Bhosale as being rich in diversity of mangrove species. Making varied use of estuarine ecosystems is not only possible, but also essential because they are the meeting point between terrestrial and marine life.

In chapter three a listing of mangrove species from the entire world, separated for the main regions of biogeographical interest is presented. The term 'mangrove' has been distinguished into: (a) mainly woody *plant formations*, distributed in the tropical or sub-tropical, tidal zone between the lowest and highest tide water mark; and (b) plant species of this formation. Barth calls 'mangal', 'mangrove community', or 'mangrove formation' when they mean the first, and 'mangrove' only, when pertaining to the plant species.

In chapter four, Sen, Rajpurohit and Wissing discuss saline areas in the Indian desert. Salt basins of the Rajasthan desert, are of the inland type and differ greatly from coastal salines. The soil's osmotic potential and its chloride ions and their relationship to certain halophytes has been discussed. Adaptability of plants in this region to saline environment is an important property which induces morphological, anatomical, eco-physiological and phenological changes.

In chapter five, Kelley, Goodin and Miller discuss the contributions made to the biological understanding of the genus *Atriplex* since 1969. This genus occupies important niches in many (and particularly arid) ecosystems. This is due to the ease with which they adapt to many diverse and generally harsh environments. This discussion centers around the physiology of salt tolerance, growth habit, transpiration and water relations, and salt effects on ultrastructure. It has been highlighted that with proper management *Atriplex* can make a significant contribution to mankind and to world productivity.

In the first chapter of part two interpretations as to why mangroves are zoned – all into two contrasting categories – are given by Snedaker. These views have been critically reviewed with reference to the western hemisphere, the Caribbean mangroves.

In the second chapter Smart studies the distribution and environmental control of productivity and growth forms of *Spartina alterniflora*.

A review of recent literature dealing with seed germination of halophytes is presented by Ungar in chapter three. An attempt has been made to point out some ecological strategies to provide successful establishment of halophytes in saline environments.

In the fourth chapter Sharma highlights indigenous Australian members of Chenopodiaceae, which are halophytic xerophytes occupying only about 6% of the land surface of that continent but constituting a major source of high protein forage. Because of their unique anatomical features, they possess a very efficient photosynthetic apparatus. High stomatal resistance to water vapour and low mesophyll resistance to CO_2 lead to their efficient use of water. An attempt is made here to synthesize information on various aspects of salinity and water relations of the chenopods.

Karmarkar and Joshi give an account of germination behaviour, seasonal changes in mineral composition, free amino acids, senescence and abscission of leaves in chapters five and six. Photosynthetic efficiency, total polyphenols, activities of certain enzymes and certain related metabolic processes in leaves of mangroves and some tropical salt marsh halophytes are described.

In chapter seven, attention has been given to the structure and function of salt glands, to the mechanism of salt excretion and to its ecological significance. This has been discussed in relation to the families Acanthaceae, Aviceniaceae, Combretaceae, Convolvulaceae, Frankeniaceae and Gramineae.

In the last chapter of part two the role of bladders for salt removal is discussed. The peculiar trichome-type of bladders is known from some genera in Chenopodiaceae. Whereas in *Salsola* species leaf succulence is the main morpho-physiological strategy coping with salinity stress, their bladders appear without ecological significance. Bladders in *Atriplex* (inclusive *Obione*) and in *Halimione*, in contrast, play an important role in the desalinization of young plant tissues and organs. In some halophytic species several layers of bladders are found, densely covering the leaves. It is concluded that salt efflux from leaf lamina is an ecological advantage, too. In ruderal, non-halophytic and xerophytic *Atriplex* species the ecological significance is much lower. Bladders seem to be an old evolutionary structure, which in several chenopod genera is developed into a salt secretion system.

Chapter one of part three shows that salt-affected soils hinder the development of agricultural production. Thus, irrigation and the implementation of chemical amendments – by planting salt tolerance species (biological desalination) – become necessary if one is to utilize such lands. The economic potentialities of certain halophytes as a source of raw material in paper (*Juncus* spp.), drug (*Salsola tetrandra*) and fodder (*Kochia* spp.) has been illustrated.

DAVID N. SEN

Contents

PART ONE

Biology and biogeography of halophytic species and salinity controlled ecosystems

DAVID N. SEN and KISHAN S. RAJPUROHIT

Introduction

It was early in nineteenth century that the name 'halophyte' was given to that group of higher plants by Pallas (see Schrader, 1809) which can grow under saline conditions. Great variations have come to our knowledge in the plant world with regard to responses to salinity. Various adaptive mechanisms have appeared in halophytes during the course of evolution. These include the elucidation of structural and functional adaptations which enable these plants to tolerate the saline environment. However, it is open for discussion whether they tolerate the saline environment or actually flourish there. Ecologists, geographers and environmentalists are interested in the inter-relationship between these plants and the environment, from the climatic, edaphic and biotic points of view.

Waisel (1972) feels that not very much is known of halophytes. He is of the opinion that in most cases we are ignorant of the metabolic adaptations and direct physiological processes that enable plants to survive under the saline conditions. It is well agreed that sodium chloride is the dominant factor in the 'halophytism', yet the mechanism of uptake of such ions, germination regulation, salt retention and expulsion, growth and such other processes in plants under saline conditions, are relatively little known. Salinity also plays an important role in the distribution of these plants. In the modern agricultural practices, the increasing use of water of poor quality, the continuous addition of waste salts to soil, as well as an increasing contamination of underground water sources lead to gradual soil salinization.

Salinity is a common phenomenon and one of the basic features of the arid and semi-arid regions throughout the world. The three major sources: marine, lithogenic, and anthropogenic contribute their share, whether small or large to salinization of soils and underground water. In nature salts undergo numerous cycles of dissolution, transport, and precipitation before accumulation occurs at certain sites. Kovda (1961) recognised five major salt cycles, each having separate and different patterns.

1. Continental salt cycles. Movement and accumulation of chloride, sulphate and carbonate salts in arid regions, where only small amounts of run-off of percolating water exist. This has been further divided into two sub-cycles.
(a) Primary cycles: Movement of salts originating by weathering or from salt containing primary rocks.
(b) Secondary cycles: Movement of salts which originated by weathering or sedimentary rocks.

2. Marine cycles. Movements and accumulation of salts near sea coasts. Such salts eventually return into the ocean.

Tasks for vegetation science, Vol. 2 ed. by D.N. Sen and K.S. Rajpurohit
© 1982, Dr W. Junk Publishers, The Hague. All rights reserved. ISBN 90 6193 942 9

3. Delta cycles. Precipitation and accumulation of salt in deltas of large rivers.

4. Artesian cycles. Accumulation of salts from underground saline water sources on the land surface. These cycles are of long term basis, and the salts which take part in such cycles are usually fossil.

5. Anthropogenic cycles. Movement and accumulation of salts produced and released by human activity.

Coastal salines are important in many ways to those countries which have connection with seas. Salt composition of such salines closely resembles with that of sea water, although their salt content varies with specific local conditions. Coastal salines are formed all over the world under semi-arid and arid regions as well as under maritime temperate climates and can be classified as: (1) salt water swamps; (2) salt marshes; and (3) salt bogs.

Inland salines are typical and an integral part of arid and semi-arid regions. It is very rarely that these are found in humid regions, and when they occur they are mostly on fossil salt depositions, or around salty springs. Although sodium chloride is the dominant component of such salines, it is not uncommon that salt composition and its concentration may vary greatly. Inland salt marshes and salt flats have developed in the seepage basins of deserts; and in arid regions such as in Thar desert, sites with a basin-shaped topography form typical concentric salines, because of the inflow of saline water from round about, or as a result of a rise of saline underground water, are common. In Indian desert these remain in existence in the rainy season only, and for the rest of the year they form a dry salty surface. Soil thickness, as well as plant cover of such salines are usually arranged in concentric rings (Rajpurohit et al., 1979).

Mangrove is a term given to a formation of trees and shrubs inhabiting the coasts of tropical and subtropical seas. It consists of a heterogenous group of plants which are mostly limited in their distribution due to tidal zone. Mangroves are distributed along most of the warm water oceans and are thus included in the following phytogeographical regions: the Caribbean region, the Brazilian region, the tropical rain forest region of Western Africa, the coasts of Savanna region of East Africa, and the Malaysian phytogeographical region (Waisel, 1972). The zonation of highly developed mangrove forests is limited between the spring and the neap tides. These habitats usually comprise a narrow strip of shallow coastal habitats, deltas, estuaries or lagoons and they are highly affected by the climate, tides, substrate, and water salinity (Chapman, 1960; Clarke and Hannon, 1967, 1969; Macnae, 1968). Coastal habitats are most saline under arid climates than under humid ones. Thus plants in the high coasts of tropical as well as temperate regions are less halophytic than those of arid regions.

Literature cited

Chapman, V.J. 1960. *Salt Marshes and Salt Deserts of the World.* Leonard Hill Ltd., London.

Clarke, L. and Hannon, N.J. 1967. The mangrove swamp and salt marsh communities of the Sydney district. I. Vegetation, soil and climate. *J. Ecol.* 55: 753–771.

Clarke, L. and Hannon, N.J. 1969. The mangrove swamp and salt marsh communities of the Sydney district. II. The holocoenotic complex with particular reference to physiography. *J. Ecol.* 57: 231–234.

Kovda, V.A. 1961. Principles of the theory and practice of reclamation and utilization of saline soils in the arid zones.

UNESCO Arid Zone Res. Salinity problems in Arid Zones. Proc. Tehran Symp. 14: 201–213.

Macnae, W. 1968. A general account of the fauna and flora of mangrove swamps and forests in the Indo-West-Pacific region. *Advan. Mar. Biol.* 6: 73–270.

Rajpurohit, K.S., Charan, A.K. and Sen, D.N. 1979. Microdistribution of plants in an abandoned salt pit at Pachpadra salt basin. *Ann. Arid Zone* 18: 122–126.

Schrader, H.A. 1809. Uber Palla's Halophyta mit besonderer Rucksicht auf die Gattungen *Salsola* und *Suaeda*. *Schrad Neues J. Bot.* 3: 58–92.

Waisel, Y. 1972. *Biology of Halophytes.* Academic Press, New York.

CHAPTER 1

Ecology of the halophytic vegetation of Egypt

M.A. ZAHRAN*

Botany Department, Faculty of Science,
Mansoura University, Mansoura, Egypt

1. Introduction

Egypt is divided into seven phytogeographical regions, namely: Mediterranean, Red Sea, Nile Valley, Oases, Sinai, Desert and Gebel Elba (Hassib, 1951; Montasir, 1954). Each of these regions is characterized by various ecosystems, e.g. desert wadis, desert plateaus, gravel desert, mountains, sandy formations, salt marshes, etc. The environmental conditions that prevail in each region (climate, soil, geomorphology, underground water, etc.) limit the number and extent of its ecosystems. The salt marsh ecosystem, however, is present in almost all of the seven phytogeographical regions of Egypt.

The present chapter is an ecological survey of the halophytes inhabiting the salt marsh ecosystem with special reference to their classification, biogeography and floristic composition of their community types.

I. Location of saline areas

The salt marsh ecosystem is associated with land that is liable to be flooded with saline water, as this water evaporates, leaving its salt behind creating salt marsh habitat. In Egypt, the salt marsh ecosystem comprises: (i) littoral salt marshes; and (ii) inland salt marshes.

* Present address: Institute of Meteorology and Arid Lands Studies, King Abdulaziz University, Post Box 1540, Jeddah, Saudi Arabia.

(i) Littoral salt marshes (LSM)

These are lands subjected to maritime influences, i.e. periodic flooding with seawater, seawater spray, seawater seepage, etc. Their formation takes place through the silting up of lagoons or shore-line areas protected by sand or shingle bars. They can be found in general, if any of the following physiographic conditions is fulfilled: the presence of estuaries, the shelter of spits, off-shore barrier islands, and large or small protected bays with shallow water (Chapman, 1938, 1964).

The littoral salt marshes of Egypt are represented by the marsh lands along the coasts of the Mediterranean Sea (MLSM, northern Egypt), Red Sea (RLSM, eastern Egypt), and also by the marsh lands associated with the northern lakes (El Borullus, Manzala and Mariut) (Fig. 1).

(ii) Inland salt marshes (ISM)

These are salt marshes far from the reach of the maritime influences. Being lower in level than the surrounding territories, the inland salt marshes are characterized by shallow underground water table. In certain localities, e.g. in the oases and depressions of the Western Desert of Egypt, the underground water is exposed forming lakes of brackish or saline water (Zahran and Girgis, 1970; Zahran, 1972).

In Egypt, the inland salt marshes may be found in the following localities:

(a) Sabkhas (saline areas) circum the lakes, springs and wells of the oases, e.g. Siwa, Moghra,

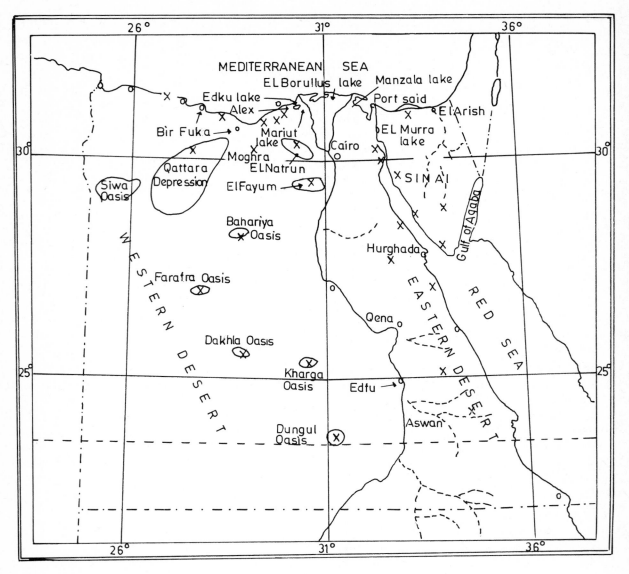

Fig. 1. Map of Egypt showing the littoral and inland salt marshes.

Dakhla, Kurkur, Dungul, etc. and of the depressions, e.g. Qattara, Wadi El Natrun, El Fayum, etc. of the Western Desert (Fig. 1).

(b) Saline patches at the feet of the high mountains of the Eastern Desert and Sinai, where the orographic precipitation usually takes place, and comparable patches are formed where run-off water collects. Evaporation of the accumulated water creates such saline habitats (playas).

II. Climate

The climatic conditions of the littoral and inland salt marshes of Egypt are not identical (Migahid, 1962; Montasir, 1954). Semi-arid climate prevails in the Mediterranean region and in the region of the northern lakes, whereas arid or extremely arid climate prevails in the Red Sea and in the inland salt marshes.

The general climate of the Mediterranean region

is characterized by a long, dry warm (summer) period and a short rainy winter extending from late autumn to early spring. This is the most rainy part in Egypt. Rainfall is greatly variable from year to year at the same place and during the same season from one part of the region to another (mean annual rainfall ranges between 100 mm to about 300 mm). Oliver (1930) states that the rain often falls in heavy downpours, extends six months of the year from October to March but attaining its maximum intensity from December to February.

Temperatures fluctuate between 10° C to 20° C in winter and 21° C to 33° C in summer. The daily fluctuations are not as notable as they are in the interior of the country.

Relative humidity is high all the year round and it reaches 80% or more in summer months and 60–70% in winter months. The prevailing winds are north-easterlies but from April to September south-western winds are not infrequent.

The pluviothermic quotient (Emberger, 1952) of the Egyptian Mediterranean region ranges between 17 and 22.

The Red Sea coastal land of Egypt is situated within a region of arid climate: hot and dry. The mean annual rainfall ranges between 25 mm in Suez and 3 mm in Hurghada (400 km south of Suez). The main bulk of rain (80–100%) occurs in the winter months, i.e. Mediterranean affinity, and summer is rainless. Variability of annual rainfall is recurrent and years of exceptionally low or exceptionally high rainfall are not unusual (Kassas and Zahran, 1962). In Suez, for example, the years 1949 and 1958 were very dry (total annual rainfall 2 mm (1949) and 3.1 mm (1958)) while the years 1952 and 1956 were relatively wet (rainfall 56.8 mm (1952) and 55 mm (1956)).

Temperature is relatively higher than that of the Mediterranean coastal land and it ranges between 14–21.7° C in winter and 23.1–36.1° C in summer. Relative humidity is generally lower than that of the Mediterranean region and it ranges between 43% in summer to 65% in winter. The Piche evaporation is higher in summer (13.7–21.5 mm/day) than in winter (5.2–10.4 mm/day).

The climate of the inland salt marshes is comparable to that of the Red Sea littoral salt marshes.

The pluviothermic quotient (Emberger, 1952) for the RLSM ranges between 2.5 and 5.5, i.e. arid or extremely arid climate.

III. Edaphic characteristics

The climatic conditions of Egypt have pronounced effects on the edaphic characteristics of the salt marshes. Aridity of climate, especially in the inland and in the Red Sea littoral salt marshes, increases the rate of evaporation of the accumulated water. As precipitation is low, there is insufficient leaching and salts accumulate in the form of surface crusts. Soil is generally alkaline (pH = 7.5–8.9) (Kassas and Zahran, 1967; Montasir, 1943).

Tables 1 and 2 include the chemical analyses of soil samples collected from: (1) areas dominated by one of four halophytes, namely: *Halocnemum strobilaceum*, *Arthrocnemum glaucum*, *Zygophyllum album* and *Nitraria retusa* in the Red Sea and Mediterranean littoral salt marshes; and (2) from areas dominated by *Cressa cretica*, *Juncus rigidus*, *Alhagi maurorum*, *Imperata cylindrica* and *Typha elephantina* in the inland salt marshes. The data in these tables elucidate the following:

The total amount of soluble salts in the salt marshes of Egypt is generally high, but it is higher in the inland and in the Red Sea littoral salt marshes than in the Mediterranean littoral salt marshes.

The surface layers, usually, contain the highest proportions of soluble salts (salt crusts). In a few localities, e.g. in a stand of *Arthrocnemum glaucum* community of the RLSM, there is accumulation of aeolian sand (sample 8, 0–60 cm, Table 1) covering the actual surface layer of the soil (sample 9, 60–65 cm). Here, the surface layer of sand cover has less salt content (3.4%) than the actual surface layer (60.5%).

The amount of soluble salt drops abruptly in the subsurface and bottom layers, e.g. in *Cressa cretica* community of the ISM (Table 2) the percentage of the soluble salts drops from 65.88% in the surface layer (sample 2, 0–25 cm) to 2.86% and 2.72% in the subsurface and bottom layers, respectively.

The least amount of soluble salts (0.39%) were recorded in the soil of the swampy habitat dominated by *Typha elephantina* of Wadi El Natrun

Table 1. Chemical analyses of soil samples collected from four community types representing both the Red Sea and the Mediterranean littoral salt marshes of Egypt.

Community type	Locality	Sample No.	Depth (cm)	Water soluble salts		
				Total	Cl	SO$_4$
Halocnemum strobilaceum	a) RLSM	1	0–2	67.37	34.0	4.4
		2	2–14	5.86	0.8	3.59
		3	14–24	0.65	0.25	tr
		4	24–40	0.62	0.23	tr
	b) MLSM	5	0–5	4.63	2.70	0.95
		6	5–25	1.34	0.82	0.40
		7	25–40	0.32	0.25	tr
Arthrocnemum glaucum	a) RLSM	8	0–60	3.4	1.35	0.5
		9	60–65	60.5	24.0	1.1
		10	65–90	5.2	1.15	0.33
		11	90–120	1.1	0.90	0.04
	b) MLSM	12	0–15	4.63	3.15	0.62
		13	15–30	2.20	1.12	0.33
		14	30–60	0.80	0.40	0.15
Zygophyllum album	a) RLSM	15	0–10	26.30	8.25	5.3
		16	10–30	1.20	0.6	0.1
		17	30–85	1.91	0.56	0.09
	b) MLSM	18	0–20	1.82	0.92	0.45
		19	20–25	0.55	0.34	0.12
		20	25–70	0.27	0.12	0.04
Nitraria retusa	a) RLSM	21	0–8	26.2	11.26	4.48
		22	8–20	22.5	8.93	3.09
		23	20–45	5.38	1.8	1.2
		24	45–70	2.28	0.3	1.2
	b) MLSM	25	0–20	2.97	1.4	0.8
		26	20–35	0.81	0.5	0.12
		27	35–80	0.43	0.2	0.05

tr = traces.

Table 2. Chemical analyses of soil samples collected from five community types of the inland salt marshes of Egypt.

Community type	Locality	Sample No.	Depth (cm)	Water soluble salts		
				Total	Cl	SO$_4$
Cressa cretica	Siwa Oasis	1	0–25	65.88	54.2	7.05
		2	25–35	2.86	1.20	0.94
		3	35–50	2.72	1.20	1.15
Juncus rigidus	Siwa Oasis	4	0–10	46.25	40.15	5.32
		5	10–25	3.18	0.92	2.17
		6	25–60	3.16	1.72	0.5
Alhagi maurorum	Siwa Oasis	7	0–15	48.02	40.91	5.62
		8	15–45	4.81	2.10	2.17
		9	45–70	1.24	0.51	0.5
Typha elephantina	Wadi El-Natrun	10	0–40	0.39	0.15	0.11
Imperata cylindrica	Dungul Oasis	11	0–10	44.11	25.0	8.3
		12	10–40	2.0	0.7	0.40
		13	40–65	1.8	0.4	0.28

(Table 2). The soil is permanently inundated with water.

The soluble salts are mainly chlorides and partly sulphates.

2. Halophytic vegetation

The halophytes, or the vegetation of the salt marshes, may be considered as highly specialized plant growth, the component of which have great tolerance to salts (Keith, 1958). The salt tolerance of the halophytes can be appraised in at least three ways: (1) their ability to survive in saline soils; (2) their absolute yield on saline soils; and (3) their relative high yield on saline soils compared to non-saline soils. In contrast to the glycophytes, the presence of certain amount of salt in soil creates more favourable conditions for the growth of the halophytes. The salt tolerance of the halophytes increases both during their growth and development and from generation to next. This adaptation is possible due to those features and properties which arise during evolution (Strogonov, 1965).

3. Classification

Classification of the halophytes of Egypt is based on: I. Adaptability to saline soils; and II. Life-forms.

I. Adaptability to saline soils

Walter (1961) classified the halophytes into: Facultative halophytes and Euhalophytes. Facultative halophytes are plants that have an optimum development in non-saline soils, but tolerate a certain amount of salts in the soil. The euhalophytes, on the other hand, are those plants which show an optimum growth on soil with certain salts but they do not flourish.

The root of the halophytes absorb the soil solution in a diluted form. As water is transpired and the salt remains in the transpiring leaves, accumulation of salts in the leaves may take place in the long run. The halophytes overcome this in different ways, viz. with or without regulating mechanism (Steiner, 1934) as follows:

(i) With regulating mechanism

Here, the halophytes are either excretives or succulents. The excretives possess glandular cells capable of excreting the excess salts. The succulents often lack the ability to secrete salts but they thwart the rise of salt concentration by an increase of their water content, and they become more and more succulent during their development.

(ii) Without regulating mechanism

In this type, the halophytes lack a regulating mechanism. The salt concentration rises continuously during the growing season but when a certain level is reached the plant dies. These are called the cumulative halophytes.

Accordingly, the halophytic plants dominating 38 community types of the littoral and inland salt marshes of Egypt are classified into: 11 succulents, 13 excretives and 14 cumulatives (Table 3). They belong to 15 families. Chenopodiaceae is represented by ten dominant halophytes, nine succulents (*Halocnemum strobilaceum, Arthrocnemum glaucum, Salicornia fruticosa, Suaeda monoica, S. fruticosa, S. pruinosa, S. vermiculata, Halopeplis perfoliata* and *Salsola tetrandra*) and one excretive (*Halimione portulacoides*). Poaceae is represented by seven dominants, three are excretives, viz: *Aeluropus* sp., *Sporobolus spicatus* and *S. virginicus* and four cumulatives, viz: *Phragmites australis, Imperata cylindrica, Lygeum spartum* and *Halopyrum mucronatum*. Four dominants (cumulatives) represent Cyperaceae, these are: *Cyperus laevigatus, Cladium mariscus, Schoenus nigricans* and *Scirpus litoralis*. Plumbaginaceae is represented by three excretive halophytes, viz: *Limonium pruinosum, L. axillare* and *Limoniastrum monopetalum*. Three families each with two dominants, are Tamaricaceae (*Tamarix mannifera* and *T. passerinoides*, excretives), Typhaceae (*Typha elephantina* and *T. domengensis*, cumulatives) and Juncaceae (*Juncus rigidus* and *J. acutus*). There are six families to each of which belong one dominant halophyte viz: *Nitraria retusa* (Nitrariaceae, excretive), *Frankenia revoluta* (Frankeniaceae, excretive), *Cressa cretica*

Table 3. Classification and biogeography of the 38 dominant halophytes of Egypt.

Dominant halophytes	Life-form	Biogeography			Families
		LSM		ISM	
		M	R		
A. Succulent type					
Halocnemum strobilaceum	Suf	+	+	+	Chenopodiaceae
Arthrocnemum glaucum	Suf	+	+	+	Chenopodiaceae
Salicornia fruticosa	Suf	+	+	+	Chenopodiaceae
Suaeda monoica	Nan	−	+	+	Chenopodiaceae
S. fruticosa	Suf	+	+	+	Chenopodiaceae
S. pruinosa	Suf	+	−	+	Chenopodiaceae
S. vermiculata	Suf	+	+	+	Chenopodiaceae
Halopeplis perfoliata	Suf	−	+	−	Chenopodiaceae
Salsola tetrandra	Suf	+	−	+	Chenopodiaceae
Inula crithmoides	Suf	+	−	+	Asteraceae
Zygophyllum album	Suf	+	+	+	Zygophyllaceae
B. Excretive type					
Avicennia marina	Mic	−	+	−	Avicenniaceae
Limonium pruinosum	Suf	+	+	+	Plumbaginaceae
L. axillare	Suf	−	+	−	Plumbaginaceae
Limoniastrum monopetalum	Suf	+	−	−	Plumbaginaceae
Aeluropus spp.	Ge	+	+	+	Poaceae
Sporobolus spicatus	Ge	+	+	+	Poaceae
S. virginicus	Ge	+	−	−	Poaceae
Nitraria retusa	Nan	+	+	+	Nitrariaceae
Tamarix mannifera	Nan	+	+	+	Tamaricaceae
T. passerinoides	Nan	−	+	−	Tamaricaceae
Frankenia revoluta	Cu	+	−	+	Frankeniaceae
Cressa cretica	Cu	+	+	+	Convolvulaceae
Halimione portulacoides	Nan	+	−	−	Chenopodiaceae
C. Cumulative type					
Rhizophora mucronata	Mic	−	+	−	Rhizophoraceae
Typha elephantina	He	−	−	+	Typhaceae
T. domengensis	He	+	+	+	Typhaceae
Phragmites australis	He	+	+	+	Poaceae
Imperata cylindrica	Ge	+	+	+	Poaceae
Lygeum spartum	Ge	+	−	−	Poaceae
Halopyrum mucronatum	Ge	−	+	−	Poaceae
Juncus rigidus	Ge	+	+	+	Juncaceae
J. acutus	Ge	+	−	+	Juncaceae
Cladium mariscus	Ge	−	−	+	Cyperaceae
Cyperus laevigatus	Ge	+	+	+	Cyperaceae
Schoenus nigricans	Ge	+	−	+	Cyperaceae
Scirpus litoralis	Ge	+	+	+	Cyperaceae
Alhagi maurorum	Sur	+	+	+	Leguminosae

Suf = Suffruticose; Nan = Nanophanerophyte; Mic = Microphanerophyte; Ge = Geophyte; He = Helophyte; Cu = Cushion chamaephyte; LSM = Littoral salt marshes; M = Mediterranean; R = Red Sea; ISM = Inland salt marshes.

(Convolvulaceae, excretive), *Zygophyllum album* (Zygophyllaceae, succulent) *Inula crithmoides* (Asteraceae, succulent) and *Alhagi maurorum* (Fabaceae, cumulative).

Sholander et al. (1962) classified the mangroves into: salt-secreting and salt non-secreting plants. We may accordingly add *Avicennia marina* (Avicenniaceae) to the excretive halophytes and *Rhizophora mucronata* (Rhizophoraceae) to the cumulative halophytes.

II. Life-forms

According to Raunkiaer's life-form classification (1934) the halophytes of Egypt are grouped under phanerophytes, chamaephytes and cryptophytes as follows:

(i) Phanerophytes

(a) Microphanerophytes are represented by the mangrove plants: *Avicennia marina* and *Rhizophora mucronata*.

(b) Nanophanerophytes are represented by *Tamarix mannifera*, *T. passerinoides*, *Nitraria retusa*, *Suaeda monoica* and *Halimione portulacoides*.

(ii) Chamaephytes

(a) Suffruticose chamaephytes are represented by *Halocnemum strobilaceum*, *Arthrocnemum glaucum*, *Salicornia fruticosa*, *Suaeda fruticosa*, *S. pruinosa*, *S. vermiculata*, *Halopeplis perfoliata*, *Salsola tetrandra*, *Inula crithmoides*, *Zygophyllum album*, *Limonium pruinosum*, *L. axillare*, *Limoniastrum monopetalum* and *Alhagi maurorum*.

(b) Cushion chamaephytes are represented by: *Cressa cretica* and *Frankenia revoluta*.

(iii) Cryptophytes

(a) Helophytes are represented by *Typha elephantina*, *T. domengensis*, *Phragmites australis* and *Scirpus litoralis*.

(b) Geophytes are represented by *Halopyrum mucronatum*, *Aeluropus* spp., *Sporobolus spicatus*, *S. virginicus*, *Juncus rigidus*, *J. acutus*, *Cladium mariscus*, *Cyperus laevigatus* and *Schoenus nigricans*.

4. The community types

The halophytic vegetation of Egypt comprises 38 community types inhabiting the following habitats: I. The mangrove swamps; II. The reed swamps; III. The salt marshes; and IV. The saline sand formations.

The general ecological characteristics of each habitat and its community types is discussed below.

I. The mangrove swamps

The mangrove swamps fringe the shore line in many parts of the Red Sea coast of Egypt but does not extend northward to the coast of the Gulf of Suez (Kassas and Zahran, 1967; Zahran, 1964; Zahran and Negm, 1973). The northernmost locality is the bay of Hyos Hormos, about 380 km south of Suez. *Avicennia marina* (excretive) dominates these swamps and usually form pure stands. Within a stretch of 40 km in the most southern part of the Red Sea coast of Egypt and between Lat. 23°40′N, *Rhizophora mucronata* (cumulative) may co-dominate with *A. marina* or it may form pure community. The foot layer of the mangrove is formed of the associate weeds, e.g. *Cymodocea ciliata*, *C. rotundata*, *Diplanthera uninervis*, *Halophila ovalis* and *H. stipulacea*. The mangrove vegetation is absent from the coasts of the Mediterranean Sea (Tackholm, 1974; Zahran, 1964).

II. The reed swamps

The reed swamp vegetation is common in various parts of Egypt, wherever there is neglected shallow water (saline, brackish or fresh) the reeds predominate (Plate 1). Four cumulative dominants are recorded, namely (*Typha elephantina*, *T. domengensis*, *Phragmites australis* and *Scirpus litoralis*. *T. elephantina* is present only in the swamps associated with the lakes of Wadi El Natrun (Boulos, 1962; Tackholm, 1974; Zahran and Girgis, 1970), elsewhere it is not recorded. *T. domengensis* and *Phragmites australis*, on the other hand, are the most abundant species of the reed swamps of Egypt. *T. australis* usually inhabits areas where soil is relatively less saline and water is less shallow.

9

Plate 1. General view of the swampy habitat of the shallow-watered lake (Siwa oasis) with *Phragmites australis* in abundance, *Juncus rigidus* dominates the wet soil adjacent to the lake (Sedge-meadow) and surface white-salt crusts are also seen.

Phragmites australis grows in the swamps closer to the dry land often with higher soil salt content.

Scirpus litoralis prevails in swamps associated with the northern lakes, e.g. Mariut lake, occupying a zone in between those of *T. domengensis* and *P. australis*. *S. litoralis* frequently forms patches of pure populations (Tadros, 1953).

The common associates of the reed swamps are water-loving species, including *Eichhornia crassipes*, *Berula erecta*, *Samolus valerandii*, *Cyperus articulatus*, *Lemna gibba*, *Spirodela polyrrhiza*, *Wolffia hyalina*, etc.

III. The salt marshes

The littoral and inland salt marshes of Egypt are dominated by thirty halophytic plants.

1–3. *Halocnemum strobilaceum*, *Arthrocnemum glaucum* and *Salicornia fruticosa* community types.

These are principal communities of the salt marshes of Egypt as they are present everywhere and their plant cover is higher than that of the other communities of the salt marshes. Their habitats are comparable, but it seems that their climatic requirements are dissimilar and accordingly their geographical distribution varies. *Halocnemum strobilaceum* is very abundant in the MLSM and in the ISM, but in the RLSM it is common in the northern region only, i.e., within the salt marshes of Gulf of Suez (western and eastern sides) but not in the region further south (Kassas and Zahran, 1967; Zahran, 1967). The plant growth often forms pure stands of the dominant species but it may be associated with some 13 halophytic species (Table 4).

Arthrocnemum glaucum and *Salicornia fruticosa*, on the other hand, are difficult to distinguish from each other in their vegetative forms. 'It is only at their time of flowering that they can be separated. *A. glaucum* flowers in April while *S. fruticosa* flowers much later' (Tadros and Atta, 1958). *A. glaucum* occurs throughout the whole marshes of Egypt, though, unlike *H. strobilaceum*, less common in the northern part of RLSM. *S. fruticosa* is also common in the inland and in the MLSM, but it is recorded only in a very limited area (Mallaha, 260 km south of Suez) in the RLSM. 'Here, *S. fruticosa* forms patches of pure growth with plant cover ranges from 50–100%. The ground is covered with a thick crust of salts (total soluble salts = 80.50%)' (Kassas and Zahran, 1967). It is of interest to note that with a few exceptions, the associate species of *A. glaucum* and *S. fruticosa* are similar (Table 4).

Salicornia herbacea is a succulent ephemeral halophyte which is abundant in seasonally flooded salt marshes of lakes in the ISM and MLSM. Its presence is not as extensive as that of *S. fruticosa*.

4. *Halopeplis perfoliata* community type

H. perfoliata is a succulent halophyte of narrow ecological range in Egypt. Its rich growth is restricted to the southern 950 km stretch of RLSM. Northward it is absent except in a locality 195 km south of Suez. In the eastern side of the Gulf of Suez (Sinai side, Fig. 1), *H. perfoliata* is recorded also, 'in one locality at the coast of Abu Zeneima about 124 km south of El-Shatt' (Zahran, 1967). *H. perfoliata* is rarely recorded in the MLSM but is absent from ISM (Tackholm, 1974). The community type dominated by *H. perfoliata* is formed either of pure stands or associated with *Arthrocnemum glaucum*, *Halocnemum strobilaceum*, *Zygophyllum album*, *Suaeda monoïca*, etc. (Table 4).

Halopeplis amplexicaulis is an ephemeral-succulent halophyte very abundant in the seasonally flooded salt marshes associated with the sandy shore-zones of the northern lakes, e.g. Manzala (Montasir, 1937) and rarely recorded in the MLSM and in the oases, but absent in the RLSM (Tackholm, 1974).

5. *Inula crithmoides* community type

I. crithmoides (succulent) predominates in the MLSM, marshes of the northern lakes and of the northern oases of the Western Desert (Montasir, 1937; Tackholm, 1974; Zahran, 1972). It is neither recorded in the RLSM nor in the salt marshes of the southern oases and depressions of the Western Desert (Boulos, 1966; Kassas and Zahran, 1967; Migahid, 1960; Zahran, 1966; Zahran and Girgis, 1970). The associate species of this community type are mainly halophytes (Table 4) and the plant cover is high (60–70%) contributed mainly by the dominant plants.

6–8. *Suaeda fruticosa, S. pruinosa* and *S. vermiculata* community types

These are three closely related succulent halophytes that seem to be comparable in their growth forms, morphology and habitat requirements. *S. fruticosa* and *S. vermiculata* are common throughout the littoral and inland salt marshes, *S. pruinosa* is abundant in the MLSM, occasionally present in the inland salt marshes and not recorded in the RLSM. The floristic composition of the communities dominated by these halophytes vary. *S. pruinosa* community includes 18 associate halophytes, *S. fruticosa* (10 associates) and *S. vermiculata* (8 associates) come next (Table 4). It is observed that *Zygophyllum album, Cressa cretica, Arthrocnemum glaucum, Halocnemum strobilaceum* and *Limonium pruinosum* are the most common associates within the three communities.

9. *Limonium pruinosum* community type

L. pruinosum (excretive) is a species with two distinctly different habitats: salt marshes (Kassas and Zahran, 1967; Tadros, 1953; Tadros and Atta, 1958; Zahran and Girgis, 1970) and desert limestone cliffs (Kassas and Girgis, 1964). It is very likely that it comprises two ecotypes, one halophytic and the other chaemophytic. In the RLSM, *L. pruinosum* dominates a common community type within

the region of the Gulf of Suez (west and east) but not farther south. Here, it is associated with halophytic species, e.g. *Limonium axillare, Suaeda vermiculata, Nitraria retusa, Zygophyllum album* and *Tamarix mannifera*, and a number of desert species, e.g. *Launaea spinosa, Convolvulus lanatus, Pituranthus tortuosus* and *Lasiurus hirsutus*. In the MLSM, *L. pruinosum* is very common and its community is associated with *Limoniastrum monopetalum, Halocnemum strobilaceum, Arthrocnemum glaucum, Salicornia fruticosa*, etc. In the ISM, *L. pruinosum* is occasionally recorded in the northern oases but absent in southern ones.

10. *Limonium axillare* community type

Unlike *L. pruinosum, L. axillare* (excretive) is not widely distributed in Egypt. It is absent from both the MLSM and the ISM. In the RLSM, *L. axillare* is only occasionally present in the northern 560 km, then it becomes very abundant dominating a characteristic community type southwards (Kassas and Zahran, 1967). The associate species of *L. axillare* community type comprise a mixed population of the halophytes (Table 4) and desert species, e.g. *Zygophyllum coccineum, Sphaerocoma hookeri, Panicum turgidum* (perennials), *Zygophyllum simplex* and *Lotononis platycarpa* (ephemerals), etc.

11. *Zygophyllum album* community type

In Egypt, *Z. album* (succulent) is an omnipresent species. It is a plant of wide ecological amplitude. In the LSM, *Z. album* community may be present within any of the salt marsh zones (Kassas and Zahran, 1967). In the inland desert, *Z. album* community is recorded in the wadis of the limestone country (Kassas and Girgis, 1964). In the ISM, e.g. in Siwa Oasis (Zahran, 1972) and Wadi El Natrun (Zahran and Girgis, 1970), *Z. album* community abounds in the sand formation associated with the lakes where the deep deposits are mainly sandy with relatively high content of soluble materials (3.3–6.5%). It is also recorded in water runnels lined with sand deposits and dissecting the gravel deposits where soil salinity is low (0.6–1.3%). The growth form of *Z. album* varies according to the habitat conditions. In the water runnels the individuals are stunted and build no sand mounds

Table 4. Floristic composition of 30 community types of the halophytic vegetation of Egypt.

The associate halophytes	The dominant halophytes Succulents											Excretives					
	Hs	Ag	Sr⁻	Sm	Sf	Sp	Se	Hp	St	In.C	Za	Av.m	Lp	La	Lm	A sp.	Ss
Zygophyllum album	+	+	+	+	+	+	+	+	+	D+	+	−	+	+	+	+	+
Arthrocnemum glaucum	+	D+	+	+	+	+	+	+	−	+	+	−	+	+	+	+	−
Salicornia fruticosa	+	+	D+	+	+	+	+	−	−	+	+	−	+	+	+	−	−
Juncus rigidus	+	+	+	+	−	−	−	−	−	−	−	−	−	−	−	−	+
Cressa cretica	+	+	+	−	+	+	+	−	−	−	+	−	−	−	+	+	−
Nitraria retusa	+	+	+	−	+	−	−	−	−	−	−	−	+	−	−	+	−
Halocnemum strobilaceum	D+	+	+	−	+	+	−	−	−	−	−	−	+	−	−	−	−
Phragmites australis	+	+	+	−	−	−	−	−	−	−	−	−	−	−	−	−	−
Limonium pruinosum	+	+	+	−	+	−	−	−	−	−	−	−	D+	+	−	−	−
Alhagi maurorum	+	+	+	−	−	−	−	−	−	−	−	−	−	−	−	−	−
Limonium axillare	−	+	+	−	−	−	−	−	−	+	+	−	+	−	D+	+	+
Tamarix mannifera	−	−	−	+	−	−	−	−	−	−	−	−	+	−	−	+	−
Inula crithmoides	+	+	+	−	−	−	−	−	−	D+	−	−	−	−	−	−	−
Suaeda monoica	−	+	+	D+	−	−	−	+	−	−	−	−	+	−	−	−	−
Aeluropus spp.	−	−	−	−	−	−	−	−	−	−	−	−	−	−	−	D+	+
Sporobolus spicatus	−	−	−	−	−	−	−	+	−	−	−	−	+	−	+	+	D+
Cyperus laevigatus	−	−	−	−	−	−	−	−	−	−	−	−	−	−	−	−	+
Halopeplis perfoliata	−	+	+	+	−	−	−	D+	−	−	+	−	−	+	−	−	+
Suaeda vermiculata	−	−	−	−	+	+	D+	−	−	−	+	−	−	+	+	−	−
S. fruticosa	−	−	+	+	D+	+	−	−	+	+	−	−	−	−	−	−	−
Scirpus litoralis	−	−	−	−	−	+	−	−	−	+	−	−	−	−	−	−	−
Sphenopus divaritus	−	−	+	−	−	+	−	−	−	+	−	−	−	−	−	−	−
Limoniastrum monopetalum	−	−	−	−	−	+	−	−	+	+	+	−	+	−	D+	−	−
Sporobolus virginicus	−	−	−	−	−	+	−	−	+	−	−	−	+	−	−	−	−
Frankenia revoluta	−	−	−	−	−	+	−	−	+	−	−	−	−	+	+	−	−
Juncus acutus	−	−	−	−	−	+	−	−	−	+	−	−	+	−	−	+	−
Frankenia pulverulenta	−	−	−	−	−	+	−	−	+	−	−	−	+	−	−	−	−
Mesembryanthemum nodiflorum	+	−	−	−	−	+	−	−	−	+	−	−	+	−	−	−	−
Reaumuria hirtella	−	−	−	−	−	−	−	−	−	−	+	−	−	−	−	−	−
Suaeda pruinosa	−	−	−	−	−	D+	+	−	+	−	−	−	−	−	−	−	−
Salsola tetrandra	−	−	−	−	−	+	−	−	D+	−	+	−	−	−	−	−	−
Imperata cylindrica	−	−	−	−	−	−	−	−	−	+	+	−	−	−	−	−	−
Samolus valerandii	−	−	−	−	−	−	−	−	−	−	−	−	−	−	−	−	−
Berula erecta	−	−	−	−	−	−	−	−	−	−	−	−	−	+	−	+	+
Panicum repens	−	−	−	−	−	−	−	−	−	−	−	−	−	−	−	−	−
Phoenix dactylifera	−	−	−	−	−	−	−	−	−	−	−	−	−	−	−	−	−
Cyperus conglomeratus	−	−	−	−	−	−	−	−	−	−	−	−	−	−	+	−	−
Salicornia herbacea	+	+	+	−	−	−	−	−	−	−	+	−	−	−	+	−	−
Carex extensa	−	−	−	−	−	−	−	−	−	−	−	−	−	−	+	−	−
Mesembryanthemum forskalei	+	−	−	−	−	−	−	−	−	+	−	−	+	−	−	−	−
Reaumuria mucronata	−	−	−	−	−	−	−	−	−	−	−	−	−	−	−	−	−
Halimione portulacoides	−	−	−	−	−	−	−	+	−	−	−	−	−	−	−	−	−
Typha elephantina	−	−	−	−	−	−	−	−	−	−	−	−	−	−	−	−	−
T. australis	−	−	−	−	−	−	−	−	−	−	−	−	−	−	−	−	−
Paspalidium geminatum	−	−	−	−	−	−	−	−	−	−	−	−	−	−	−	−	−
Parapholis incurva	−	−	−	−	−	−	−	−	−	−	+	−	−	−	−	−	−
Tamarix passerinoides	−	−	−	−	−	−	−	−	−	−	−	−	−	−	−	−	−
T. arborea	−	−	−	−	−	−	−	−	−	−	−	−	−	−	−	−	−
Atriplex halimus	−	−	−	−	−	−	−	−	−	−	+	−	+	−	−	−	−
Limonium tubiflorum	−	+	−	−	−	−	−	−	−	−	−	−	−	−	−	−	−
Halopeplis amplexicaulis	−	+	+	−	−	−	−	−	−	−	−	D+	−	−	−	−	−
Avicennia marina	−	−	−	−	−	−	−	−	−	−	−	+	−	−	−	−	−
Rhizophora mucronata	−	−	−	−	−	−	−	−	−	−	+	−	−	−	−	−	−
Lygeum spartum	−	−	−	−	−	−	−	−	−	−	−	−	−	−	−	−	−
Cladium mariscus	−	−	−	−	−	−	−	−	−	−	−	−	−	−	−	−	−
Schoenus nigricans	−	−	−	−	−	−	−	+	−	−	+	−	−	−	−	−	−
Atriplex farinosa	−	−	−	−	−	−	−	+	−	−	+	−	−	−	−	−	−
A. leucoclada	−	−	+	−	−	−	−	−	−	−	−	−	−	−	−	−	−
Telephium sphaerospermum	−	−	−	−	−	−	−	−	−	−	−	−	−	−	−	−	−
Halopyrum mucronatum	−	−	−	−	−	−	−	−	−	−	+	−	−	−	−	−	−
Silene succulenta	−	−	−	−	−	−	−	−	−	−	+	−	−	−	−	−	−
NS	14	17	17	7	10	15	8	7	9	13	29	1	11	7	8	9	7

NC = Number of the halophytic community types in which the halophytic species is present, P = presence, NS = Number of the associate halophytes in each community type, Hs = *Halocnemum strobilaceum*, Ag = *Arthrocnemum glaucum*, Sr = *Salicornia fruticosa*, Sm = *Suaeda monoica*, Sf = *Suaeda fruticosa*, Sp = *Suaeda pruinosa*, Se = *Suaeda vermiculata*, Hp = *Halopeplis perfoliata*, St = *Salsola tetrandra*, In.C = *Inula crithmoides*, Za = *Zygophyllum album*, Av.m = *Avicennia marina*, Lp = *Limonium pruinosum*, La = *Limonium axillare*, Lm = *Limoniastrum monopetalum*, A.sp = *Aeluropus* spp., Ss = *Sporobolus spicatus*, Sv = *Sporobolus virginicus*, Nr = *Nitraria retusa*, Tm = *Tamarix mannifera*, Tp = *Tamarix passerinoides*, Fr = *Frankenia revoluta*, Cc

					Cumulatives																	
Sv	Nr	Tm	Tp	Fr	Cc	Ht	Rm	Te	Ta	Pc	Im.C	Ls	Hm	Jr	Ja	Cm	Cl	Sn	Sl	Al.m	NC	P%
−	+	+	+	−	+	+	−	−	−	+	+	+	+	+	−	−	−	−	−	+	26	67.6
−	+	+	−	+	+	−	−	−	−	+	−	−	−	+	−	−	−	−	−	−	20	52.0
−	+	+	−	+	+	−	−	−	−	+	−	−	−	+	−	−	−	−	−	+	19	49.4
+	+	+	−	−	−	+	−	−	−	+	+	−	−	D+	+	+	+	+	+	+	18	46.8
−	+	+	−	+	D+	+	−	−	−	+	−	−	−	−	−	−	−	−	−	+	18	46.8
−	D+	+	+	+	+	−	−	−	−	−	−	−	−	−	−	−	−	−	−	+	18	46.8
−	+	−	−	−	−	−	−	−	−	+	−	−	−	−	−	−	−	−	−	+	15	39.0
−	−	−	−	−	−	−	−	+	+	D+	−	−	−	+	+	+	−	+	−	+	13	33.8
−	+	−	−	−	−	−	−	−	−	−	−	−	−	−	−	−	−	−	−	−	12	31.2
−	+	+	−	−	−	−	−	−	−	+	+	−	−	+	−	+	−	−	−	D+	12	31.2
−	−	−	−	−	−	−	−	−	−	−	+	−	−	−	−	−	−	−	−	−	9	23.4
−	+	D+	+	−	−	−	−	−	−	−	−	−	−	−	−	−	−	−	−	+	9	23.4
−	−	−	−	−	−	−	−	−	−	−	+	+	−	−	−	−	−	−	−	−	8	20.8
−	+	+	−	−	−	−	−	−	−	−	−	−	−	−	−	−	−	−	−	−	8	20.8
+	−	+	−	−	−	−	−	−	−	−	−	−	−	−	−	−	−	−	−	−	7	18.2
+	−	−	−	−	−	−	−	−	−	+	−	−	−	+	−	−	−	−	−	+	7	18.2
+	−	−	−	−	−	−	−	−	−	−	−	−	−	−	−	D+	−	−	−	−	7	18.2
+	−	−	−	−	−	−	−	−	−	−	−	−	−	−	−	−	−	−	−	−	7	18.2
−	+	−	−	−	−	−	−	−	−	−	−	−	−	−	−	−	−	−	−	−	7	18.2
−	−	−	−	−	−	−	−	−	−	−	−	−	+	−	+	−	−	+	D+	−	6	15.6
−	−	−	−	−	−	−	−	−	−	−	−	+	−	−	−	−	−	−	−	−	6	15.6
−	−	−	−	−	−	−	−	−	−	−	−	−	+	−	−	−	−	−	−	−	6	15.6
D+	−	−	−	−	−	−	−	−	−	−	−	−	+	−	−	−	−	−	−	−	6	15.6
−	−	−	−	D+	−	+	−	−	−	−	−	−	−	−	−	−	−	−	−	−	5	13
−	−	−	−	−	+	+	−	−	−	−	−	−	−	−	D+	−	+	−	−	+	5	13
−	−	+	−	−	−	+	−	−	−	−	−	−	−	−	−	−	−	−	−	−	5	13
−	+	−	−	+	−	−	−	−	−	−	−	−	−	−	−	−	−	−	−	−	5	13
−	+	−	−	−	−	−	−	−	−	−	−	−	−	−	−	−	−	−	−	−	5	13
−	+	−	−	−	−	−	−	−	−	−	−	−	−	−	−	−	−	−	−	−	4	10.4
−	−	−	−	−	−	−	−	−	−	−	−	−	−	−	−	−	−	−	−	+	4	10.4
+	+	−	−	−	−	−	−	−	−	D+	−	−	−	−	+	−	−	−	−	−	4	10.4
−	−	−	−	−	−	−	−	+	+	−	−	−	−	−	+	−	−	−	−	−	4	10.4
−	−	−	−	−	−	−	−	−	−	−	−	−	+	+	+	−	−	−	−	−	4	10.4
−	−	−	−	−	−	−	−	−	−	−	−	−	−	−	+	−	−	−	−	−	4	10.4
−	−	−	−	−	−	−	−	−	−	−	−	−	−	−	+	−	−	−	−	−	4	10.4
−	−	−	−	−	−	−	+	+	−	+	−	−	−	−	−	−	−	−	−	−	4	10.4
+	+	−	−	−	−	−	−	−	−	−	−	−	−	−	−	−	−	−	−	−	3	7.8
−	−	−	−	−	−	+	−	−	−	−	−	−	−	−	−	−	−	−	−	−	3	7.8
−	−	−	−	−	D+	−	−	−	−	−	−	−	−	+	−	−	−	−	−	−	3	7.8
−	−	−	−	−	−	−	D+	+	−	+	−	−	−	−	−	−	−	−	−	−	3	7.8
−	−	−	−	−	−	−	+	D+	+	−	−	−	−	−	−	−	−	−	−	−	3	7.8
−	+	−	−	−	−	−	−	+	+	−	−	−	−	−	−	−	−	−	−	−	3	7.8
−	+	+	D+	−	−	−	−	+	+	−	−	−	−	−	−	−	−	−	−	−	3	7.8
−	+	−	−	−	−	−	−	−	−	−	−	−	−	−	−	−	−	−	−	−	2	5.2
−	−	−	−	−	+	−	−	−	−	−	−	−	−	−	−	−	−	−	−	−	2	5.2
−	−	−	−	+	−	−	−	−	−	−	−	−	−	−	−	−	−	−	−	−	2	5.2
−	−	−	−	D+	−	−	−	−	−	−	−	−	−	−	−	−	−	−	−	−	2	5.2
−	−	−	−	−	−	−	−	−	−	−	D+	−	−	−	−	−	−	−	−	−	2	5.2
−	−	−	−	−	−	−	−	−	+	−	−	−	−	−	−	−	D+	−	−	−	2	5.2
−	−	−	−	−	−	−	−	−	−	−	−	−	−	−	−	+	−	D+	−	−	2	5.2
−	−	−	−	−	−	−	−	−	−	−	−	−	−	−	+	−	−	−	−	−	2	5.2
−	−	−	−	−	−	−	−	−	−	−	−	−	D+	−	−	−	−	−	−	−	2	5.2
−	−	−	−	−	−	−	−	−	−	−	−	+	−	−	−	−	−	−	−	−	2	5.2
6	20	11	3	6	7	6	1	7	8	14	7	7	5	13	9	4	6	2	6	12	12	

= *Cressa cretica*, Ht = *Halimione portulacoides*, Rm = *Rhizophora mucronata*, Te = *Typha elephantina*, Ta = *Typha domengensis*, Pc = *Phragmites australis*, Im.C = *Imperata cylindrica*, Ls = *Lygeum spartum*, Hm = *Halopyrum mucronatum*, Jr = *Juncus rigidus*, Ja = *Juncus acutus*, Cm = *Cladium mariscus*, Cl = *Cyperus laevigatus*, Sn = *Schoenus nigricans*, Sl = *Scirpus litoralis*, Al.m = *Alhagi maurorum*, D = Dominant plant, + = associate species, − = absent.

while in the salt marsh habitats, *Z. album* forms luxuriant growth and the individual bushes build phytogenic mounds and hummocks.

The floristic composition of *Z. album* community includes the highest number of the associate species among the other halophytes communities of Egypt. These associates are about 30 halophytes (Table 4) and a number of desert plants, e.g. *Artemisia monosperma, Polycarpaea repens, Calligonum comosum, Aristida plumosa, Pituranthus tortuosus, Cornulaca monacantha* (perennials), *Monsonia nivea, Neurada procumbens, Zygophyllum simplex, Mesembryanthemum nodiflorum* (ephemerals), etc.

12. *Alhagi maurorum* community type

A. maurorum (cumulative) is a widespread plant in Egypt as it occurs in all of the phytogeographical regions (Kassas, 1955). Kassas and Zahran (1967) considered that *A. maurorum* is an alien plant to the salt marsh habitats. Zahran (1967) classified it under what he called 'apparently salt tolerant plants', *A. maurorum* has a long root system that may extend several meters deep (Kassas, 1955). It reaches soil layers that are less saline and permanently wet. But, as *A. maurorum* is an abundant species dominating a characteristic community of the ISM (Migahid et al., 1960; Zahran, 1966; Zahran, 1972) and of the LSM (Kassas and Zahran, 1962) we may consider it as a cumulative halophyte. It shows a range of morphological variations of leaf-spine relationships, and may comprise a number of ecotypes. This needs further studies.

The floristic composition of *A. maurorum* community includes 14 halophytes (Table 4) and desert species, e.g. *Panicum turgidum, Lycium europaeum, Thymelaea hirsuta*, etc.

13. *Salsola tetrandra* community type

S. tetrandra (succulent) community is a characteristic feature of the MLSM. It is rarely recorded in the ISM and absent from the RLSM. In the MLSM, *S. tetrandra* community occupies a continuous strip of the saline land parallel to the sea shore. Its area extends farther inland than the general limit of the downstream-ends of wadis crossing the area from south to north (Migahid et al., 1971; Zahran and Negm, 1973). The presence of the associate halophytes, e.g. *Halocnemum strobilaceum, Limonium pruinosum, Frankenia revoluta, Suaeda fruticosa*, etc. (Table 4) shows its halophytic affinity, whilst the occasional appearance of desert species, e.g. *Thymelaea hirsuta, Pituranthus tortuosus, Astragalus spinosus*, etc. give the impression of a non-halophytic (glycophytic) community.

14. *Halimione portulacoides* community type

This is a frequent and luxuriant excretive halophyte in the MLSM (Tadros, 1953) but not recorded elsewhere in Egypt. Its associates are totally halophytes (Table 4).

15. *Limoniastrum monopetalum* community type

L. monopetalum is an excretive halophyte that seems to be intolerant to arid climate (like the preceding one) as it flourishes only in the semi-arid region of Egypt. Its community type is one of the principal communities of the MLSM (Tadros, 1956) and of the northern lakes (Montasir, 1937). The total plant cover of *L. monopetalum* community ranges between 25–35% contributed by the dominant plant. The common associates are halophytes (Table 4) but occasional glycophytes may occur, e.g. *Thymelaea hirsuta, Pituranthus tortuosus, Anabasis articulata, Haloxylon articulatum*, etc.

16. *Cressa cretica* community type

This is a cushion-chamaephyte excretive halophyte that is common in the salt marshes of Egypt. Its domination is observed in two habitats: (a) in the LSM, e.g. Delta of Wadi Hommath of the Gulf of Suez where the soil is formed of saline sand flats with total soluble salts = 6.0% (Kassas and Zahran, 1964). Here, the associates include: *Zygophyllum album, Arthrocnemum glaucum, Halocnemum strobilaceum, Nitraria retusa, Limonium pruinosum, Alhagi maurorum*, etc. (Table 4) with total plant cover = 10.20%; (b) in the ISM, e.g. salt marshes of Siwa Oasis, *Cressa cretica* is restricted to dry saline areas 'at the far inner edge of the salt marsh system the depth of the surface salt crust is 25–40 cm (total soluble salts = 65.88%). In this habitat *C. cretica* forms pure community with very thin plant cover (<2%)' (Zahran, 1972).

14

17. *Frankenia revoluta* community type

Unlike *C. cretica*, *F. revoluta* (cushion-chamaephyte, excretive) is not very widely present in Egypt (Tackholm, 1974). It is abundant in the MLSM and its domination is limited to the dry salt marsh habitat where the salt crusts accumulate on the soil surface. *F. revoluta* is absent from the RLSM but rare in the ISM. The associates of this community are mainly salt tolerant species (Table 4) and the total plant cover is low (5–10%).

18. *Aeluropus* spp. community type

A. lagopoides and *A. brevifolius* (excretives) are similar ecologically and morphologically, but *A. lagopoides* is most common in the LSM and ISM (Tackholm, 1974). *A. brevifolius* is restricted to the RLSM where it co-dominates with *A. lagopoides* in the southern part from 700 km south of Suez and in the eastern side of the Gulf of Suez (Zahran, 1967). 'The growth of *Aeluropus* spp. is usually that of creeping but in one locality (Mersa Alam area) the plants form peculiar cone-like tussocks of interwoven roots rhizomes and sand' (Kassas and Zahran, 1967). The associates of *Aeluropus* spp. community include 11 halophytes (Table 4) and desert plants, e.g. *Salsola baryosma*, *Sphaerocoma hockeri*, *Launaea cassiniana*, *Suaeda volkensi* (perennials), *Zygophyllum simplex*, *Astragalus eremophulous*, *Lotononis platycarpa*, *Aristida meccana*, *A. funiculata*, *Polycarpaea spicata* (ephemerals), etc.

19. *Sporobolus spicatus* community type

S. spicatus is a salt marsh (excretive) grass commonly present in Egypt. It dominates a well developed community in the ISM (Hassib, 1945; Migahid et al., 1960; Zahran and Girgis, 1970) as well as in the salt marshes of the northern lakes (Montasir, 1937). In the RLSM, *S. spicatus* is abundant in the southern 100 km but not northward (Kassas and Zahran, 1967). *S. spicatus* is also common in the MLSM and it co-dominates with *S. virginicus* a characteristic community. The list of the associate plants include 43 species: 28 perennials (7 halophytes and 21 desert plants) and 15 ephemerals. The desert plants include: *Panicum turgidum*, *Suaeda schimperi*, *Salsola baryosma* (perennials), *Zygophyllum simplex*, *Launaea cassiniana*, *Eremopogon foveolatus* (ephemerals), etc.

20. *Sporobolus virginicus* community type

S. virginicus (excretive) dominates the transitional habitat between the littoral sand dunes and the salt marsh habitats of the Mediterranean region (Tadros, 1953). It forms almost pure community sometimes covering quite large patches of the line of separation between the dunes and the marshes (Rezk, 1970). *S. virginicus* is rarely recorded or absent in the other salt marshes of Egypt (Tackholm, 1974).

21. *Lygeum spartum* community type

The most suitable habitat for the domination of *L. spartum* seems to be the saline sand sheets of the MLSM. Here the soil has high content of sand (>60%) and low content of soluble salts (1–1.8%). The plant growth forms a quite distinct homogeneous community, the species itself forming luxuriant tufts (Zahran, 1967). *L. spartum* usually forms pure stands (Ayyad, 1969), but its common associates are: *Limoniastrum monopetalum*, *Nitraria retusa*, *Zygophyllum album*, *Atriplex halimus*, etc. (Table 4). *L. spartum* is not recorded either in the ISM or in the RLSM.

22. *Cyperus laevigatus* community type

C. laevigatus (cumulative) is a most common halophyte dominating the wet salt marshes especially around the northern lakes and in the northern oases of the Western Desert. In Wadi El Natrun, for example, *C. laevigatus* is very abundant and makes with *Juncus acutus* the *Cyperus-Juncus* complex the phytocoenosis of which is composed of two strata: an upper stratum (restricted) dominated by *J. acutus* and a lower stratum (extensive) dominated by *C. laevigatus* (Zahran and Girgis, 1970). The total plant cover is usually high (80–100%) and the associates are the water loving types (Table 4).

23. *Cladium mariscus* community type

The occurrence of *C. mariscus* (cumulative) is rare in Egypt (Tackholm, 1974). 'Its domination is recorded only in Siwa Oasis where it flourishes in the marshy land about the springs where the water level is very shallow (5 cm depth) or exposed' (Zahran, 1972). Its associates are few which include, *Phragmites australis*, *Juncus rigidus*, *Cyperus laevigatus* and *Alhagi maurorum*.

15

24. *Schoenus nigricans* community type

In the MLSM *S. nigricans* (cumulative) is frequently met with scattered between patches of *J. rigidus* community but occupying slightly lower level apparently with a higher moisture content. 'From a distance it is difficult to separate the two communities owing to the similarity in appearance between small bushes of *J. rigidus* and *S. nigricans*' (Tadros, 1953). The associated halophytes include: *Cressa cretica, Juncus rigidus, Phragmites australis*, etc. (Table 4), while the associated glycophytes include: *Thymelaea hirsuta, Koeleria phleoides*, etc. *S. nigricans* is rare in the ISM and not recorded in the RLSM.

25. *Juncus rigidus* community type

J. rigidus (cumulative) is strongly tolerant to the increased soil salinity as well as to the aridity of the climate. It dominates a widespread community in various parts of the salt marshes of Egypt (Zahran et al., 1972). In the oases of the Western Desert, *J. rigidus* community covers more than 60% of the salt marshes with plant cover up to 80–90% (Zahran, 1972). Also in the MLSM *J. rigidus* is very abundant (Tadros, 1953; Tadros and Atta, 1958) while in the RLSM it covers extensive areas at both sides of the Gulf of Suez, e.g. Ein Sokhna area (about 50 km south of Suez) where its cover reach 100% (Kassas and Zahran, 1962; Zahran, 1967). The associates are mostly halophytes (Tadros, 1953).

26. *Juncus acutus* community type

J. acutus (cumulative), unlike *J. rigidus*, seems to be intolerant to the aridity of the climate or to the dryness of soil. It abounds in the habitats where climate is mild and soil is highly moistened. Accordingly, its abundance is observed in the wet salt marshes of the northern lakes MLSM, Wadi El Natrun and Fayum Depression (Montasir, 1937; Tadros, 1953, 1956; Zahran and Girgis, 1970). *J. acutus* is not recorded in the RLSM and in the salt marshes of the southern oases. The associates of *J. acutus* community include: *Samolus valerandii, Berula erecta, Panicum repens* (water loving plants), *Inula crithmoides, J. rigidus, Cyperus laevigatus, Aeluropus* spp., etc. (Table 4).

27–28. *Nitraria retusa* and *Suaeda monoica* community types

N. retusa (excretive) and *S. monoica* (succulent) are salt marsh bushes that are comparable in habit and habitats. The two species have an ecological range that extends beyond the limits of the salt marshes to the fringes of the deserts. In Egypt, they vary biogeographically. *N. retusa* is abundant along the northern stretch of the RLSM. It is neither recorded in the southern oases of the Western Desert nor in the southern part of the Red Sea marshes (Boulos, 1966; Migahid et al., 1960; Zahran, 1966; Zahran and Girgis, 1970; Zahran, 1972). *N. retusa* is also common within the wadis of the limestone desert of Egypt (Kassas and Girgis, 1965).

The domination of *S. monoica* is, however, restricted to the southern part of RLSM. It gradually replaces *N. retusa* within the 300–650 km south of Suez, further southward *N. retusa* is absent and *S. monoica* is the abundant species. *S. monoica* is rarely recorded in the MLSM and in the ISM (Tackholm, 1974; Tadros, 1953).

29–30. *Tamarix mannifera* and *T. passerinoides* community types

These excretive species represent the climax stage of the halosere succession in Egypt (Kassas and Zahran, 1967; Tadros and Atta, 1958; Zahran and Girgis, 1970) (Plate 2).

T. mannifera grows in a variety of habitats and in various forms. It may form thickets in the dry salt marshes and may also form sand hillocks in the sand-choked deltaic parts of certain wadis that

Plate 2. *Tamarix mannifera* scrubland representing the climax stage of the salt marshes of Egypt.

drain in the Red Sea and Mediterranean Sea. *T. mannifera* is abundant in the ISM (Zahran and Girgis, 1970; Zahran, 1972).

The morphologically and ecologically comparable, *T. passerinoides*, is a rare species in Egypt. Its presence is limited to a narrow stretch (El Mallaha, 260–280 km south of Suez) in the RLSM (Kassas and Zahran, 1964). It is rarely recorded in the other salt marshes of Egypt (Tackholm, 1974). The associates of *T. mannifera* community are mostly perennials and include: 13 salt marsh plants (Table 4) and 17 glycophytes, e.g. *Zygophyllum coccineum, Haloxylon salicornicum, Anabasis articulata, Tamarix aphylla, Acacia raddiana, A. tortilis, Calotropis procera, Leptadenia pyrotechnica,* etc. The floristic composition of *T. passerinoides* includes 3 halophytes (Table 4).

IV. The saline sand formation

These are the sand sheets, terraces and embankments formed mainly of aeolian sand deposits parallel to the Red Sea and Mediterranean Sea and about the lakes of the oases and depressions of the Western Desert. They are under the direct influence of sea (or lake) water spray and/or seepage and accordingly they are relatively saline. Two ammophilous grasses namely: *Imperata cylindrica* and *Halophyrum mucronatum* predominate these formations.

Imperata cylindrica* is widespread in Egypt. In the ISM it may cover sandy-bars or terraces about the lakes, e.g. Wadi El Natrun lakes (Zahran and Girgis, 1970). In the RLSM, *I. cylindrica* predominates in the sand sheets of the downstream parts of certain wadis, e.g. Wadi Hommath (Kassas and Zahran, 1962). The associates are mainly a few halophytes (Table 4). *I. cylindrica* is also common in the MLSM.

Halophyrum mucronatum is an ammophilous grass of a very limited occurrence. It dominates maritime sand embankments in an area (about 2 km long extending to the south of Mersa Abu Ramad, 1030 km south of Suez) in the RLSM. It is

---

* *Imperata cylindrica* and *Desmostachya bipinnata* are two similar grasses and without the inflorescence it is not easy to difrentiate between them.

a new record to the flora of Egypt (Tackholm, 1974; Zahran, 1964). The floristic composition of this community includes 5 halophytes (Table 4) and 7 desert and sand dune species, e.g. *Panicum turgidum, Indigofera argentea, Heliotropium pterocarpum,* etc.

5. Conclusion

'Plants growing on saline soils are considered from the point of view of form, structure and function as xerophytes. Both xerophytes and halophytes are physiologically and anatomically adapted to inadequate amount of water in soil' (Strogonov, 1965). The range of tolerance of the halophytes is relatively wide, but when soil salinity and/or climatic aridity becomes higher than the tolerance of the plants, their growth will be impaired and they may die. Barren patches, where soil salinity exceeds the range of the abundant halophytes, are a common feature of the salt marshes of Egypt (Plate 3).

The ecological studies of the halophytic vegetation of Egypt elucidate that 90% of these plants are not only highly tolerant to saline soils but also to arid climate. Most (61%) of the succulent halophytes (*Halocnemum strobilaceum, Arthrocnemum glaucum, Salicornia fruticosa, Suaeda fruticosa, S. vermiculata* and *Zygophyllum album*), 46% of the excretive halophytes (*Limonium pruinosum, Aeluropus* spp., *Sporobolus spicatus, Nitraria retusa, Tamarix mannifera* and *Cressa cretica*) and 45% of the cumulatives (*Typha domengensis, Phragmites australis, Juncus rigidus, Imperata cylindrica, Scirpus litoralis, Cyperus laevigatus* and *Alhagi maurorum*) are widely distributed and dominant in the inland salt marshes and in both the Red Sea and Mediterranean littoral salt marshes of Egypt. The other dominant halophytes are not as widely distributed but are geographically restricted as follows.

The mangroves, *Avicennia marina* (excretive) and *Rhizophora mucronata* (cumulative), being tropical formation, are recorded in the southern part of the RLSM.

The succulents, *Halopeplis perfoliata* and *Suaeda monoica,* the excretives, *Limonium axillare* and *Tamarix passerinoides,* the cumulative, *Halopyrum*

Plate 3. Salt marsh dominated by *Arthrocnemum glaucum* (total plant cover upto 80–90%) and in the barren patches, thick salt crusts kill the plants, preventing the growth of new generation.

mucronatum are dominant in the RLSM, *S. monoica* is rarely recorded in the ISM.

Limoniastrum monopetalum, Frankenia revoluta, Sporobolus virginicus, Halimione portulacoides (excretives) and *Lygeum spartum* (cumulative) are dominant in the MLSM.

Typha elephantina and *Cladium mariscus* (cumulatives) are dominant in the ISM. Each is restricted to one locality: *T. elephantina* in Wadi El Natrun (Boulos, 1962; Zahran and Girgis, 1970) while *C. mariscus* in Siwa Oasis (Zahran, 1972).

Suaeda pruinosa, Salsola tetrandra, Inula crithmoides (succulents); *Juncus acutus* and *Schoenus nigricans* (cumulatives) are dominant in the inland and MLSM.

The community types dominated by halophytic plants are usually either pure or associated with a limited number of species. In Egypt, except for a few annuals, e.g. *Juncus bufonius, Mesembryanthemum nodiflorum, Atriplex littoralis, Scirpus supinus, Crypsis aculeata, Frankenia pulverulenta, Salicornia herbacea, Halopeplis amplexicaulis*, etc., the associates of the halophytic communities are mainly perennial halophytes. Glycophytes (perennials and ephemerals) may associate certain halophytic communities the extent of which extends towards

inland into the desert country, e.g. *Salsola tetrandra, Nitraria retusa, Suaeda monoica* and *Tamarix mannifera* communities.

The mangrove communities of the Red Sea (*Avicennia marina*, and *Rhizophora mucronata*) are associated with the sea weeds, e.g. *Cymodocea ciliata, C. serrulata, C. rotundata, Diplanthera uninervis, Halophila ovalis* and *H. stipulacea*, etc. In the inland and littoral reed swamps dominated by *T. elephantina, T. domengensis, Phragmites australis* and *Scirpus litoralis* where soil salinity is relatively low (0.39%), water loving species, e.g. *Berula erecta, Samolus valerandii, Cyperus articulatus, C. mundtii, C. dives, Scirpus tuberosus, S. mucronata, Lemna gibba, Spirodela polyrrhiza, Wolffia hyalina*, etc. are the common associates.

Apart from the associate species mentioned in Table 4, the other associate perennial halophytes in the moist and/or dry salt marshes of Egypt include: *Frankenia hispida, Reaumuria palestina, Juncus inflexus, J. punctorius, J. effusus, J. fontansii, Schangenia hortensis, Cyperus capitatus, C. bulbosus, Carex extensa, Silene succulenta, Fimbristylis bis-umbellata*, etc. recorded as rare or occasional associates.

With respect to the life-form of the dominant

18

halophytes of Egypt, it is quite obvious that the main bulk of the plants (about 68.2%) are of the suffruticose (36.7%) and of the geophytes (31.5%). The nanophanerophytes represent about 13.3%, the helophytes 7.9% while each of the microphanerophytes and the cushion chamaephytes represents 5.3% of the total halophytes.

Chenopodiaceae includes the greatest number of the dominant halophytes (about 26.3% of the total, 92.2% of the succulents and 7.7% of the excretives). Poaceae comes next, to which belongs 18.2% of the total halophytes (22.1% of the excretives and 28.3% of the cumulatives). To Cyperaceae belongs 4 halophytes, all are cumulatives (10.4% of the total and 28.2% of the cumulatives). Plumbaginaceae includes 3 excretive halophytes, i.e. 7.8% of the total

and 18.1% of the excretives. To Tamaricaceae belongs about 5.2% of the total and 14.4% of the excretive halophytes. Each of Typhaceae and Juncaceae include 14.2% of the cumulative (about 5.2% of the total halophytes). There are seven families to each of which belongs one halophyte, i.e. 2.6% of the total halophytes, these are: Avicenniaceae, Nitrariaceae, Frankeniaceae and Convolvulaceae (include excretive species), Zygophyllaceae and Asteraceae (include succulent species) and Rhizophoraceae and Leguminosae (include cumu·lative species).

Acknowledgements. The author wishes to express his sincere thanks to Professor M. Kassas, Faculty of Science, Cairo University, for reading the manuscript.

Literature cited

Ayyad, M. 1969. An edaphic study of habitats at Ras El-Hikma. *Bull. Inst. Deserte d'Egypte* 19: 245–259.

Boulos, L. 1962. *Typha elephantina* Roxb. in Egypt. *Candollea* 18: 129–135.

Boulos, L. 1966. A natural history study of Kurkur Oasis, Libyan Desert, Egypt. IV - The Vegetation. *Postillea* 100: 1–21.

Chapman, V.J. 1938. Studies on salt marsh ecology, sections I-III. *J. Ecol.* 26: 144–179.

Chapman, V.J. 1960. *Salt Marshes and Salt Deserts of the World.* Leonard Hill, New York.

Chapman, V.J. 1964. *Coastal Vegetation.* Pergman Press, London.

Emberger, L. 1952. Report on the arid and semi-arid regions of North-West Africa UNESCO/US/AZ, 89. Paris.

Girgis, W., Zahran, M.A., Reda, Kamelia A. and Shams, H. 1971. Ecological notes on Moghra Oasis. *UAR J. Bot.* 14: 147–155.

Hassib, M. 1945. An ecological study of a salty district caused by a drainage from the sewage disposal farm. *Cairo, Bull. Fac. Sc. Fouad I Univ.* 25: 59–76.

Hassib, M. 1951. Distribution of plant communities in Egypt. *Bull. Fac. Sc. Fouad I Univ.* 29: 59–258.

Kassas, M. 1955. On the distribution of *Alhagi maurorum* in Egypt. *Egyptian Academy of Science Proc.* 3: 140–151.

Kassas, M. and Girgis, W. 1964. Habitat and plant communities of the Egyptian desert. V - The limestone plateau. *J. Ecol.* 53: 107–199.

Kassas, M. and Girgis, W. 1965. Habitat and plant communities of the Egyptian desert. VI - The unit of a desert ecosystem. *J. Ecol.* 53: 725–729.

Kassas, M. and Zahran, M.A. 1962. Studies on the ecology of the Red Sea coastal land. I - The district of Gebel Ataqa and El-Galala El-Bahariya. *Bull. Soc. Geog. d'Egypte* 35: 129–175.

Kassas, M. and Zahran, M.A. 1964. Studies on the ecology of the Red Sea coastal land. II - The district from El-Galala El-Qibliya to Hurghada. *Ibid.* 38: 156–193.

Kassas, M. and Zahran, M.A. 1967. On the ecology of the Red Sea littoral salt marsh, Egypt. *Ecol. Monog.* 37: 297–315.

Keith, L.B. 1958. Some effects of increasing soil salinity on plant communities. *Can. J. Bot.* 36: 79–89.

Long, G.A. 1956. Phyto-ecological map of Ras El-Hekma. FAO/9/7130, Rome.

Migahid, A.M. 1962. The drought resistance of Egyptian desert plants. *Proc. Madrid Symp.* UNESCO/AZ/213–233.

Migahid, A.M., Abdel Rahman, A.A., El-Shafei, Ali, M. and Hammouda, M.A. 1955. Types of habitat and vegetation at Ras El-Hekma. *Bull. Inst. Deserte d'Egypte* 5: 107–190.

Migahid, A.M., Abdel Rahman, A.A., El-Shafei Ali, M. and Hammouda, M.A. 1960. An ecological study of Kharga and Dakhla Oases. *Bull. Soc. Geog. d'Egypte* 33: 279–309.

Migahid, A.M., El-Batanouny, K.H. and Zaki, M.A.F. 1971. Phytosociological and ecological study of a sector in the Mediterranean coastal region, Egypt. *Vegetatio* 23: 113–134.

Montasir, A.H. 1937. Ecology of Lake Manzala. *Bull. Fac. Sci. Egyptian Univ.* 12: 1–50.

Montasir, A.H. 1943. Soil structure in relation to plants at Mariut. *Bull. Inst. d'Egypte* 15: 205–236.

Montasir, A.H. 1954. Habitat factors and plant communities in Egypt. *Bull. Soc. Geog. d'Egypte* 27: 115–143.

Oliver, F.W. 1930. The Egyptian desert. *Transections of the Norfolk and Norwick Naturalists Soc.* 13: Part I.

Raunkiaer, C. 1934. *The Life Form of Plants and Statistical Plant Geography.* Translated by Carter Fausboll and Tansley. Oxford Univ. Press, London.

Rezk, R.M. 1970. Vegetation change from a sand dune community to a salt marsh as related to soil characters in Mariut district, Egypt. *Oikos* 21: 341–343.

Sholander, P.E., Hammel, H.T., Hemmingsen, E. and Garey, W. 1962. Salt balance in mangroves. *Plant Physiol.* 37: 722–729.

Steiner, M. 1934. Zur Ökologie der Salzmarschen der nordöstlichen Vereinigten Staaten von Nordamerika. *Fb. Wiss. Bot.* 81: 94–202.

Stocker, O. 1927. Der Wadi Natrun. In *Vegetationsbilder*, eds. G. Karsten et K. Schenk, 18 Reihe I.

Strogonov, B.P. 1965. *Physiological Basis of Salt Tolerance of Plants*. Oldourne Press, London.

Tackholm, Vivi. 1974. *Student's Flora of Egypt*. Cooperative Printing Company, Beirut.

Tadros, T.M. 1953. A phytosociological study of halophilous communities from Mareotes (Egypt). *Vegetatio* 4: 102–104.

Tadros, T.M. 1956. An ecological survey of the semiarid coastal strip of the Western Desert of Egypt. *Bull. Inst. Deserte d'Egypte* 6: 28–56.

Tadros, T.M. and Atta, A. Berlanta. 1958. Further contribution to the study of the sociology and ecology of the halophilous plant communities of Mareotes, Egypt. *Vegetatio* 8: 137–160.

Tadros, T.M. and El-Sharkawi, M. El-Hassanein. 1960. Phytosociological and ecological studies on the vegetation of Fuka Ras El-Hekma area. *Bull Inst. Deserte d'Egypte* 10: 37–59.

Tansley, A.G. 1949. *Britain's Green Mantle*. G. Allen and Unwin, London.

Walter, H. 1961. The adaptation of plants to saline soils. *AZ Proc. Teheran Symp.* UNESCO 14: 129–134.

Zahran, M.A. 1964. Contributions to the study on the ecology of the Red Sea coasts. Ph.D. Thesis Fac. Sc. Cairo Univ., Cairo.

Zahran, M.A. 1964. Distribution of mangrove vegetation in UAR (Egypt). *Bull. Inst. Deserte d'Egypte*. 15: 7–11.

Zahran, M.A. 1966. Ecological study of Wadi Dungul. *Ibid.* 16: 127–143.

Zahran, M.A. 1967. On the ecology of the east coast of the Gulf of Suez. I - Littoral salt marshes. *Ibid.* 17: 225–261.

Zahran, M.A. 1972. On the ecology of Siwa Oasis. *Egyptian J. Bot.* 25: 223–242.

Zahran, M.A. 1974. Biogeography of mangrove vegetation along the Red Sea coasts. *Int. Symp. on The Biology and Management of Mangrove Vegetation*, Honolulu, pp. 43–51.

Zahran, M.A. and Girgis, W. 1970. On the ecology of Wadi El-Natrun. *Bull. Inst. Deserte d'Egypte* 18: 229–267.

Zahran, M.A., Kamal El-Din, H. and Boulous, S. 1972. Potentialities of the fibrous plants of Egyptian flora in national economy. I - *Juncus rigidus* C. Mey and paper industry. *Bull. Inst. Deserte d'Egypte* 22: 193–203.

Zahran, M.A. and Negm, S.A. 1973. Ecological and pharmacological studies on *Salsola tetrandra* Forsk. *Bull Fac. Sci. Mansoura Univ.* 1: 67–75.

CHAPTER 2

Estuarine ecosystem of India

G.V. JOSHI and LEELA J. BHOSALE
Department of Botany, Shivaji University,
Kolhapur – 416004, India

1. Introduction

An estuary is considered as a partly enclosed body of water that is formed when a river joins the sea. The water of the former is salinated by sea water. Such an estuarine environment is obviously to be found at the mouth of a river and will extend to as far as sea water can reach. The river, flowing over a considerable distance, brings a large amount of silt with it which is deposited on the banks of the river and, which is at the same time responsible for island formations in the river. These islands, which are barren at first, are soon subjected to the process of succession and develop into stable ecosystems.

The salinity of the soil, partial coverage by tidal waters, poor aeration and shifting soil conditions create such an environment that only a particular type of ecosystem can thrive under tropical and subtropical conditions. This ecosystem consists of salt-tolerant plants of which mangroves constitute the dominant part. According to Pannier (1979) mangroves represent an ecosystem typical of tropical coasts and estuaries. They are of exceptional biological importance. As nutrient filters and synthesizers of organic matter, mangroves create a living buffer between land and sea. Being highly dependent upon the inorganic nutrients contributed by rivers, they play a primary role in supporting the productivity of the associated marine environment. The SCOR/UNESCO panel (1976) has stated: 'Scientists are agreed that mangrove areas, far from being waste lands, are highly productive and comparable to good agricultural land. Apart from forest products they produce fish, prawns, edible crabs and shell fish. The outflow from mangroves carries foods for fisheries upon which many coastal communities depend. Unfortunately mangrove ecosystems are also sensitive and easily destroyed by activities such as poorly executed logging operations, alluvial mining, drainage canals and the building of harbours and marinas. Nevertheless the rapid growth of human population makes it necessary that such developments must proceed in selected areas; sufficient knowledge however already exists to immediately implement conservation measures'.

According to Teas (1977) mangroves are valuable for shore-line protection and stabilization, as well as habitat for wild life and sources of photosynthetic productivity. Hence he has stressed the need of restoration of shore lines with proper mangrove species (Teas et al., 1976). McGill (1958) estimated that 75% of the coast lines between 25° north and south latitudes are dominated by mangroves. Mangroves reach their maximum development and greatest luxuriance in Southeast Asia, Indonesia and Borneo (Macnae, 1968). Even though India does not fall in this region it has several estuarine areas where mangrove vegetation dominates. Andman-Nicobar islands and Gangetic Sundarbans are the largest mangrove areas in Southeast Asia while the deltas of the Krishna and Godavari rivers exhibit luxuriant growth of mangrove forests on the eastern coast of India.

Tasks for vegetation science, Vol. 2 ed. by D.N. Sen and K.S. Rajpurohit

The environment of an estuary is unique as it harbours a unique ecosystem, where the temperature is warm, tidal onslaught is at low intensity, the soil is rich in nutrients, saline, muddy and shifting, and the dominant component and main producer is mangrove forest. It is followed by mud flat halophytes like *Sesuvium portulacastrum*, mangrove grass like *Aeluropus lagopoides*, sea grass like *Halophila beccarii*, and borderline salt-tolerant plants like *Clerodendrum inerme* and *Salvadora persica*. On the sandy shores *Spinifex squarrosus* and *Ipomoea pes-caprae* dominate.

In the ecosystem the food chain may be observed. (Fig. 1.)

Even though there was very scanty literature till 1970 on the estuarine ecosystems and one had to rely on reviews by Adriani (1958) and Walter (1973), during the last decade a good number of treatises have appeared on this subject. To mention a few, Biology of Halophytes (Waisel, 1972), Ecology of Halophytes (Reimold and Queen, 1974), The Mangroves of India (Blasco, 1975), Biology and management of Mangroves (Walsh et al., 1975), Mangrove Vegetation (Chapman, 1976), Wet-coastal Ecosystems (Chapman, 1977) and Eco-geographical studies in Vashishti and Terekhol river (Joshi and Shinde, 1978).

In view of the abundance of literature on this subject it does not seem useful to repeat the same information here. Hence an attempt is made in this chapter to give information not given earlier on estuarine ecosystems, and then specifically, for plant components only. The information is taken from the few estuaries of Western India. It will not vary, however, from other estuarine ecosystems in India except in minor details.

2. Estuarine regions in India

India has a coast line which runs up to 5100 km and embraces both tropical and subtropical zones. The area covered by the Indian mangroves may be 3, 50, 500 hectares (Blasco, 1977). This figure is approximate and may be slightly more when actual estimates are made. The mangroves occur on the coast as well as on the estuaries. The Gangetic, Krishna-Godavari and Kaveri estuaries have rich mangrove vegetation. On the western coast, only Tapi and Narmada run over a long distance and the rest of the rivers are having a short distance. The estuarine vegetation on the rivers of Konkan and Goa though less extensive is compact and well defined. Some of the estuaries (Plate 1) in the Ratnagiri district were studied extensively and the results are summarized here. The values in other estuaries in India may show a similar trend upon investigation.

3. Structure of vegetation

The estuaries of western India are formed by rivers which have a narrow run of 40–50 km from source to mouth. Hence, delta formation is not seen. This results in development of vegetation only on the banks and islands. The estuarine ecosystem is seen on the banks and extends for 100 meters to 1/2 km only. At some places it is less than 100 meters. Due to the industrial and urban development of the region the estuaries are being subjected to various stresses. The effluents from the cities and villages are left in the estuarine environment and this changes its composition considerably. The nitrogen

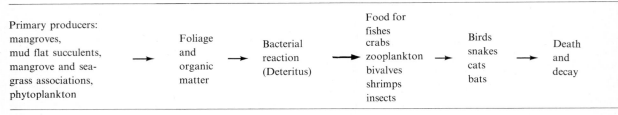

Fig. 1

Plate 1. Estuary with mangrove vegetation on the banks.

Plate 2. *Sonneratia alba* with pneumatophores.

dynamics of the ecosystems are affected greatly. This does not so much alter the number of mangrove species as it does their relative dominance. The direct effect of pollution in the form of more organic matter is reflected by *Acanthus ilicifolius* stands; when the nitrogen balance is disturbed in the mangrove swamp, *Halophila beccarii*, a sea grass also appears in the mangrove mud.

The structure of estuarine vegetation at various places is shown in Table 1. As many as fifteen mangrove species are seen in the estuary. This shows that the estuaries are rich in diversity of mangrove species. *Rhizophora apiculata, Bruguiera gymnorrhiza* and *Avicennia alba, Sonneratia alba* (Plate 2) and *Kandelia candel* are found in few places as their distribution is restricted.

The structure of the estuarine community in Table 1 shows that the dominant component is the mangrove group. This component is rich in variety. There may be less density and extent of cover in the Indian estuarine mangroves but various families are represented. From the data presented in Table 1, it is clear that the mangroves are a botanically diverse group, all of which have close relatives among ordinary land plants. Teas (1976) has stated that the mangroves should be considered by botanists to be land plants that have adapted to the salty or brackish-water habitat, rather than plants that have evolved in the sea. Even though these plants originally grew on land or in fresh water areas, they have, in the course of time, come to depend on salinity for full development. Some of their enzyme

systems have become salt-dependent (Larsen, 1967; Shomer-Ilan and Waisel, 1973).

Macnae (1968) has stated that in spite of their ability of many mangrove species to grow under non-saline conditions, not many mangrove species are found to be successful competitors in upland or fresh water environments. This is further confirmed by Joshi and Shinde's observations (1978) on the estuarine ecosystem at Chiplun which has undergone change due to fresh water influx and which has resulted in the replacement mangroves by fresh water marsh plants.

It is clear from Table 1 that representatives from divergent and unrelated families of mangroves have migrated from land towards estuarine environments. They have established themselves as a successful species after acquiring modifications on divergent lines. This is clear from the divergent root modifications, stem structure and germination adaptations. Only *Avicennia* and *Sonneratia* show development of pneumatophores while many Rhizophoraceae members have stilt roots; *Kandelia* has buttress roots. Vivipary is restricted to a few plants. Cryptovivipary is seen in a few being altogether absent in others. It shows that each species has gone for its own line of adaptations in estuarine environment and this has resulted in the development of a common ecosystem with heterogeneous modifications.

The plants listed in Table 1 are usually found in the estuaries in India. However a short account of *R. apiculata, B. gymnorrhiza, K. candel, A. marina,*

Table 1. Structure of estuarine vegetation at different stations in Ratnagiri district.

Species	Occurrence at					Form and root system	Flowering time
	R	G	D	M	B		
Mangroves							
Rhizophora mucronata	C	C	C	C	LC	Tree with prop roots	Aug–Dec
Rhizophora apiculata	—	—	—	—	C	Tree with prop roots	Aug–Jan
Ceriops tagal	C	C	C	C	C	Shrub	July–Sept
Bruguiera gymnorrhiza	—	C	C	—	—	Tree	Dec–Feb
Kandelia candel	—	—	—	—	C	Small tree with buttress roots	March
Avicennia alba	LC	LC	—	—	—	Shrub	March–May
Avicennia officinalis	—	C	C	C	C	Tree	April–June
Avicennia marina	C	—	C	C	C	Tree	Aug–Sept
Sonneratia alba	VC	C	VC	VC	VC	Tree	Feb–July
Sonneratia apetala	—	—	—	LC	—	Tree	Feb
Excoecaria agallocha	C	C	LC	C	C	Tree/Shrub	July–Aug
Lumnitzera racemosa	C	C	LC	C	C	Shrub	Jan–Apr
Derris heterophylla	C	C	C	C	C	Climbing shrub	Feb–Mar
Aegiceras corniculatum	C	C	C	C	C	Shrub	Feb
Acanthus ilicifolius	C	C	C	C	C	Herb/Shrub	Apr–May
Mangrove grass							
Aeluropus lagopoides	C	C	C	C	C	Herb	Nov–Dec
Sea grass							
Halophila beccarii	C	—	C	C	C	Small herb	July–Sept
Mudflat plants							
Sesuvium portulacastrum	C	C	—	—	—	Herb	Nov–Dec
Brackish water fern							
Acrosticum aureum	—	—	C	C	—	Herb	Nov–Jan (sporing period)
Borderline mangroves							
Salvadora persica	C	C	—	—	—	Shrub/Tree	Nov–Feb
Clerodendrum inerme	C	C	C	C	C	Shrub	Nov–Jan
Premna integrifolia	C	—	LC	LC	—	Shrub	June–July
Thespesia populnea	C	—	—	—	—	Tree	cold season
Stenophyllus barbata	C	C	C	C	C	Herb	July
Cyperus sp.	—	—	—	—	C	Herb	Dec
Sand binders							
Ipomoea pes-caprae	—	C	C	C	—	Herb	throughout year
Spinifex squarrosus	C	C	—	—	?	Herb	Nov

C = common, VC = very common, LC = less common, R = Ratnagiri, G = Ganapatipule, D = Deogad, M = Mumbra, B = Bhatya.

S. apetala and *H. beccarii* is essential. *R. apiculata* is found in restricted localities in Ratnagiri upstream in Bhatya creek towards Chinchkhari. It flowers from August to January (Plate 3). *B. gymnorrhiza* also has a very restricted distribution and is uncommon to the estuaries. *K. candel* is seen only after a long journey upstream – at Bhatya, Terekhol and Chiplun. It is a mangrove with a very fine canopy and a majestic, reddish coloured, buttressed trunk. The flowers here are seen in March. *A. marina* is a component of the mangrove system newly identified by the Pondicherry group and later on confirmed by us. It is a tall, black-barked *Avicennia* which flowers from August to September. *Sonneratia apetala* is

Plate 3. A small plant of *Rhizophora apiculata*.

Plate 4. A tree of *Avicennia officinalis*.

another rare component and is seen only at Mumbra and Chiplun.

The only sea grass found in the mangroves is *H. beccarii*. Its presence in Goa has been reported by Untawale and Jagtap (1977). It has, however, been found by us to be very common in the mangroves throughout the Ratnagiri district and more dominant where conditions of pollution exist. The flowers from this sea grass were collected in July–September.

Acrosticum aureum is the mangrove fern which is not uniformly distributed but occurs in restricted places. It is seen extensively at Goa as well as at Deogad and Mumbra. It is considered to be representative where the mangrove forest has been cut down; regeneration starts with this fern. The only climber seen from mangroves is *Derris heterophylla* which was formerly known as *D. uliginosa*. It flowers from February to March. It differs from *D. uliginosa* in having more than seven leaflets (up to 11) and stalks that are not grooved. According to Untawale (1979) *D. trifoliata* and *D. pentaphylla* are found in the mangrove swamps of Goa and elsewhere.

Most of the mangroves start flowering only after the onslaught of the monsoon is over, sometime around August. This enables them to prepare seeds ready for disposal and establishment by the next monsoon period. The only monsoon flowering estuarine plants are *E. agallocha* and *H. beccarii*. The summer flowering mangroves are *Bruguiera*, *Avicennia* (Plate 4), *Sonneratia*, *Lumnitzera*, *Acanthus*, *Derris* and *Aegiceras*.

4. Zonation

It is known that the mangroves are usually found between the levels of equinoctial high tides and slightly below mean low tide. If the area available is large, the mangroves can be seen in zones or bands. However, occurrence of a species is not limited to only one zone. In the estuaries it was found that there were no distinct zonations. In fact no satisfactory explanation is yet available regarding zonation in mangroves. According to Macnae (1968) zonation is attributed to such factors as frequency and depth of tidal inundation, salinity of the water and soil, or soil maturity along the shore line. He further stated that zonation does not occur where rainfall is less than 1778 mm/year. Watson (1928) has stated that the zonation in the mangroves is most marked in areas such as Malaysia where the average tidal range is great and where there is high, relatively non-seasonal rainfall. Such conditions do not exist in the estuaries studied here and hence there is no definite zonation. However *Avicennia, Rhizophora* and *Sonneratia* grow towards the seaside while *Acanthus* and *Lumnitzera* always grow away from it. *Acanthus* forms pure and dominant patches where tides are not high and the area is covered by the influx of fresh water and industrial or city waste.

The associate tree species, the grasses and mud flat plants are shown in Table 1. *Suaeda nodiflora, S. fruticosa* and *Arthrocnemum indicum* seen at Bombay are not very common in the estuaries here. These succulent herbs are found in fields where sea water reaches them only at very high tide, when it converts them into salty ones. The only member of this region which is common in the two estuaries is *Sesuvium portulacastrum*.

The new areas in the form of islands or denudation of original stands provide for study of succession. The pioneers on newly formed islands are ribbon-like marine algae and *Cyperus* sp. which ultimately make way either for *Avicennia, Sonneratia* or *Excoecaria* which act as the pioneers. The dominance of *Sonneratia alba* on the fringes of estuaries at Ratnagiri and Chiplun is a noteworthy feature. Similarly such dominance of *R. mucronata* at Terekhol estuary and of *A. marina* at Bhatya are also worth mentioning. However in general the estuaries have a mixed mangrove vegetation.

5. Mineral constituents

A good review of the work done on physiology of mangroves has been given by Walsh (1974). In this work he has also covered mineral metabolism and chemical composition. He has stated that there is a paucity of literature concerning the elemental composition of the mangroves. This situation did not change very much in 1979. Our laboratory, however, has been engaged in the study of this aspect of the mangrove; our results are given in Tables 2, 3 and 4. These cover concentrations of major and minor elements and the seasonal variations in them. The estuaries in India are subject to distinct seasonal variations. This is reflected in the composition of the water inundating the estuaries, as well as the elements in the leaves of the estuarine plants.

Table 2 shows the concentrations of Ca, Mg, Na, K and Cl ions in the saline plants. The results have been reported in an earlier communication (Joshi, 1975). The present table, however, has additional information and new species included in it. The dominance of Na and Cl ions is obvious. This is accompanied by lesser amounts of Mg. The capacity of the saline plants to develop salt tolerance is due to their capacity to increase uptake of Ca and K ions in spite of richness of Na and Cl ions in the soil. This is explained in the works of Larsen (1967) and Rains and Epstein (1967) as well as in our results on several saline plants. The accumulating as well as the excreting types have higher Na and Cl ions. No general statement, however, can be made in this regard.

Seasonal variations in Na, K, Ca, Cl, total chlorophylls and N are shown in Table 3. Part of this data has been published (Jamale and Joshi, 1977); the other part is yet to be published. In all the plants, the lowest values are recorded in monsoon and the highest in the summer months. These observations are similar to those of Karmarkar (personal communication) for the halophytes and the mangroves in the Creeks of Bombay. The increase in K and Ca contents in the mangroves in summer may be for the development of salt tolerance. In *Acanthus*, however, high Ca values are recorded in winter and not in summer which has the lowest. This is difficult to explain.

The seasonal variations in the chlorophylls and N in three mangrove species have been investigated. In the mangroves the chlorophyll contents are higher during the winter and lower during the summer and monsoon. This may be due to light and temperature fluctuations. In summer there is bright light and high temperature while in monsoon the light is poor and temperatures are high. In summer salinity of sea water is high. These factors may

Table 2. Inorganic constituents in the leaves of mangroves and associate halophytes of estuaries in Maharashtra.

Species	Calcium	Magnesium	Sodium	Potassium	Chloride	Na/K
Salt excluding						
Rhizophora mucronata	1.28	0.76	2.57	1.26	4.47	2.04
Ceriops tagal	1.06	0.79	1.57	1.22	2.57	1.29
Bruguiera gymnorrhiza	1.23	0.89	4.13	1.33	3.87	3.11
Kandelia candel	1.28	0.729	1.20	1.80	2.67	0.67
Salt excreting						
Avicennia officinalis	1.18	1.15	5.30	1.63	3.88	3.25
Avicennia alba	1.18	0.88	3.11	1.65	5.124	1.88
Aegiceras corniculatum	0.46	0.32	2.35	0.35	3.66	6.71
Salt accumulating						
Acanthus ilicifolius	0.83	0.94	3.46	2.49	4.27	1.39
Lumnitzera racemosa	2.17	0.57	5.54	2.99	11.73	1.85
Sonneratia alba	1.58	0.73	5.05	0.78	9.94	6.47
Excoecaria agallocha	1.25	0.41	3.26	1.36	6.39	2.40
Associate mangrove spp.						
Clerodendrum inerme	2.48	0.69	5.49	3.02	6.24	1.82
Salvadora persica	5.88	1.43	2.75	1.96	9.29	1.40
Aeluropus lagopoides	0.24	1.64	2.75	1.60	3.64	1.12
Thespesia populena	0.16	0.04	2.02	0.62	2.03	3.25
(saline habitat)						
Succulent halophytes						
Sesuvium portulacastrum	0.37	0.54	9.9	2.97	8.64	3.33
Suaeda sp.	0.56	0.50	8.55	1.94	—	4.41
Glycophyte						
Zea mays	0.23	0.18	—	0.92	0.14	—
(value of Miller, 1938)						
Saline fern						
Acrosticum aureum	0.05	0.17	2.10	1.29	2.52	1.63
Substratum						
Mangrove soil	0.13	0.06	0.51	0.05	0.63	10.2
Sea water	0.83	0.67	10.81	0.38	18.11	28.44

Values expressed as g per 100 g of dry tissue/air dry soil in case of sea values per litre.

inhibit chlorophyll synthesis while moderate light and temperatures will promote chlorophyll synthesis. N contents are low in summer when light intensity, temperature and salinity are high and chlorophyll contents are low. It appears that salinity and temperature control nitrogen metabolism in the estuarine plants.

The literature on trace elements in the mangrove is still lacking. Walsh (1974) has given some values. Ramadas et al. (1975) have reported trace elements from the mangrove ecosystem and stated that the trace elements have effect on mangrove productivity. Table 4 records the value reported by Bhosale (1979). Mn and Mo are quite high while Co is less. Zn and Cu are within the normal ranges mentioned by Epstein (1972). It appears that there is no deficiency of trace elements in the mangroves.

6. Effect of fresh water on mangroves

In the Ratnagiri district the two estuaries, namely Terekhol and Vashishti, provide excellent areas where effect of fresh water on the mangroves can be studied. Terekhol estuary has an undisturbed estuarine ecosystem, while the construction of the Koyna dam and the emptying of its tail waters in the estuary of Vashishti has a considerable effect on the ecosystem. The mangroves from Chiplun onwards have been exposed to fresh water conditions

Table 3. Seasonal variations in Na, K, Ca, Cl, chlorophylls and nitrogen in mangroves.

Species	Season	Sodium	Potassium	Calcium	Chlorides	Total chlorophylls	Nitrogen
Sonneratia alba							
	Monsoon	4.32	0.77	0.81	6.75	70.81	0.93
	Winter	4.18	0.94	1.23	7.50	90.78	0.74
	Summer	5.61	0.94	1.47	9.85	87.77	0.60
Avicennia officinalis							
	Monsoon	1.91	0.35	0.44	1.77	65.00	2.51
	Winter	3.37	1.49	1.22	2.85	57.00	2.59
	Summer	5.30	1.63	1.18	3.88	90.00	2.77
Avicennia alba							
	Monsoon	2.82	0.82	0.38	3.24	100.26	1.04
	Winter	3.09	1.36	0.90	4.26	114.12	0.83
	Summer	3.11	2.01	1.19	4.78	108.92	0.79
Aegiceras corniculatum							
	Monsoon	1.27	0.84	0.39	1.42	126.00	2.45
	Winter	2.34	0.70	0.57	2.75	71.00	1.06
	Summer	2.65	0.35	0.46	3.66	77.00	2.50
Excoecaria agallocha							
	Monsoon	1.44	1.57	1.15	3.20	112.27	1.24
	Winter	3.34	1.33	1.25	5.66	139.36	0.99
	Summer	3.55	1.87	1.36	6.57	135.58	0.51
Rhizophora mucronata							
	Monsoon	3.35	0.60	0.71	3.80	—	—
	Winter	3.36	0.61	1.00	4.26	—	—
	Summer	3.60	0.66	1.02	5.51	—	—
Acanthus ilicifolius							
	Monsoon	2.19	2.17	1.74	1.28	75.00	6.27
	Winter	3.42	2.81	1.95	4.27	37.00	3.80
	Summer	3.47	2.49	0.83	4.85	50.00	4.17

Values expressed as g per 100 g of dry tissue.
Chlorophyll values expressed as mg per 100 g fresh tissue.

Table 4. Distribution of trace elements in the leaves of mangroves.

Species	Mn	Mo	Zn	Cu	Co
Rhizophora mucronata	32.2	11.5	11.5	4.1	2.3
Bruguiera gymnorrhiza	46.5	11.6	12.8	6.0	2.3
Ceriops tagal	82.2	11.1	11.1	5.3	2.2
Avicennia officinalis	158.7	15.9	17.5	9.2	3.2
Lumnitzera racemosa	29.4	29.4	29.4	4.1	5.9
Aegiceras corniculatum	25.8	8.3	8.3	4.3	1.7
Acanthus ilicifolius	96.7	16.7	24.7	9.0	3.3
Aeluropus lagopoides	54.3	7.1	107.1	6.6	1.4
Clerodendrum inerme	133.3	15.9	36.5	10.8	3.2

Values expressed as ppm (mg/1000 g dry wt).

Table 5. Major inorganic constituents in mangroves and water samples at Terekhol and Chiplun.

Species	Terekhol					Chiplun (Vashishti)				
	Na	K	Ca	Cl	Na/K	Na	K	Ca	Cl	Na/K
Water sample	2.76	0.14	0.15	5.31	19.7	0.016	0.001	0.001	0.04	16.0
Sonneratia alba	2.04	1.44	1.56	4.12	1.4	0.96	0.92	1.18	2.41	1.1
Rhizophora mucronata	2.44	0.76	1.14	4.41	3.2	1.6	0.48	0.75	2.9	3.3
Kandelia candel	2.0	0.84	0.68	2.84	2.4	0.72	1.24	0.85	2.27	0.6
Aegiceras corniculatum	1.72	0.9	0.5	2.27	1.9	0.48	0.4	0.27	0.71	1.2
Avicennia officinalis	2.76	1.12	0.84	3.2	2.5	1.6	1.08	0.44	2.12	1.5
Acanthus ilicifolius	2.68	1.64	1.12	5.11	1.6	0.68	2.4	1.28	3.33	0.3
Excoecaria agallocha	0.8	1.9	1.72	2.56	0.4	0.1	0.52	1.8	0.57	0.2

Values in water sample are expressed as g per litre and in leaves as g per 100 g of dry tissue.

for the last 12 years. The analyzed results have been reported by Joshi (1976), Joshi and Jamale (1975) and Joshi and Shinde (1978), and are shown in Table 5.

The results show that when mangroves grow in non-saline areas they increase their uptake of potassium and calcium which are supposed to play an important role in salt tolerance. Walsh (1974) and McMillan (1975) have summarized the views on the growth of mangroves under saline and non-saline conditions. Stern and Voight (1959), Connor (1969) and McMillan (1971) have stressed the importance of salinity for a good development of mangroves. However, many investigators (Bowman, 1917; Warming and Vahl, 1925; Rosevear, 1947; Egler, 1948; Daiber, 1960) are of the opinion that the mangroves grow equally well under fresh water conditions. It appears that there are two views regarding the salinity requirements of the mangroves. It was observed that under fresh water conditions there is an invasion and competition from fresh water marsh plants. At Chiplun these plants have outclassed the mangroves in competition. The areas occupied by *Rhizophora* and *Sonneratia* are gradually occupied by *Acanthus* and *Clerodendrum* first and by fresh water sedge like *Cyperus rotundus* and *Asteracantha longifolia* later. Waisel (1972) has stated that genera such as *Sonneratia* disappear in habitats which are frequently eluted with fresh water. Our observations are on similar lines and suggest that for good growth of an estuarine ecosystem a definite or minimum amount of salinity is essential.

7. Photosynthesis and productivity

Waisel (1972) has reported the effect of salinity on CO_2 assimilation. The succulent members like *Atriplex* and *Suaeda* show a deviated C_4 pathway. However this information on mangroves is meagre. In his work, Walsh (1974) reviewed energy relationships in mangroves. It shows that most of the work done is on productivity and a little on initial products of CO_2 fixation. Webb and Burley (1965) investigated the dark $^{14}CO_2$ fixation in obligate and facultative salt marsh halophytes. They found that in the dark, $^{14}CO_2$ incorporation in the halophytes is more in the amino acids. Shomer-Ilan and Waisel (1973) have found that in *Aeluropus litoralis*, aspartate is the initial product of light CO_2 assimilation and PEPCase is the salt dependent enzyme. It was found by Waghmode and Joshi (1979) that even in *A. lagopoides*, aspartate is the initial product, PEP-Case is the dominant enzyme and Kranz syndrome with pyruvate-pi-dikinase is very much present. Ramadas and Raghvendra (1977) have reported C_4 syndrome and the aspartate former type of photosynthesis in saline sandy grass such as *Spinifex squarrosus*.

The process of photosynthesis in the estuarine plants has been investigated in our laboratory for the last ten years (Joshi et al., 1974 and 1975a; Bhosale and Karadge, 1975; Joshi 1976; Waghmode and Joshi, 1979). The results are shown in Table 6. It was found that the stomata are open during early hours of the day and CO_2 assimilation takes place before the light intensity increases.

Table 6. Products of short-term photosynthetic $^{14}CO_2$ fixation in saline plants. Period of fixation 10 sec unless mentioned otherwise.

Species	Aspartate	Alanine	Glutamate	Malate	Citrate	Succinate	PEP + PGA	Origin + others	Glycollate	Glycerate
Sonneratia alba (5 sec)	48.58	32.55	—	4.72	—	—	14.15	—	—	—
Excoecaria agallocha	22.86	48.58	—	11.43	11.43	5.7	—	—	—	—
Aegiceras corniculatum (5 sec)	36.36	56.82	—	—	—	—	6.82	—	—	—
Ceriops tagal (5 sec)	31.74	33.33	14.28	—	—	—	—	20.63	—	—
Lumnitzera racemosa (5 sec)	15.78	58.42	2.10	—	—	—	—	23.68	—	—
Aeluropus lagopoides	45.00	12.49	21.93	16.98	1.53	—	0.36	—	—	—
Halophilla beccarii	11.30	18.60	—	5.80	—	—	29.80	—	18.9	14.40

Values are expressed as percentage of total radioactivity counted on chromatogram.

There is no Kranz anatomy but PEPCase is more active than RuBPCase. The initial products of photosynthesis are aspartate and alanine. Pyruvate-pi-dikinase, an enzyme of C_4 pathway is also present (Joshi et al., 1980). The path of photosynthesis in sea grass, *Halophilla beccarii*, has also been investigated (Ghevade and Joshi, 1980). It has C_3 and C_4 features of photosynthesis and hence no definite pathway can be attributed.

The values for productivity of the mangroves are given by Walsh (1974). The non-mangrove component of the estuarine ecosystem contributes only up to 15%. Teas (1977) has reported 1.28 to 10.7 tonnes/ha/yr (dry wt of organic matter in metric tonnes per hectare per year) for the Florida mangroves. Golley and Lieth (1972) estimated the average NPP for tropical forests to be 25.3 tonnes/ha/yr. Hence Teas remarks that the mangrove forests are not highly productive in comparison with other tropical forests. Information on the productivity of estuarine vegetation in India is scanty (Untawale, 1979). According to him the average rate of growth of the mangrove species is 30–35 cm in height 1–2 cm in diameter per year. He has given a figure of 4.2 to 10.05 gc/m^3/day[1] for phytoplankton in Pichavarum mangroves worked out by Krishnamurthy (1975) while the values for Untawale's laboratory are 2.24 to 2.9 gc/m^2/day[1]. Bhosale and Karadge (1975) reported photosynthetic efficiency and productivity

of mangroves of western India. They found *Avicennia* and *Sonneratia* to be more efficient CO_2 assimilators than the rest. The results show that the mangroves are not a highly productive ecosystem. However, this is only in terms of dry matter production. If we think in terms of input and long-term benefits, the estuarine ecosystem is the best system under those natural ecological conditions. This is confirmed by a recent report by Glenn (University of Arizona; personal communication) in which he states that *Atriplex* spp., *Batis maritima* and *Salicornea europaea* produce 895–1365 g/m^2 of dry matter per year and are as good as any crop system.

8. Exploitation of estuarine ecosystem

Walsh (1974) has given details about the silviculture of the mangroves in several parts of Southeast Asia for exploitation of their commercial products. Teas (1979) has also given good data on the silviculture of mangroves. Walsh (1977) has again given a detailed account of exploitation of the mangroves, while Queen (1977) has covered economic uses of salt marshes.

When population and industrial pressures were not present the estuarine ecosystem used to adjust itself to the small pressures exerted by human intervention. These ecosystems have, however, in

Table 7. Monthly variations in total polyphenols in mangroves.

Species	July/Aug	Sept/Oct	Nov/Dec	Jan/Feb	Mar/Apr	May/June
Sonneratia alba						
Leaves	1.914	3.000	1.305	2.390	1.460	2.672
Stem	4.750	3.219	1.925	2.200	3.290	4.042
Bark	6.210	4.690	5.500	5.175	7.020	5.500
Rhizophora mucronata						
Leaves	1.568	3.420	6.000	5.200	9.300	7.680
Stem	3.570	3.844	3.135	5.720	5.406	4.950
Bark	5.483	11.390	7.770	9.120	13.300	10.395
Avicennia alba						
Leaves	1.275	3.120	0.969	2.040	2.010	1.980
Stem	0.500	2.000	0.480	0.868	0.882	0.666
Bark	0.600	2.200	1.540	1.716	1.785	1.600
Excoecaria agallocha						
Leaves	3.913	6.200	3.730	2.560	5.712	3.888
Stem	4.860	4.420	1.739	1.352	1.650	0.988
Bark	5.312	3.659	8.140	7.030	8.540	6.820

Values expressed as g per 100 g of fresh weight.

the last few years come under tremendous pressure, especially in developing countries. The estuaries have become the cities' drainages, areas for shrimp and other marine cultures, agriculture, industries and mining. In bigger cities like Bombay and Calcutta these are the areas where slums develop and stabilize. The Dacca Seminar (1979) has given considerable thought to this problem while a UNESCO Report (1976) has given the direction into which constructive steps can be taken.

Untawale (1979) has prepared a report describing possibilities for human use of the mangrove environment and the management implications of this for India. One of the uses is the exploitation of the bark of certain mangrove species as a source for tannin. The monthly variations in total polyphenols in the mangroves were studied in our laboratory (Jamale and Joshi, 1978) and shown in Table 7. *Rhizophora* sp. appears to be the best one and October and March appear to be the best for harvest.

In the case of an estuarine ecosystem, the matter of multipurpose usage is both possible and essential because this ecosystem is a meeting point between terrestrial and marine life. At a time when attempts are being made to restore coastlines in Florida (Teas, 1977) by serial planting (Teas and Jurgens, 1978) and when countries like Saudi Arabia are eager to restore the mangrove vegetation which existed there a few centuries ago, it would be worthwhile for a developing country like India – with scientific manpower and technical know-how at its disposal – to have a strong, constructive policy for the care and use of its large number of estuarine ecosystems.

Acknowledgements. The results included here form part of Ph.D. theses written by several students working in this laboratory. The help rendered by Shri A.P. Waghmode and S.Y. Kotmire in preparing the manuscript is gratefully acknowledged. We are also grateful to the Department of Science and Technology of the Indian Government for their generous grant to the Project (No. 7(15)/76 SERC) entitled 'Photosynthesis and marine environment'. Through them our work was made possible.

Literature cited

Adriani, M.J. 1958. Halophytes. In *Encyclopedia of Plant Physiology* 4: pp. 709–736. Springer-Verlag, Berlin.

Bhosale, L.J. 1979. Distribution of trace elements in the leaves of mangroves. *Indian J. Mar. Sci.* 8: 58–59.

Bhosale, L.J. and Karadge, B.A. 1975. Photosynthetic efficiency and productivity of mangroves of western India. *Bull. Dept. Mar. Sci. Univ. Cochin* 7: 205–212.

Blasco, F. 1975. *The Mangroves of India*. French Institute, Pondichery, India.

Blasco, F. 1977. Outlines of ecology, botany and forestry of the mangles of the Indian subcontinent. In *Ecosystems of the World*. I. *Wet Coastal Ecosystems*, ed. V.J. Chapman, pp. 241–260. Elsevier Scientific Publ. Co., Amsterdam.

Bowman, H.H.M. 1917. Ecology and physiology of the Red mangrove. *Proc. Am. Philos. Soc.* 56: 589–672.

Chapman, V.J. 1976. *Mangrove Vegetation*. Strauss & Cramer, Germany.

Chapman, V.J. 1977. *Ecosystems of the World*. Vol. 1. *Wet Coastal Ecosystems*. Elsevier Scientific Publ. Co., Amsterdam.

Connor, D.J. 1969. Growth of grey mangrove (*Avicennia marina*) in nutrient culture. *Biotropica* 1: 36–40.

Daiber, F.C. 1960. Mangroves: The tidal marshes of the tropics. *Univ. Delaware Estuarine Bull* 5: 10–15.

Egler, F.E. 1948. The dispersal and establishment of red mangrove in Florida. *Carbibb. For.* 9: 299–310.

Epstein, E. 1972. *Mineral Nutrition in Plants – Principles and Perspectives*. John Wiley and Sons, New York.

Ghevade, K.S. and Joshi, G.V. 1980. Photosynthetic and photorespiratory carbon metabolism in the sea grass, *Halophila beccarii* Ascher S. *Indian J. exp. Biol.* 18: 1344–1345.

Golley, F.B. and Lieth, H. 1972. The bases of tropical production. In *Tropical Ecology*, eds. F.B. Golley and R. Misra, pp. 1–26. Univ. Georgia, Athens, GA.

Jamale, B.B. and Joshi, G.V. 1977. Seasonal variations in organic and inorganic constituents in mangroves. *J. Shivaji Univ. (Science)* 17: 131–138.

Jamale, B.B. and Joshi, G.V. 1978. Effect of age on mineral constituents, polyphenols, polyphenol oxidase and peroxidase in leaves of mangroves. *Indian J. exp. Biol.* 6: 117–119.

Joshi, G.V. 1975. Physiology of salt tolerance in plants. *Biovigyanam* 1: 21–39.

Joshi, G.V. 1976. Early products of photosynthesis in plants exposed to salt and water stresses. *The Biochem. J.* 3: 33–43.

Joshi, G.V., Bhosale, L.J., Jamale, B.B. and Karadge, B.A. 1975a. Photosynthetic carbon metabolism in mangroves. In *Proc. Int. Symp. Biol. and Management of Mangroves* Vol. II, eds. G.E. Walsh, S.C. Snedaker and H.J. Teas, pp. 579–594. Int. Food Agric. Sci., Univ. Florida, Gainesville.

Joshi, G.V. and Jamale, B.B. 1975. Ecological studies in mangroves of Terekhol and Vashishti rivers. *Bull. Dept. Mar. Sci. Univ. Cochin* 7: 751–760.

Joshi, G.V., Jamale, B.B. and Bhosale, L.J. 1975b. Ion regulation in mangroves. In *Proc. Int. Symp. Biol. and Management of Mangroves*, Vol. II, eds. G.E. Walsh, S.C. Snedaker

and H.J. Teas, pp. 595–607. Int. Food Agric. Sci., Univ. Florida, Gainesville.

Joshi, G.V., Karekar, M.D., Gowda, C.A. and Bhosale, L.J. 1974. Photosynthetic carbon metabolism and carboxylating enzymes in algae and mangrove under saline conditions. *Photosynthetica* 8: 51–52.

Joshi, G.V. and Shinde, S.D. 1978. *Ecogeographical Studies in Terekhol and Vashisti Rivers*. Shivaji University Publication, Kolhapur.

Joshi, G.V., Sontakke, S.D. and Bhosale, L.J. 1980. Studies in photosynthetic enzymes from mangroves. *Bot Marina* 23: 745–747.

Karmarkar, S.M. 1978. (Personal communication).

Krishnamurthy, K. 1975. Aspects of an Indian mangrove forest. In *Proc. Int. Symp. Biol. and Management of Mangroves*, Vol. II, eds. G.E. Walsh, S.C. Snedaker and H.J. Teas, pp. 88–95. Int. Food Agric. Sci., Univ. Florida, Gainesville.

Kunstadter, P. and Snedaker, S.C. (eds.). 1978. *UNESCO Seminar on Human Uses of the Mangrove Environment and Management Implications*. Held in Dacca, Bangladesh, December, 1978. (In preparation.).

Larsen, H. 1967. Biochemical aspect of extreme halophilism. *Adv. In Microbiol.* 1: 97–132.

Macnae, W. 1968. A general account of the fauna and flora of mangrove swamps and forests in the Indo-West-Pacific region. *Adv. Mar. Biol.* 6: 73–270.

McGill, J.T. 1958. Map of coastal landforms of the world. *Geog. Rev.* 48: 402–405.

McMillan, C. 1971. Environmental factors affecting seedling establishment of the black mangrove on the central Texas coast. *Ecology* 52: 927–930.

McMillan, C. 1975. Interaction of soil texture with salinity tolerances of *Avicennia germinans* and *Laguncularia racemosa* from North America. In *Proc. Int. Symp. Biol. Management of Mangroves*, Vol. II, eds. G.E. Walsh, S.C. Snedaker and H.J. Teas, pp. 561–569. Int. Food Agric. Sci., Univ. of Florida, Gainesville.

Pannier, F. 1979. Mangroves impacted by human-induced disturbances: A case study of the Orinoco delta mangrove ecosystem. *Environmental Management* 3: 205–216.

Queen, W.H. 1977. Human uses of salt marshes. In *Wet Coastal Ecosystems*, ed. V.J. Chapman, pp. 363–368. Elsevier Scientific Publishing Co., Amsterdam.

Rains, D.W. and Epstein, E. 1967. Preferent absorption of potassium by leaf tissue of the mangrove *Avicennia marina*: an aspect of halophytic competence in coping with salt. *Aust. J. Biol. Sci.* 20: 847–857.

Ramadas, V.S. and Raghvendra, A.S. 1977. Kranz leaf anatomy and C_4 dicarboxylic acid pathway of photosynthesis in *Spinifex squarrosus* L. *Indian J. exp. Biol.* 15: 645–648.

Ramadas, V.S., Rajendran, A. and Venugopalan, U.K. 1975. Studies on trace elements in Pichavaram mangroves (south India). In *Proc. Int. Symp. Biol. and Management of Mangrove*, Vol. I, eds. G.E. Walsh, S.C. Snedaker and H.J. Teas, pp. 96–114. Int. Food Agric. Sci., Univ. Florida, Gainesville.

Reimold, R.J. and Queen, W.H. 1974. *Ecology of Halophytes*.

Academic Press, New York.

Rosevear, D.R. 1947. Mangrove swamps. *Farm For.* 8: 23–30.

Shomer-Ilan, A. and Waisel, Y. 1973. The effect of sodium chloride on balance between the C_3 and C_4 carbon fixation pathways. *Physiol. Plant.* 29: 190–193.

Stern, W.L. and Voight, G.K. 1959. Effect of salt concentration on growth of red mangrove in culture. *Bot. Gaz.* 121: 36–39.

Teas, H.J. 1976. Productivity of Biscayne bay mangroves. *Biscayne Bay Symposium* I. Special report No. 5: 103–111.

Teas, H.J. 1977. Ecology and restoration of mangrove shorelines in Florida. *Environmental Conservation* 4: 51–58.

Teas, H.J. 1979. Silviculture with saline water. In *The Biosaline Concept*, ed. Alexander Hollaender, pp. 117–161. Plenum Publ. Corporation, U.S.A.

Teas, H.J. and Jurgens, W. 1978. Aerial planting of Rhizophora mangrove propagules in Florida. *Proc. of fifth Annual Conference on Restoration of Coastal Vegetation in Florida*, Florida. pp. 1–25.

Teas, H.J., Wanless, H.R. and Chardon, R. 1976. Effects of man on the shore vegetation of Biscayne bay. *Biscayne Bay Symposium* I. Special report No. 5: 133–156.

UNESCO, Panel Report (Mangroves). 1976. Meeting held at Phucket, Thailand.

Untawale, A.G. 1979. Country report for India. Regional Seminar on human uses of the mangrove environment and management implications. *UNESCO Dacca Seminar*: 1–22.

Untawale, A.G. and Jagtap, T.G. 1977. A new record of *Halophilla beccarii* (Ascher). *Mahasagar* 10: 91–94.

Waghmode, A.P. and Joshi, G.V. 1979. Kranz leaf anatomy and C_4 dicarboxylic acid pathway of photosynthesis in *Aeluropus lagopoides* L. *Indian J. exp. Biol.* 17: 606–607.

Waisel, Y. 1972. *Biology of Halophytes*. Academic press, London.

Walsh, G.E. 1974. Mangroves: a review. In *Ecology of Halophytes*, eds. R.J. Reimold and W.H. Queen, pp. 51–174. Academic press, London.

Walsh, G.E. 1977. Exploitations of mangal. In *Wet Coastal Ecosystems*, ed. V.J. Chapman, pp. 347–362. Elsevier Scientific Publishing Co., Amsterdam.

Walsh, G.E., Snedaker, S.C. and Teas, H.J. 1975. *Proc. Int. Symp. Biol. and Management of Mangroves*, Vols. I and II, Int. Food. Agric. Sci., Univ. of Florida, Gainesville.

Walter, H. 1973. *Vegetation of the Earth in Relation to Climate of the Eco-physiological Conditions*. Springer-Verlag, New York Science Library, 15.

Warming, E. and Vahl, M. 1925. *Ecology of Plants*. Oxford Univ. Press, London.

Waison, J.E. 1928. Mangrove forests of the Malay Peninsula *Malay. For. Rec.* 6: 1–275.

Webb, K.L. and Burley, J.W.A. 1965. Dark fixation of $^{14}CO_2$ by obligate and facultative salt marsh halophytes. *Can. J. Bot.* 43: 281–285.

CHAPTER 3

The biogeography of mangroves

HARTMUT BARTH

FB 5 Biologie/Chemie Arbeitsgruppe Ökologie,
Universität Osnabrück, D-4500 Osnabrück, FRG

1. Introduction and definition

The determination of a number of mangrove species in a particular location of the world requires first the definition of the word 'mangrove': This term has a double meaning: 1) it is used for a mainly woody *plant formation*, distributed in the tropical or subtropical tidal zone between the lowest and highest tide water mark; 2) it is used for a *plant species* of this formation. These species are usually woody.

Some authors distinguish between the 'mangal', 'mangrove community', or 'mangrove formation', when they mean the first, and 'mangrove' only, when pertaining to the plant species. I shall adhere to this definition and use in the following text 'mangal' and 'mangrove' as described above.

The definition given above for the mangal is not precise, since typical mangrove trees (as e.g. *Avicennia marina*) do not only grow along the coast or along the estuaries of tropical rivers, but also as so-called inland mangrove. On the other hand, plants of the tropical rainforest, normally growing in fresh water stands, may sometimes extend into brackish water or even into the saltwater region of the mangal. Such a mixture of trees is often found in the landward fringe of the mangal. This fact makes it difficult to draw the border line between the mangal and neighbouring wetland plant communities. Some authors consider for example a number of the brackish water tolerating *Barringtonia* species to be

a mangrove while others consider them as associate species.

Many authors differentiate between the typical mangrove on the one hand and its accompanying flora on the other hand (e.g. 'principal' or 'typical mangrove' or 'core mangrove' and 'associate mangrove'). Others list the mangrove species alphabetically and/or in the order of 'zones' dominated by particular mangrove species.

The subdivision into typical mangroves and associates is treated very differently in literature. Undisputed as typical mangroves are the genera *Bruguiera, Ceriops, Kandelia,* and *Rhizophora* of the Rhizophoraceae, the genus *Avicennia* of the Avicenniaceae (formerly included in the Verbenaceae) as well as the genus *Sonneratia* of the Sonneratiaceae. The Combretaceae *(Conocarpus, Laguncularia,* and *Lumnitzera)* are partly considered typical mangroves and partly to be associates.

In a number of species lists and distribution maps of mangroves one particular species, the Bignoniaceae *Dolichandrone spathacea* (L.f.) K.Sch., growing on the banks of tidal rivers and estuaries, is left out. Van Steenis (1977, p. 142) in the Flora Malesiana presents the following distribution of this tree of 5–20 m height: 'From the coast of Malabar throughout tropical SE Asia and the whole of Malesia to New Guinea, Micronesia (W. Carolines: Korror; Yap; Tomil I.), the Solomons, the New Hebrides and New Caledonia, not found in Australia and Polynesia.'

The main problem with this chapter was to decide which associates had to be included as mangroves. Should, for example, *Acrostichum aureum*, the great pantropical fern (but not a tree fern!), which occurs in nearly all mangals, be listed and enumerated, or should only the trees and woody shrubs be taken into account? If *Acrostichum*, as a typical mangal associate, is enumerated, where then should one draw the line between mangal associates and plants of other formations of the salt marsh or dune vegetation?

A similar problem exists with the epiphytes and climbers. Therefore we have to leave it open as to what should be included at the end of the species list and what should be left out, until better classification criteria become available.

At least there are some taxonomic problems of particular interest regarding the species distribution. One important example is the taxonomy and distribution of the genus *Rhizophora* in the western Pacific, since there seems to be an overlap between the group of taxa with an 'Indo-Pacific' distribution ('eastern mangroves', i.e. *R. apiculata* Bl., *R. mucronata* Lamk., *R. stylosa* Griff., and *R. x lamarckii* Montr.) and the group with an 'Atlantic' distribution ('western mangroves', i.e. *R. mangle* L., *R. harrisonii* Leechm. and *R. racemosa* G.F.W. Mey.) (Ding Hou, 1960; Tomlinson, 1978). As Tomlinson (1978) pointed out, a distinct form of *R. mangle* recognized and named as either a variety (*R. mangle* var. *samoensis* (Hochr.) Salv.) or a separate species (*R. samoensis* (Hochr.) Salv.) produces in some islands of Australasia (found in Fiji and New Caledonia) a hybrid population with *R. stylosa* which he recognizes as *R. x selala* (Salv.) Toml. But the taxonomic status of *R. mangle* or *R. x samoensis* is still disputed (Breteler, 1977).

2. Distribution of mangal

The occurrence of mangal around the world can be divided into two main distribution groups, the species-rich 'Indo-Pacific' or 'eastern' mangal and the 'Atlantic' or 'western' mangal. The Cape of Good Hope in South Africa seems to be a strict barrier between these two distribution areas. Some mangrove genera of the East African mangal are indeed present in West Africa, but represented by well distinguishable species. For example, from the genus *Avicennia* the East African species *A. marina* is replaced by *A. germinans* (*A. africana*) in West Africa. Other East African mangrove genera are absent in West Africa, e.g. *Bruguiera*, *Ceriops*, *Sonneratia* and *Xylocarpus*.

The Indo-Pacific region can be subdivided into six distribution zones, East Africa, Red Sea/Persian Gulf, India, Malaysia/Guinea, Micronesia/Macronesia and Australia/New Zealand, the Atlantic region into three subgroups respectively, Pacific America, Atlantic America and West Africa (Table 3). The mangal of Hawaii is not native, all four mangrove species there have been introduced (Walsh, 1974).

The geographical distribution of mangal is presented in Fig. 1, which contains also the number of mangrove species, including the main mangal associates, from the various growth localities (Table 2). A detailed list of the species composition of the mangal at the locations listed in Table 2 is given in Table 4. A taxonomic review of the mangrove species and associates is presented in Table 1.

Also drawn in the distribution map of mangal (Fig. 1) are sea surface temperatures in form of the 'calendar month sea surface isotheres' of 24°C (= 75°F), 27°C (= 80°F) and 30°C (= 85°F) according to Hutchins and Scharff (1947). The calendar month sea surface isothere means the isotherm of maximum warming over all 12 months of the year. Polewards of the e.g. 24°C isotheres the sea surface temperatures are everywhere and every time colder than 24°C.

As is shown in Fig. 1, over nearly the whole range of mangal distribution the sea surface calendar month isotheres of 24°C correspond very well to the latitudinal limits of mangrove growth except the north Atlantic coasts, where the 27°C isothere corresponds better to the distribution limits.

East Africa

This distribution zone includes the East-African coast from Tanzania to Natal (no records of mangroves for the southeast coast of Somalia), Madagascar and Seychelles. Good descriptions of the

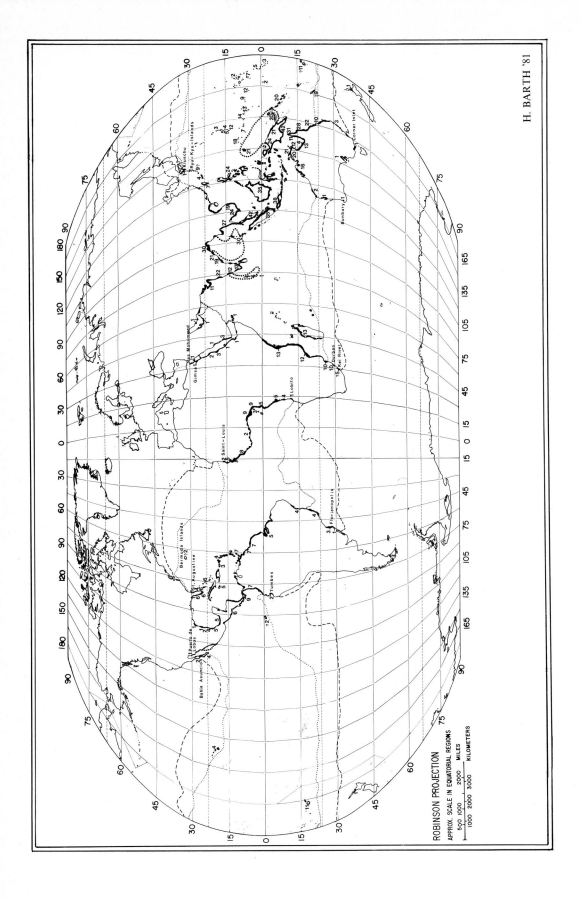

Fig. 1. World distribution of mangal. The numbers refer to the number of mangrove taxa reported in the literature from the respective area. - - - - 24° C; ——— 27° C; .... 30° C calendar month sea surface isotheres from Hutchins and Scharff (1947).

Table 1. Taxonomy of mangrove species including the main mangal associates (+) after several authors, e.g. Breteler (1977), Chapman (1976), Ding Hou (1958, 1960, 1978), Exell (1954), Graham (1964), Lecompte (1969), Jonker (1959), Keay (1953), Savory (1953), Stearn (1958), Tomlinson (1978), Tomlinson and Womersley (1976) and Van Steenis (1958, 1977).

Acanthaceae
Acanthus ebracteatus Vahl.
   (syn. *A. ilicifolius* Lour.)
Acanthus ilicifolius L.
   (syn. *A. doloarin* Blanco)

Apocynaceae
? *Cerbera floribunda* K. Sch.
Cerbera manghas L.
   (syn. *C. lactaria* Buch.-Ham)
   (syn. *Elcana seminuda* Blanco)
Cerbera odollam Gaertn.
   (syn. *C. forsteri* Seem.)
   (syn. *C. laurifolia* Lodd.)
   (syn. *Excoecaria ovatifolia* Noronha)

Avicenniaceae
Avicennia alba Bl.
   (syn. *A. acuminata* Cornw.)
   (syn. *A. albida* Bl. var. *latifolia* Mold.)
   (syn. *A. marina* var. *alba* Bakh.)
   (syn. *A. officinalis* Lam.)
   (syn. *A. officinalis* var. *alba* Clarke)
   (syn. *A. resinifera* Griff. (non Forst.))
   (syn. *A. spicata* Kuntze)
   (syn. *A. tomentosa* Bl.)

? *Avicennia balanophora* Stapf & Mold.

Avicennia eucalyptifolia Zipp. ex Miq.
   (syn. *A. alba* Karsten)
   (syn. *A. alba* var. *acuminatissima* Merr.)
   (syn. *A. officinalis* var. *acuminata* Domin.)
   (syn. *A. officinalis* var. *eucalyptifolia* Valet)

Avicennia germinans (L.)Stearn.
   (? syn. *A. africana* P. Beauv.)
   (syn. *A. americana* Nutt.)
   (syn. *A. angustifolia* Hornem.)
   (syn. *A. germinalis* Soukup.)
   (syn. *A. germinans* L.)
   (syn. *A. lamarckiana* Presl.)
   (syn. *A. nitida* Jacq.)
   (syn. *A. nitida* var. *africana* Chapm.)
   (syn. *A. nitida* Beebe)
   (syn. *A. nitida* var. *angustata* Forst.)
   (syn. *A. oblongifolia* Nutt.)
   (syn. *A. officinalis a nitida* Kuntze)
   (syn. *A. officinalis β lanceolata* Kuntze)
   (syn. *A. papulosa* Orst.)

   (syn. *A. tomentosa* var. *acutifolia* Blanchet)
   (syn. *A. tomentosa* var. *owarensis* Walp.)
   (syn. *Hilairanthus nitidus* van Tiegh.)
   (syn. *Hilairanthus tomentosus* van Tiegh.)

Avicennia lanata Ridl.
   (syn. *A. officinalis* var. *spathulata* Kuntze)
   (syn. *A. officinalis* var. *spathulata* f. *tomentosa* Kuntze)

Avicennia marina (Forsk.) Vierh.
   (syn. *A. alba* Wight)
   (? syn. *A. intermedia* Griff., Watson)
   (syn. *A. marina* f. *typica* Bakh.)
   (syn. *A. marina* var. *acutissima* Stapf & Mold.)
   (syn. *A. marina* var. *anomala* Mold.)
   (syn. *A. marina* var. *australasica*
   (syn. *A. marina* var. *intermedia* Bakh.)
   (syn. *A. marina* var. *marina* (Forsk.) Bakh.)
   (syn. *A. marina* var. *resinifera* (Forst.) Bakh.)
   (syn. *A. mindanaense* (Elm.) Prain)
   (syn. *A. nitida* Blanco)
   (syn. *A. officinalis* Miq.)
   (syn. *A. officinalis* Maxim)
   (syn. *A. officinalis* var. *nigra* Cowan)
   (syn. *A. officinalis* var. *ovatifolia* Kuntze)
   (syn. *A. officinalis* var. *ovatifolia* f. *flaviflora* Kuntze)
   (syn. *A. officinalis* var. *ovatifolia* f. *tomentosa* Kuntze)
   (syn. *A. officinalis* var. *resinifera* Biswas)
   (syn. *A. officinalis* (L.) var. *rumphiana* (Hall.)Bakh.)
   (syn. *A. resinifera* Forst)
   (syn. *A. resinosa* Forsk.)
   (syn. *A. rumphiana* Hallier)
   (syn. *A. sphaerocarpa* (Stapf.)Ridl.)
   (syn. *A. tomentosa* Blume)
   (syn. *A. tomentosa* var. *arabica* Walp.)
   (syn. *A. tomentosa* Blanco)
   (syn. *A. tomentosa* var. *australasica* Walp.)
   (syn. *A. tomentosa* R.Br.)
   (syn. *Racka ovata* Roem. & Schult.)
   (syn. *Racka torrida* J.F. Gmel.)
   (syn. *Sceura marina* Forsk.)
   (syn. *Trichorhiza lechenaultii* Miq.)

Avicennia officinalis L.
   (syn. *A. obovata* Griff.)
   (syn. *A. obtusifolia* Wall.)
   (syn. *A. oepata* Hamilton)
   (syn. *A. officinalis* var. *tomentosa* Cowan)
   (syn. *A. tomentosa* Willd.)
   (syn. *A. tomentosa* var. *asiatica* Walp.)

Table 1 *(continued)*.

Avicennia schaueriana Stapf & Leechm.
(syn. *A. nitida* var. *trinilensis* Mold.)
(syn. *A. tomentosa* Schauer)

Avicennia tonduzii Mold.
(syn. *A. tonduzzi* Cuatrecasas)

Bignoniaceae (after Van Steenis, 1977)
Dolichandrone spathacea (L.f.)K.Schn.
(syn. *Bignonia javanica* Thunb.)
(syn. *Bignonia longiflora* Willd. ex DC.)
(syn. *Bignonia longissima* Lour., non Jacq.)
(syn. *Bignonia spathacea* L.f.)
(syn. *Dolichandrone crispa* Seem.)
(syn. *D. longissima* K. Sch.)
(syn. *D. rheedii* Seem.)
(syn. *Lignum equinum* Rumph.)
(syn. *Pongelia longiflora* Rafin)
(syn. *Spathodea diepenhorstii* Miq.)
(syn. *Spathodea longiflora* Vent. ex DC.)
(syn. *Spathodea loureiriana* DC.)
(syn. *Spathodea luzonica* Blanco)
(syn. *Spathodea rheedii* Spreng.)

+ **Bombacaceae** (after Chapman, 1976)
Camptostemon philippinense (Vidal)Becc.
Camptostemon schultzii Mast.

Combretaceae (Terminaliaceae)
(after Chapman, 1976; Exell, 1954; Graham, 1964;
Lecompte, 1969)
Conocarpus erecta L.
(syn. *C. erectus* var. *arborea* Griseb.)
var. *argentea* Millsp.)
var. *procumbens* Jacq.)
var. *sericea* Griseb.)
(syn. *C. sericea* Forst.)
Laguncularia racemosa (L.) Gaertn.
(syn. *Conocarpus racemosa* L.)
Lumnitzera littorea (Jack)Voigt
(syn. *Laguncularia purpurea* Gaud.)
(syn. *Lumnitzera coccinea* W.&A.)
(syn. *Lumnitzera littoralis* Ogura)
(syn. *Lumnitzera purpura* Presl.)
(syn. *Pyrrhanthus littoreus* Jack)

Lumnitzera racemosa Willd. var. *lutea* (Gaud.) Exell
(syn. *Laguncularia lutea* Gaud.)
(syn. *Lumnitzera lutea* Presl.)

Lumnitzera racemosa Willd. var. *racemosa* Willd.) Van Steenis
(syn. *Laguncularia rosea* Gaud.)
(syn. *Lumnitzera racemosa* Willd.
(syn. *Lumnitzera racemosa* var. *pubescens* Koord.&Val.)
(syn. *Lumnitzera rosea* Presl.)
(syn. *Petaloma alba* Blanco)

Euphorbiaceae
Excoecaria agallocha L.
(syn. *E. affinis* Endl.)
(syn. *E. camettia* Willd.)
(syn. *E. ovalis* Endl.)

+ **Filices**
Acrostichum aureum L.
Acrostichum speciosum Willd.

+ **Leguminosae** (Caesalpiniaceae)
Cynometra iripa Kostel
Cynometra ramiflora L. sub sp.
bijuga (Span.) Prain.
(syn. *Cynometra ramiflora* MacNae)
Derris heptaphylla (L.)Merr.
(syn. *D. macroloba* Miq.)
(syn. *D. sinuata* Thw.)
Derris heterophylla (Willd.)Back.
(syn. *D. uliginosa* (Roxb.) Benth.)
Intsia bijuga (Colebr.)O.Ktze.
(syn. *Afzelia bijuga* (Colebr.) Gray)
(syn. *Intsia retusa* (Kurz)O.Ktze.)

+ **Lythraceae**
Barringtonia racemosa (L.)Spreng.
Barringtonia speciosa (L.f.)
(syn. *B. asiatica* (L.) Kurz.)
Pemphis acidula J.R. & G. Forst.

+ **Malvaceae**
Hibiscus tiliaceus L.
Thespesia acutiloba (E.G. Bak.) Exell & Mend.
Thespesia populnea (L.) Soland. ex Correa
(syn. *Bupariti pupulnea* (L.) Roth.)
(syn. *Thespesia macrophylla* Bl.)
(syn. *T. populneoides* Kostel)

Meliaceae
Xylocarpus granatum (L.) Koenig
(syn. *Amoora salomoniensis* DC.)
(syn. *Carapa moluccensis* Oliver)
(syn. *Carapa indica* Juss.)
(syn. *Carapa obovata* Bl.)
(syn. *Granatum obovatum* (Dl.) Kuntze)
(syn. *Milnea austro-caledonica* Jeann.)
(syn. *Monosoma littorata* Griff.)
(? syn. *Xylocarpus benadirensis* Mattei)
(syn. *Xylocarpus littorata* Griff.)
(? syn. *Xylocarpus minor* Ridley)
(syn. *Xylocarpus obovatus* Grewe)

Xylocarpus mekongensis Pierre
(syn. *Carapa borneensis* Becc.)
(syn. *Carapa mekongensis* (Pierre) Pellegrin)

39

Table 1 *(continued)*.

(syn. *Carapa moluccensis* Watson)
(syn. *Carapa moluccensis* var. *gangetica* Prain)
(syn. *Xylocarpus gangeticus* (Prain) Parking)
(syn. *Xylocarpus moluccensis* var. *ellipticus* Koord. & Val.)
(syn. *Xylocarpus parvifolius* Ridl.)

Xylocarpus moluccensis (Lamx.) Roem.
(syn. *Carapa indica* A. Juss.)
(syn. *Carapa moluccensis* Lamx.)
(syn. *Carapa rumphii* Kostel)
(syn. *Granatum moluccensis* (Lamx.) O. Ktze.
(? syn. *Xylocarpus australasicus* Ridl.)
(syn. *Xylocarpus carnulosus* Zoll. & Mor.)
(syn. *Xylocarpus forsterii* Miq.)

Amoora cucullata Roxb.

+ **Myristicaceae**
Myristica hollrungei Warb.

Myrsinaceae
Aegiceras corniculatum (L.) Blanco
(syn. *Aegiceras majus* Gaertn.)
(syn. *Aegiceras malaspinaea* DC.)
(syn. *Malaspinaea laurifolia* Presl.)
(syn. *Rhizophora corniculatum* Stickman)
Aegiceras floridum Roem. & Schult.
(syn. *A. ferreum* Bl.)

+ **Myrtaceae**
Osbornia octodonta F. v. Muell.

Palmae
+ *Oncosperma filamentosum* Bl.
+ *Oncosperma horridum* Scheff.
Nypa fruticans Wurmb.
(syn. *Nipa fruticans* Thunb.)
(syn. *Nypa arborescens* Wurmb.)
(syn. *Nypa littoralis* Blanco)

+ *Phoenix paludosa* Roxb.)
(syn. *P. siamensis* Miq.)
+ *Phoenix reclinata* Jacq.
(syn. *P. equinoxialis* Boj.)
(syn. *P. leonensis* Lodd ex Kunth)
(syn. *P. senegalensis* van Houtte)
(syn. *P. spinosa* Schum. & Thorn)

+ *Raphia vinifera* Beauv.
(syn. *R. taedigera* Mart.)
(syn. *R. nicaraguensis* Oerst.)

+ **Pandanaceae**
Pandanus candelabrum
Pandanus livingstoniana Rendle
Pandanus tectorius Soland.

Plumbaginaceae
Augialitis annulata R.Br.
(syn. *Aegianilitis annulata* Presl.)
Aegialitis rotundifolia Roxb.
(syn. *Aegialitis annulata* var. *rotundifolia* Boiss.)
(syn. *Aegianilitis rotundifolia* Presl.)

Rhizophoraceae (after Chapman, 1976; Ding Hou, 1958, 1960, 1978; Jonker, 1959; Keay, 1953; Savory, 1953; Stearn, 1958; Toml 1978; Tomlinson and Womersley, 1976; Van Steenis, 1958)

Bruguiera cylindrica (L.)Bl.
(syn. *B. caryophylloides* Bl.)
(syn. *B. malabarica* Arn.)
(syn. *Kanilia caryophylloides* Bl.)
(syn. *Rhizophora caryophylloides* Burm.)
(syn. *Rhizophora ceratophylloides* Gmel.)
(syn. *Rhizophora cylindrica* L.)

Bruguiera exaristata Ding Hou

Bruguiera gymnorrhiza (L.) Lamk.
(syn. *B. capensis* Bl.)
(syn. *B. conjugata* Merr.)
(syn. *B. cylindrica* (non Bl.) Hance)
(syn. *B. rheedii* Bl.)
(syn. *B. rumphii* Bl.)
(syn. *B. wightii* Bl.)
(syn. *B. zippelii* Bl.)
(syn. *Rhizophora gymnorrhiza* (L.)Willd.)
(syn. *Rhizophora palun* DC.)
(syn. *Rhizophora rheedii* Stend.)
(syn. *Rhizophora tinctoria* Blanco)

Bruguiera hainesii C.G. Rogers
(syn. *Rhizophora caryophylloides* (non Burm.) Griff.)

Bruguiera parviflora (Roxb.) W. & A. ex Griff.
(syn. *B. ritchiei* Merr.)
(? syn. *Rhizophora cylindrica* (non L.) Roxb.)
(syn. *Rhizophora parviflora* Roxb.)
(syn. *Kanilia parviflora* Bl.)

Bruguiera sexangula (Lour.)Poir.
(syn. *B. 10-angulata* Griff.)
(syn. *B. australis* A. Cunn.ex Arn.)
(syn. *B. eriopetala* W. & A. ex Arn.)
(syn. *B. malabarica* (non Arn.) F.-Vill.)
(syn. *B. oxyphylla* Miq.)
(syn. *B. parietosa* Griff.)
(syn. *B. sexangularis* Spreng.)
(syn. *Rhizophora australis* Steud.)
(syn. *Rhizophora eriopetala* Steud.)
(syn. *Rhizophora plicata* Blanco)
(syn. *Rhizophora polyandra* Blanco)
(syn. *Rhizophora sexangula* Lour.)

Table 1 *(continued)*.

Ceriops decandra (Griff.) Ding Hou
  (syn. *C. roxburghiana* Arn.)
  (syn. *C. zippeliana* Bl.)
  (syn. *Bruguiera decandra* Griff.)
  (pro syn. *Rhizophora decandra* (Roxb. ex Griff.)W.&A.)
  (syn *Rhizophora glomerulata* Zipp. ex. Bl.)

Ceriops tagal (Perr.) C.B. Rob.
  (incl. *C. tagal* var. *australis* MacNae)
  (incl. *C. tagal* var. *tagal* Mac Nae)
  (syn *C. boviniana* Tul.)
  (syn *C. candolleana* Arn.)
  (syn. *C. forsteniana* Bl.)
  (syn. *C. lucida* Miq.)
  (syn. *C. pauciflora* Bth.)
  (syn. *C. somalensis* Chiov.)
  (syn. *C. timoriensis* Domin.)
  (syn. *Rhizophora candel* (non.L.)Blanco)
  (syn. *Rhizophora tagal* Perr.)
  (syn. *Rhizophora timoriensis* D.C.)

Kandelia candel (L.) Druce
  (syn. *K. rheedii* W. & A.)
  (syn. *Rhizophora candel* (L.)

Rhizophora apiculata Bl.
  (syn. *R. candelaria* DC.)
  (syn. *R. conjugata* (non L.) Arn.)
  (syn. *R. mangle* (non L.) Blanco)

Rhizophora harrisonii Leechm.
  (syn. *R. brevistyla* Salv.)
  (syn. *R. mangle* Hooker.)
  (syn. *R. mangle* var. *racemosa* Mart.)
  (syn. *R. racemosa* (non Meyer) Mult. auct. pro parte Benth.)

Rhizophora × *lamarckii* Montr.
  (syn. *R. conjugata* var. *lamarckii* Guill.)
  (syn. *R. pachypoda* Baill.)

Rhizophora mangle L.
  (syn. *R. mangle* var. *racemosa* Hiern.)
  (? syn. *R. mangle* var. *samoensis* Hochr.)
  (? syn. *R. samoensis* (Hochr.) Salv.)

Rhizophora mucronata Lamk.
  (syn. *R. latifolia* Miq.)
  (syn. *R. longissima* Blanco)
  (syn. *R. macrorrhiza* Griff.)
  (syn. *R. mangle* (non L.) Roxb.)
  (syn. *R. mucronata* var. *typica* (Schimp.) Hochr.)

Rhizophora racemosa G.F.W. Meyer
  (syn. *R. harrisonii* Salv.)

Rhizophora × *selala* (Salv.) Toml.

  (syn. *R. mucronata* Lamk. var. *selala* Salv.)
  (syn. *R.* "*selala*" Toml. & Womersl.)

Rhizophora stylosa Griff.
  (syn. *R. mucronata* var. *stylosa* Schimp.)

+ Rubiaceae
Scyphiphora hydrophylaccea Gaertn.

Sonneratiaceae
Sonneratia alba J. Sm.
  (syn. *Blatti alba* et *leucantha* O.K.)
  (syn *Chiratia leucantha* Montr.)
  (syn. *Rhizophora caseolaris* L. pro parte)
  (syn. *Sonneratia alba* var. *iriomotensis* Yamam.)
  (syn. *S. caseolaris* Engl.)
  (syn. *S. griffithii* Watson)
  (syn. *S. iriomotensis* Masamune)
  (syn. *S. mossambicensis* Klotzsch ex Peters)

Sonneratia apetala Buch.Ham.
  (syn. *Blatti apetala* O.K.)
  (syn. *Kambala apetala* Rafin.)

Sonneratia caseolaris (L.) Engler
  (syn. *Aubletia caseolaris* Gaertn.)
  (syn. *Blatti caseolaris* O.K.)
  (syn. *Blatti pagatpat* Niedenzu)
  (syn. *Rhizophora caseolaris* (L.) Burm.)
  (syn. *Sonneratia acida* (L.) Back.)
  (syn. *S. evenia* Bl.)
  (syn. *S. lanceolata* Bl.)
  (syn. *S. neglecta* Bl.)
  (syn. *S. obovata* Bl.)
  (syn. *S. ovalis* Korth)
  (syn. *S. pagatpat* Blanco)
  (syn. *S. rubra* Oken.)

Sonneratia griffithii Kurz
  (syn. *S. acida* var. *griffithii* King)
  (syn. *S. alba* Griff. in Watson)

Sonneratia ovata Back.
  (syn. *S. alba* Watson)

Sterculiaceae
Heritiera fomes Buch.-Ham.
  (syn. *H. minor*)

Heritiera littoralis Dryand ex h. Ait.
  (syn. *H. fomes* Wall.)
  (syn. *H. tothila* Kurz.)

Theaceae (Pellicieraceae, Ternstroemiaceae)
Pelliciera rhizophorae Planch & Triana

Table 1 *(continued)*.

+ **Tiliaceae**	**Verbenaceae**
Brownlowia lanceolata Benth	+ *Clerondendrum inerme* (L.) Gaertn.
Brownlowia argentata	

Table 2. Localities of mangal distribution.

Location No.	Location	References
01	Tanzania (Tanga, Morrumbene-estuary, Rufiji-delta)	Day (1974), Grant (1938), Grewe (1941), Walter and Steiner (1936/37)
02	Madagascar	Chapman (1977), Kiener (1972)
03	Northern Moçambique (down to 23°S)	MacNae (1968, Figs. 4, 7 and p. 96), MacNae and Kalk (1962)
04	Southern Moçambique (Inhaca Island, Morrumbene-Estuary)	Day (1974), MacNae and Kalk (1962)
05	Natal (north of Kosi Bay)	Chapman (1977), MacNae (1968), Steinke (1972), Werger (1978), West (1945)
06	Natal/Transkei (south of Kosi Bay, down to Port St. Johns; Kei Rivermouth)	Chapman (1977), MacNae (1968), Steinke (1972), Werger (1978)
07	Sokotra	Vierhapper (1907)
08	French Somali and Ethiopia (Massawa)	Zahran (1975, 1977)
09	Sudan (Suakin, 19°N)	Zahran (1975, 1977)
10	Sudan/Egypt (around Marsa Halaib 22°N)	Zahran (1975, 1977)
11	Egypt and Sinai (Marsa Halaib 22°N - Hurghada 27° 14′ N and Ras Mohammed, 27° 40′ N)	Zahran (1975, 1977)
12	Saudi Arabia (18°N)	Zahran (1975, 1977)
13	Saudi Arabia (Gizan, 16°N)	Zahran (1975, 1977)
14	Jemen (El Hodeida, 13° 30′ N)	Zahran (1975, 1977)
15	Arabian Gulf (UAE - Qatif, Qeshm, Strait of Hormuz)	Basson et al. (1977), Bolay and Schedler (1978), Dickson (1955), Evans et al. (1973), Halwagy and Macksad (1972), Zahran (1977), and pers. visit
16	West India (Kathiawar-Peninsula)	Blasco (1975, 1977)
17	West India (Bombay and Goa region)	Blasco (1975, 1977), Blasco et al. (1975), Chatterjee (1957/58), Dwivedi et al. (1975), Joshi et al. (1975)
18	West India (Kerala coast)	Blasco (1975, 1977), Blasco et al. (1975)
19	East India (Cauvery delta)	Blasco (1975, 1977) Blasco et al. (1975)
20	East India (Krishna and Godavari delta)	Blasco (1975, 1977) Blasco et al. (1975), Chatterjee (1957/58), Sidhu (1963)
21	East India (Mahanodi delta)	Blasco (1975, 1977), Blasco et al. (1975)
22	East India (Ganges delta)	Blasco (1975, 1977), Blasco et al. (1975), Sen Gupta (1938), Sidhu (1975)
23	Andamans and Nicobars	Blasco (1975, 1977), Blasco et al. (1975)
24	Burma, Thailand	Ding Hou (1958, 1960), Lecompte (1969), Vu Van Cuong (1964)
25	Cambodja	Ding Hou (1958, 1960), Lecompte (1969), Vu Van Cuong (1964)
26	Viet-Nam	Ding Hou (1958, 1960), Lecompte (1969), Ross (1975), Vu Van Cuong (1964)
27	South-China	Ding Hou (1958, 1960), Lecompte (1969),
28	Taiwan	Ding Hou (1958, 1960), Hosokawa et al. (1977), Hsieh and Jang (1969)
29	Iriomote-shima (24° 20′ N)	Hosokawa et al. (1977)

Table 2 *(continued)*.

Location No.	Location	References
30	Ryukyu-Islands	Ding Hou (1960), Hosokawa et al. (1977), Walsh (1974, p. 55)
31	Kyushu	Hosokawa et al. (1977)
32	Philippines	Ding Hou (1969), Hosokawa et al. (1977)
33	Malaysia (Malay Penisula)	Carter (1959), Diemont et al. (1975), Ding Hou (1969), Exell (1954), Van Steenis (1962?), Vu Van Cuong (1964), Watson (1928)
34	Sumatra	Ding Hou (1958, 1960), Exell (1954)
35	Borneo	Ding Hou (1958, 1960), Exell (1954)
36	Java	Backer and Bakhuizen van den Brink (1963/65/68), Exell (1954)
37	Celebes	Ding Hou (1958, 1960), Exell (1954)
38	Lesser Sunda Islands	Ding Hou (1958, 1960), Exell (1954)
39	West-New Guinea	Exell (1954), Frodin (1980), Percival & Womersley (1975)
40	Northeast-New Guinea	Exell (1954), Frodin (1980), Percival & Womersley (1975)
41	Papua-New Guinea	Exell (1954), Percival & Womersley (1975), Tomlinson & Womersley (1976), Womersley (1980)
42	Bismarck-Archipelago	Percival & Womersley (1975)
43	Solomon-Islands (incl. Brit. Solomon-Island)	Percival & Womersley (1975)
44	New Caledonia	Hosokawa et al. (1977), Percival & Womersley (1975), Tomlinson (1978)
45	Fiji	Hosokawa et al. (1977), Richmond & Ackermann (1975), Tomlinson (1978)
46	Samoa	Hosokawa et al. (1977), Tomlinson (1978)
47	Palau	Fosberg (1975), Hosokawa et al. (1977)
48	Yap	Fosberg (1975), Hosokawa et al. (1977)
49	West-Caroline-Atolls	Fosberg (1975)
50	Truk	Fosberg (1975), Hosokawa et al. (1977)
51	Ponape	Fosberg (1975), Hosokawa et al. (1977)
52	East-Caroline-Atolls	Fosberg (1975)
53	Kusaie	Fosberg (1975), Hosokawa et al. (1977)
54	Guam	Fosberg (1975), Hosokawa et al. (1977)
55	Rota	Fosberg (1975)
56	Tinian	Fosberg (1975)
57	Saipan	Fosberg (1975), Hosokawa et al. (1977)
58	North-Marianas	Forsberg (1975)
59	South-Marshall Atolls	Fosberg (1975), Hosokawa et al. (1977)
60	North-Marshall Atolls	Fosberg (1975), Hosokawa et al. (1977)
61	North-Gilbert Atolls	Fosberg (1975)
62	South-Gilbert Atolls	Fosberg (1975)
63	Nauru	Fosberg (1975)
64	Cape York (Southeast-Gulf of Carpentaria, Australia)	Jones (1971), Saenger and Hopkins (1975), Saenger et al. (1977)
65	Weiba (Southeast-Gulf of Carpentaria, Australia)	Saenger and Hopkins (1975), Saenger et al. (1977)
66	Mornington Island (South-Gulf of Carpentaria, Australia)	Saenger and Hopkins (1975), Saenger et al. (1977), Woolston (1973)
67	Sweers Island (Southeast-Gulf of Carpentaria, Australia)	Saenger and Hopkins (1975), Saenger et al. (1977)
68	Tarrant Point (Southeast-Gulf of Carpentaria, Australia)	Saenger and Hopkins (1975), Saenger et al. (1977)
69	Arnhem Land (Southeast-Gulf of Carpentaria, Australia)	Saenger and Hopkins (1975), Saenger et al. (1977), Specht (1958)

Table 2 *(continued)*.

Location No.	Location	References
70	Darwin, N.T.	Chippendale (1972), Saenger and Hopkins (1975), Saenger et al. (1977)
71	Great Barrier Reef (12° 14.5′S - 16° 23′S)	Stoddart (1980)
72	Cape York (11°S) - Daintree River (16°17′S)	Jones (1971), Saenger et al. (1977)
73	Daintree R. (16°17′S) - Lucinda/Herbert River (18°32′S)	Jones (1971), MacNae (1966, 1968), Saenger et al. (1977)
74	Lucinda/Herbert R. (18° 32′S) - Sarina (21° 30′S)	Jones (1971), Saenger et al. (1977)
75	Sarina (21° 30′S) - Richmond River (28° 53′S)	Jones (1971), Saenger et al. (1977)
76	Richmond R. (28° 53′S) - Bateman's Bay (35° 42′S)	Clarke and Hannon (1967), Jones (1971), Saenger et al. (1977)
77	Bateman's Bay (35° 42′S) - Corner Inlet (38° 45′S)	Jones (1971), Saenger et al. (1977), Wells (1980)
78	Westernport Bay (Melbourne) St. Vincent Gulf / Spencer's Gulf / Bunbury	MacNae (1966)
79	Northern Territory	Chippendale (1972), Saenger et al. (1977)
80	Cambridge Gulf / Ord River - King Sound (West Australia 14° 45′S - 17° 30′S)	Saenger et al. (1977), Thom et al. (1975)
81	West Australia (21°S - 23°S)	Saenger et al. (1977)
82	West Australia, Shark Bay + Bunbury	Saenger et al. (1977)
83	Hawaii	Walsh (1974, p. 55)
84	Baja California (Ballenas Bay/ Bahia de los Angeles, Pond Lagoon, Bahia Asuncion - Puerto Chale 27°N - 24° 30′N)	Mac Donald (1977), West (1977), Wiggins (1980)
85	Baja California (Bahia Santo Domingo - Bahia de Magdalena)	Mac Donald (1977), Wiggins (1980)
86	Pacific Mexico (Puerto de Lobos, 30° 15′ N)	West (1977)
87	Mexico (Pacific coast)	Breteler (1977), West (1977)
88	Guatemala (Pacific coast)	Breteler (1977), West (1977)
89	El Salvador	Breteler (1977), Daugherty (1975), Lötschert (1960), West (1977)
90	Honduras (Pacific coast)	Breteler (1977), West (1977)
91	Nicaragua (Pacific coast)	Breteler (1977), West (1977)
92	Costa Rica (Pacific coast)	Breteler (1977), West (1977)
93	Panama (Pacific coast)	Breteler (1977), West (1977)
94	Columbia	Breteler (1977), Chapman (1970), West (1956, 1977)
95	Ecuador	Breteler (1977), West (1977)
96	Tumbes (Ecuador/Peru 3° 40′S	Peña (1971), West (1977)
97	Galapagos Islands	Chapman (1939-1945), West (1977)
98	Venezuela	West (1977)
99	Brit. Guyana	Walsh (1974), West (1977)
100	Surinam	Jonker (1959), Pons and Pons (1975), Walsh (1974), West (1977)
101	French Guyana	Boyé (1962), Walsh (1974), West (1977)
102	Brazil (Amazonas delta)	West (1977)
103	Brazil (Florianopolis, 27° 30′S)	West (1977)
104	Brazil (Aranangua River mouth; 29°S)	West (1977)
105	Leeward Islands (Barbuda)	Stoddart et al. (1973), West (1977)
106	Puerto Rico	Cintrón et al. (1978), Lugo and Cintrón (1975), West (1977)

Table 2 *(continued)*.

Location No.	Location	References
107	Haiti	West (1977)
108	Jamaica	Chapman (1939-1945)
109	Cuba	West (1977)
110	Bahamas	Britton and Millspaugh (1962), West (1977)
111	Mexico (Gulf coast, Tabasco)	Thom (1967)
112	Mexico (Gulf coast, Mandinga 18° 30' - 10° 03'N)	Chapman (1970), Lot-Helgueras et al. (1975)
113	Mexico (Gulf coast, Veracruz and Tamaulipas 19° 38' - 23° 47'N)	Lot-Helgueras et al. (1975)
114	Mexico Gulf coast, 23° 48' - 25° 45'N)	Lot-Helgueras et al. (1975)
115	Mexico (Rio Bravo region)	Lot-Helgueras et al. (1975)
116	Texas (near Corpus Christi, 18°N)	McMillan (1971)
117	Florida (Gulf coast, Cedar Keys, 29°N)	Davis (1940), West (1977)
118	Florida (Gulf coast, Tampa Bay)	Davis (1940), West (1977)
119	Florida (Gulf coast, Ten Thousand Islands, 25° 45'N)	Davis (1940), Graham (1964), West (1977)
120	Florida (Atlantic coast, St. Augustine, 29° 53'N)	Davis (1940), West (1977)
121	Bermuda Islands	West (1977)
122	Senegal (Casamance)	Grewe (1941), Keay (1953), Lawson (1966)
123	Gambia (Gambia River delta)	Giglioli and Thornton (1966), Grewe (1941), Keay (1953), Lawson (1966)
124	Guinea	Grewe (1941), Keay (1953), Lawson (1966)
125	Sierra Leone	Grewe (1941), Jordan (1964), Keay (1953), Lawson (1966)
126	Liberia	Grewe (1941), Keay (1953), Lawson (1966)
127	Ivory coast	Grewe (1941), Keay (1953), Lawson (1966)
128	Ghana (Gold coast)	Grewe (1941), Keay (1953), Lawson (1966)
129	Togo	Grewe (1941), Keay (1953), Lawson (1966)
130	Dahome	Grewe (1941), Keay (1953), Lawson (1966)
131	Nigeria	Grewe (1941), Keay (1953), Lawson (1966), Savory (1953)
132	Cameroon (Wouri estuary)	Boyé et al. (1975), Grewe (1941), Keay (1953), Lawson (1966), Walter and Steiner (1936)
133	Gabun	Grewe (1941), Keay (1953), Lawson (1966)
134	Sao Tomé	Grewe (1941), Keay (1953), Lawson (1966)
135	Congo (French Congo)	Grewe (1941), Keay (1953), Lawson (1966)
136	Cabinda (Angola)	Grewe (1941), Keay (1953), Lawson (1966)
137	Zaire (Belgian Congo)	Grewe (1941), Keay (1953), Lawson (1966)
138	Angola	Grewe (1941), Keay (1953), Lawson (1966), Walter and Steiner (1936)
139	Angola (Lobito)	Grewe (1941), Walter and Steiner (1936)
* 82a	New Zealand	Taylor (1980)

mangal of this region are given by Day (1974), Grewe (1941), Kiener (1972), MacNae and Kalk (1962), Walter and Steiner (1936/37), West (1945) and others. It is characterized by the predominance of *Bruguiera gymnorrhiza*, *Ceriops tagal*, *Rhizophora mucronata*, *Sonneratia alba*, and *Avicennia marina*.

Red Sea/Persian Gulf

In this region the mangal flora at most places is restricted to one species: *Avicennia marina*. At some localities along the coasts of the Red Sea *Rhizophora mucronata* and/or *Bruguiera gymnorrhiza* are associated.

In a lot of recent distribution maps of mangroves they are missing for the Persian Gulf (e.g. Chap-

Table 4. Distribution of mangrove species including the main mangal associates. The location numbers refer to Tables 2 and 3.
× = dominant or present; ○ = rare or very rare; ? presence not well known or disputed.

Mangrove species	01	02	03	04	05	06	07	08	09	10	11	12	13	14	15	16	17	18	19	20	21	22	23	24	25	26	27	28	29	30	31	32
Rhizophoraceae																																
Bruguiera gymnorrhiza	x	x	x	x	x	x		x					x	x		○	○	○		x	x	x	x	x	x	x	x	x		x		x
Bruguiera cylindrica																○		x	○	○	○	○		x	x	x						
Bruguiera parviflora																○					○	○	x	x	x	x						
Bruguiera sexangula																						x	?									
Bruguiera exaristata																																
Bruguiera hainesii																								○								
Ceriops tagal	x	x	x	x	x											x	x	○		?		x		x	x	x		x				x
Ceriops decandra																x			x	x	x	x	?	x	x	x						○
Kandelia candel																○	○				○	x	?	x	x	x	x	x	x	?	?	x
Rhizophora mucronata	x	x	x	x	x	x		x	x				x			○	x	○	x	○	○	x	x	x	x	x	x	x	x	x		x
Rhizophora apiculata	x	x	x	x	x	x		x	x				x			○	x	○	x	○	○	x	x	x	x	x	x	x	x	x		x
Rhizophora stylosa																○	○	○	x	○	○	x	x	x	x	x		x				○
Rhizophora x lamarckii																												x		○		
Rhizophora x selala																																
Rhizophora mangle																																
Rhizophora racemosa																																
Rhizophora harrisonii																																
Sonneratiaceae																																
Sonneratia alba	x	x	x	x	x											○			○		○											
Sonneratia apetala																x																x
Sonneratia caseolaris																x		○	x	○	x											
Sonneratia ovata																x	?	?	?	?	x	?						x				
Sonneratia griffithii																						x		x	x	x						
																						x		x								
Avicenniaceae																																
Avicennia marina	x	x	x	x	x	x	○	x	x	x	x	x	x	x	x	x	○		x	○	x	x	?	x	x	?			x			x
Avicennia alba																x	x		x	x	?	x	x	x	x	?						
Avicennia officinalis																x	x		x	x	?	x	x	x	x	?						
Avicennia eucalyptifolia																x	x	x	x	x	x	x	x	x	x	?		x				x
Avicennia schaueriana																																
Avicennia tonduzii																																
Avicennia bicolor																																
Avicennia germinans																																
Euphorbiaceae																																
Excoecaria agallocha																x		x	x	x	x	x		x	x	?					x	
Meliaceae																																
Xylocarpus granatum						x																										
Xylocarpus moluccensis	○	x																	?	x	x	x		x	x	?				x		x
Xylocarpus benadirensis	x	x	x																○	?	x	x		x	x	?				x		x
Xylocarpus australasicum																																
Xylocarpus guianensis																																
Sterculiaceae																																
Heritiera littoralis	○	○	○		x																x	x		x	x	?				x		x
Heritiera fomes																				○	○	?										
Combretaceae																																
Lumnitzera racemosa	x	x		x	x											x		x	x	x	x	x		x	?	x	x	x		x		x
Lumnitzera littorea																						x		x	x	x				x		x
Conocarpus erecta																																
Languncularia racemosa																																
Myrsinaceae																																
Aegiceras corniculatum																x	x		x	x	?	x	x	x	x	x						x
Aegiceras floridum																																
Plumbaginaceae																																
Aegialites annulata																				?	x	x										
Aegialites rotundifolia																																
Acanthaceae																																
Acanthus ilicifolius																x	x	x	x	x	x	x	x	x	x	x					x	
Acanthus ebracteatus																						x									x	
(Acanthus volubilis)																						○										

Mangrove species	01	02	03	04	05	06	07	08	09	10	11	12	13	14	15	16	17	18	19	20	21	22	23	24	25	26	27	28	29	30	31	32
Apocynaceae																																
Cerbera manghas																		x			o	x								?		
Cerbera odollam																														?		
Cerbera floribunda																																
Lythraceae																																
Pemphis acidula	x																															
Barringtonia racemosa		o	x																										x			
Barringtonia speciosa																																
Myrtaceae																																
Osbornia octodonta																									x							x
Filices (Pteridaceae)																																
Acrostichum aureum	o	o	o	o	o	o												x		x	x	o		x	x	x			x			x
Acrostichum speciosum			o																													
Palmae																																
Nypa fruticans																					x	x		x	x	x			x			x
Phoenix spinosa																																
Phoenix paludosa																				x	x	x										
Phoenix reclinata	x	x																														
Oncosperma filamentosa																																
Oncosperma tigillaria																																
Malvaceae																																
Hibiscus tiliaceus	x		x	x	x	x												x	o	x	x	x	x	x	x	x						x
Thespesia populnea		o	x	x												x																x
Bignoniaceae																																
Dolichandrone spathacea																		x	x	x	x	x	x	x	x	x						x
Pandanaceae																																
Pandanus candelabrum																																
Pandanus tectorius																	x	x	x	x	x	x										
Pandanus kanehirae																																
Rubiaceae																																
Scyphiphora hydrophyllacea																			o				x	x	x	x					?	
Salvadoraceae																																
Salvadora persica																x	x		x	o												
Salvadora oleides																x	o															
Tiliaceae																																
Brownlowia lanceolata																				o	o	x										
Brownlowia riedelii																																
Brownlowia argentata																																
Verbenaceae																																
Clerodendrum inerme																x																
Caesalpiniaceae																																
Cynometra remiflora var. bijuga																																
Intsia bijuga																																
Myristicaceae																																
Myristica hollrungei																																
Lepiniopsis trilocularis																																
Bombacaceae																																
Camptostemon schultzii																																
Theaceae																																
Pelliciera rhizophorae																																

Table 4 *(continued)*.

Location number

Mangrove species	33	34	35	36	37	38	39	40	41	42	43	44	45	46	47	48	49	50	51	52	53	54	55	56	57	58	59	60	61	62	63
Rhizophoraceae																															
Bruguiera gymnorrhiza	x	x	x	x	x		x	x	x		x	x	x	x	x	x	x	x	x	x	x			x			x	x	x	x	x
Bruguiera cylindrica	x	x	x	x	x		x	x	x																						
Bruguiera parviflora	x	x	x	x	x		x	x	x	x	x	x																			
Bruguiera sexangula	x	x	x	x	?		x	x	x	x	x	?																			
Bruguiera exaristata						o																									
Bruguiera hainesii	o																														
Ceriops tagal	x	x	x	x	?		x	x	x	x	x	?			x	x															
Ceriops decandra	o	o	o	x	o		x	x																							
Kandelia candel	o	o	o																												
Rhizophora mucronata	x	x	x	x	x				x	?	?	?	?		x	x	x	x	x	x	x	x					x		x	x	
Rhizophora apiculata	x	x	x	x	x		x	x	x	x	x	o			x	x		x	x	x	?	?									
Rhizophora stylosa	o	x	x	x	x			x	x	x	x	x	x	?								x									
Rhizophora x lamarckii								o	o	o	o	o																			
Rhizophora x selala												o	o																		
Rhizophora mangle												o	o	o																	
Rhizophora racemosa																															
Rhizophora harrisonii																															
Sonneratiaceae																															
Sonneratia alba	x	x	x	x	x		x	x	x	x	x				x	x		x	x	x	x						x		x		
Sonneratia apetala		?																													
Sonneratia caseolaris	x	x	x	x	x		x	x	x	x	x																				
Sonneratia ovata	x	x	o	x	x			?																							
Sonneratia griffithii	o								o																						
Avicenniaceae																															
Avicennia marina	x	x	x	x	?		x		x		x																				
Avicennia alba	x	x	x	x					x						x	x					x										
Avicennia officinalis	x	x	x	x	?				x			x																			
Avicennia eucalyptifolia							x		x																						
Avicennia schaueriana																															
Avicennia tonduzii																															
Avicennia bicolor																															
Avicennia germinans																															
Euphorbiaceae																															
Excoecaria agallocha	x	x	x	x			x	x	x	x	x	x	x	x	x			x	?		?	?									
Meliaceae																															
Xylocarpus granatum	x	x	x	x			x	x	x	x	x	x						x	x					x	?						
Xylocarpus moluccensis	x	x	x	x			x		x	x		x										x	x	x							
Xylocarpus benadirensis																															
Xylocarpus australasicum									x																						
Xylocarpus guianensis																															
Sterculiaceae																															
Heritiera littoralis	x	x	x	x			x	x	x	x	x	x	x	x	x			x	x		x	x				x					
Heritiera fomes																															
Combretaceae																															
Lumnitzera racemosa	x	x	x	x	x	x	x	x	x			x																			
Lumnitzera littorea	x	x	x	x	x	x	x	x	x	o	o	x			x	x	x	x	x	x	x	x			?		x		x		
Conocarpus erecta																															
Languncularia racemosa																															
Myrsinaceae																															
Aegiceras corniculatum	x	x	x	x					o																						
Aegiceras floridum			x																												
Plumbaginaceae																															
Aegialites annulata																															
Aegialites rotundifolia																															
Acanthaceae																															
Acanthus ilicifolius	x	x	x	x					x																						
Acanthus ebracteatus	x	x	x	x					x					x																	
(Acanthus volubilis)																															

Mangrove species	33	34	35	36	37	38	39	40	41	42	43	44	45	46	47	48	49	50	51	52	53	54	55	56	57	58	59	60	61	62	63
Apocynaceae																															
Cerbera manghas	x	x	x	x			x	x	x	x	x				x	x		x	x	x	x									x	
Cerbera odollam	x	x	x	x																											
Cerbera floribunda							x	x	x		x																				
Lythraceae																															
Pemphis acidula			x				x	x	x	x	x				x	x	x	x	x	x	x	x	x	x	x	x	x	x	x	x	
Barringtonia racemosa																															
Barringtonia speciosa												x	x	x	x	x	x	x	x	x	x	x	x								
Myrtaceae																															
Osbornia octodonta	x	x	x	x	x				o																						
Filices (Pteridaceae)																															
Acrostichum aureum	x	x	x	x				x				x			x	x	x	x	x			x	x	x	x	x	x	x			
Acrostichum speciosum	x							x							?	?	?														
Palmae																															
Nypa fruticans	x	x	x	x	x		x	o			x				x	x		x	x		x	x									
Phoenix spinosa																															
Phoenix paludosa																															
Phoenix reclinata																															
Oncosperma filamentosa	x																														
Oncosperma tigillaria																															
Malvaceae																															
Hibiscus tiliaceus	x	x	x	x				x				x			x	x	x	x	x	x	x	x	x	x	x	x	x				
Thespesia populnea	x	x	x	x				x							?	?		?	?		?										
Bignoniaceae																															
Dolichandrone spathacea	x	x	x	x	x	x	x	x	x	x	x				x	x∙															
Pandanaceae																															
Pandanus candelabrum																															
Pandanus tectorius																															
Pandanus kanehirae														x																	
Rubiaceae																															
Scyphiphora hydrophyllacea	x	x	x	x				x							x	x															
Salvadoraceae																															
Salvadora persica																															
Salvadora oleides																															
Tiliaceae																															
Brownlowia lanceolata	x							x																							
Brownlowia riedelii																															
Brownlowia argentata	x							x																							
Verbenaceae																															
Clerodendrum inerme			x									x		?	?			?	?		?	?			?	?					
Caesalpiniaceae																															
Cynometra remiflora var. bijuga			x			x	x	x	x	x			x																		
Intsia bijuga	x	x	x	x																											
Myristicaceae																															
Myristica hollrungei						x	x	x	x																						
Lepiniopsis trilocularis						x			x																						
Bombacaceae																															
Camptostemon schultzii																															
Theaceae																															
Pelliciera rhizophorae																															

Location number

Table 4 *(continued)*.

Mangrove species	64	65	66	67	68	69	70	71	72	73	74	75	76	77	78	79	80	81	82	83	84	85	86	87	88	89	90	91	92	93	94	95	96
Rhizophoraceae																																	
Bruguiera gymnorrhiza	x				x	x	x	x	x	x	x	x				x	x																
Bruguiera cylindrica	x							x	x	x																							
Bruguiera parviflora	x				x			x		x	x	x																					
Bruguiera sexangula																x				x													
Bruguiera exaristata	x	x					x	x	x	x	x	x																					
Bruguiera hainesii																																	
Ceriops tagal	x	x			x	x	x	x	x	x	x	x	x			x	x																
Ceriops decandra	x								x	x																							
Kandelia candel																																	
Rhizophora mucronata	x				x	x	x			x	x					x	X																
Rhizophora apiculata	x									x	x	x																					
Rhizophora stylosa	x	x	x	x	x			x	x	x	x	x	x			x	x																
Rhizophora x lamarckii																																	
Rhizophora x selala																																	
Rhizophora mangle																				x	x	x		x	x	x	x	x	x	x	?	?	?
Rhizophora racemosa																																	
Rhizophora harrisonii																												x	x	x	x	x	x
Sonneratiaceae																																	
Sonneratia alba	x							x	x	x	x					x	x																
Sonneratia apetala																																	
Sonneratia caseolaris	X									x										x													
Sonneratia ovata																																	
Sonneratia griffithii																																	
Avicenniaceae																																	
Avicennia marina	x	x	x	x	x	x	x	x	x	x	x	x	x	x	x	x	x	x	x														
Avicennia alba																																	
Avicennia officinalis																																	
Avicennia eucalyptifolia	x				x		x	x	x	x	x					x																	
Avicennia schaueriana																																	
Avicennia tonduzii																																	
Avicennia bicolor																												x	x	x	x		
Avicennia germinans																				x		x	x	x	x	x	x	x	x	x	x	x	x
Euphorbiaceae																																	
Excoecaria agallocha	x	x		x	x	x	x	x	x	x	x	x				x	x																
Meliaceae																																	
Xylocarpus granatum	x	x			x	x	x	x	x	x						x	x																
Xylocarpus moluccensis																																	
Xylocarpus benadirensis																																	
Xylocarpus australasicum	x				x	x	x	x	x							x	x																
Xylocarpus guianensis																																	
Sterculiaceae																																	
Heritiera littoralis	x							x	x	x							x																
Heritiera fomes																																	
Combretaceae																																	
Lumnitzera racemosa	x						x	x	x	x	x	x				x	x																
Lumnitzera littorea	x	x		x	x	x		x	x	x						x																	
Conocarpus erecta																				x	x	x		x	x	x	x	x	x	x			
Languncularia racemosa																					x	x		x	x	x	x	x	x	x	x	x	x
Myrsinaceae																																	
Aegiceras corniculatum	x	x	x		x	x	x	x	x	x	x	x	x			x	x	x															
Aegiceras floridum																																	
Plumbaginaceae																																	
Aegialites annulata	x	x		x	x	x	x	x	x	x	x	x				x	x																
Aegialites rotundifolia																																	
Acanthaceae																																	
Acanthus ilicifolius	x			x	x	x			x	x	x	x					x																
Acanthus ebracteatus																																	
(Acanthus volubilis)																																	

Mangrove species	64	65	66	67	68	69	70	71	72	73	74	75	76	77	78	79	80	81	82	83	84	85	86	87	88	89	90	91	92	93	94	95	96
Apocynaceae																																	
Cerbera manghas																																	
Cerbera odollam																																	
Cerbera floribunda																																	
Lythraceae																																	
Pemphis acidula	x							x	x																								
Barringtonia racemosa																																	
Barringtonia speciosa																																	
Myrtaceae																																	
Osbornia octodonta	x	x				x	x	x	x	x					x																		
Filices (Pteridaceae)																																	
Acrostichum aureum																																	
Acrostichum speciosum	x							x	x	x	x				x	x															x	x	
Palmae																																	
Nypa fruticans	x					x		x	x						x																		
Phoenix spinosa																																	
Phoenix paludosa																																	
Phoenix reclinata																																	
Oncosperma filamentosa																																	
Oncosperma tigillaria																																	
Malvaceae																																	
Hibiscus tiliaceus	x							x	x																	x							
Thespesia populnea	x							x	x																								
Bignoniaceae																																	
Dolichandrone spathacea																																	
Pandanaceae																																	
Pandanus candelabrum																																	
Pandanus tectorius																																	
Pandanus kanehirae																																	
Rubiaceae																																	
Scyphiphora hydrophyllacea	x	x				x		x	x	x					x																		
Salvadoraceae																																	
Salvadora persica																																	
Salvadora oleides																																	
Tiliaceae																																	
Brownlowia lanceolata																																	
Brownlowia riedelii																																	
Brownlowia argentata																																	
Verbenaceae																																	
Clerodendrum inerme																																	
Caesalpiniaceae																																	
Cynometra remiflora var.bijuga	x							x	x	x																							
Intsia bijuga																																	
Myristicaceae																																	
Myristica hollrungei																																	
Lepiniopsis trilocularis																																	
Bombacaceae																																	
Camptostemon schultzii	x	x						x							x	x																	
Theaceae																																	
Pelliciera rhizophorae																											x	x	x	x	x		

51

Table 4 *(continued)*.

Mangrove species	Location number																							
	97	98	99	100	101	102	103	104	105	106	107	108	109	110	111	112	113	114	115	116	117	118	119	120
Rhizophoraceae																								
Bruguiera gymnorrhiza																								
Bruguiera cylindrica																								
Bruguiera parviflora																								
Bruguiera sexangula																								
Bruguiera exaristata																								
Bruguiera hainesii																								
Ceriops tagal																								
Ceriops decandra																								
Kandelia candel																								
Rhizophora mucronata																								
Rhizophora apiculata																								
Rhizophora stylosa																								
Rhizophora x lamarckii																								
Rhizophora x selala																								
Rhizophora mangle	?	x	x	x	?	x	x		x	x	?	x	?	x	x	x	x							
Rhizophora racemosa		x	x	x	?	x																		
Rhizophora harrisonii	x	x	x	x	x																			
Sonneratiaceae																								
Sonneratia alba																								
Sonneratia apetala																								
Sonneratia caseolaris																								
Sonneratia ovata																								
Sonneratia griffithii																								
Avicenniaceae																								
Avicennia marina																								
Avicennia alba																								
Avicennia officinalis																								
Avicennia eucalyptifolia																								
Avicennia schaueriana		x	x	x	x	x	x		?															
Avicennia tonduzii																								
Avicennia bicolor																								
Avicennia germinans	x	x	x	x	x	x			x	x	?	x	?	x	x	x	x	x	x	x	x			x
Euphorbiaceae																								
Excoecaria agallocha																								
Meliaceae																								
Xylocarpus granatum																								
Xylocarpus moluccensis																								
Xylocarpus benadirensis																								
Xylocarpus australasicum																								
Xylocarpus guianensis		x	?	?																				
Sterculiaceae																								
Heritiera littoralis																								
Heritiera fomes																								
Combretaceae																								
Lumnitzera racemosa																								
Lumnitzera littorea																								
Conocarpus erecta	x			x					x			x	?	x	x	?	x	x						
Languncularia racemosa	?	?	x	x	x	x	x	x	x	x	?	x	?	x	x	x	x							
Myrsinaceae																								
Aegiceras corniculatum																								
Aegiceras floridum																								
Plumbaginaceae																								
Aegialites annulata																								
Aegialites rotundifolia																								
Acanthaceae																								
Acanthus ilicifolius																								
Acanthus ebracteatus																								
(Acanthus volubilis)																								

Mangrove species

Location number

Mangrove species	97	98	99	100	101	102	103	104	105	106	107	108	109	110	111	112	113	114	115	116	117	118	119	120
Apocynaceae																								
Cerbera manghas																								
Cerbera odollam																								
Cerbera floribunda																								
Lythraceae																								
Pemphis acidula																								
Barringtonia racemosa																								
Barringtonia speciosa																								
Myrtaceae																								
Osbornia octodonta																								
Filices (Pteridaceae)																								
Acrostichum aureum					x	?	x							?			x	x						
Acrostichum speciosum																								
Palmae																								
Nypa fruticans																								
Phoenix spinosa																								
Phoenix paludosa																								
Phoenix reclinata																								
Oncosperma filamentosa																								
Oncosperma tigillaria																								
Malvaceae																								
Hibiscus tiliaceus				x	x	?	x					?		?		x	x	x						
Thespesia populnea				x								?	x	?		x	?							
Bignoniaceae																								
Dolichandrone spathacea																								
Pandanaceae																								
Pandanus candelabrum																								
Pandanus tectorius																								
Pandanus kanehirae																								
Rubiaceae																								
Scyphiphora hydrophyllacea																								
Salvadoraceae																								
Salvadora persica																								
Salvadora oleides																								
Tiliaceae																								
Brownlowia lanceolata																								
Brownlowia riedelii																								
Brownlowia argentata																								
Verbenaceae																								
Clerodendrum inerme																								
Caesalpiniaceae																								
Cynometra remiflora var. bijuga																								
Intsia bijuga																								
Myristicaceae																								
Myristica hollrungei																								
Lepiniopsis trilocularis																								
Bombacaceae																								
Camptostemon schultzii																								
Theaceae																								
Pelliciera rhizophorae																								

No species reported

Table 4 *(continued)*.

Mangrove species	121	122	123	124	125	126	127	128	129	130	131	132	133	134	135	136	137	138	139
Rhizophoraceae																			
Bruguiera gymnorrhiza																			
Bruguiera cylindrica																			
Bruguiera parviflora																			
Bruguiera sexangula																			
Bruguiera exaristata																			
Bruguiera hainesii																			
Ceriops tagal																			
Ceriops decandra																			
Kandelia candel																			
Rhizophora mucronata																			
Rhizophora apiculata																			
Rhizophora stylosa																			
Rhizophora x lamarckii																			
Rhizophora x selala																			
Rhizophora mangle	x	x	x	x	x	x					x							x	
Rhizophora racemosa		x	x	x	x	x					x	x	x	x				x	
Rhizophora harrisonii		x	x	?	x	x	x	x	x	x	x	x	x		x	x	x		
Sonneratiaceae																			
Sonneratia alba																			
Sonneratia apetala																			
Sonneratia caseolaris																			
Sonneratia ovata																			
Sonneratia griffithii																			
Avicenniaceae																			
Avicennia marina																			
Avicennia alba																			
Avicennia officinalis																			
Avicennia eucalyptifolia																			
Avicennia schaueriana																			
Avicennia tonduzii																			
Avicennia bicolor																			
Avicennia germinans	x	x	x	x	x	x	x	x	x	x	x	x	x	x	x	x	x	x	x
Euphorbiaceae																			
Excoecaria agallocha																			
Meliaceae																			
Xylocarpus granatum																			
Xylocarpus moluccensis																			
Xylocarpus benadirensis																			
Xylocarpus australasicum																			
Xylocarpus guianensis																			
Sterculiaceae																			
Heritiera littoralis																			
Heritiera fomes																			
Combretaceae																			
Lumnitzera racemosa																			
Lumnitzera littorea																			
Conocarpus erecta				x			?	?	?	?	?	x	?	?	?				
Languncularia racemosa			o	x	x				?	?	?	?	?	x	?	?	?	**X**	?
Myrsinaceae																			
Aegiceras corniculatum																			
Aegiceras floridum																			
Plumbaginaceae																			
Aegialites annulata																			
Aegialites rotundifolia																			
Acanthaceae																			
Acanthus ilicifolius																			
Acanthus ebracteatus																			
(Acanthus volubilis)																			

Mangrove species	121	122	123	124	125	126	127	128	129	130	131	132	133	134	135	136	137	138	139
Apocynaceae																			
Cerbera manghas																			
Cerbera odollam																			
Cerbera floribunda																			
Lythraceae																			
Pemphis acidula																			
Barringtonia racemosa																			
Barringtonia speciosa																			
Myrtaceae																			
Osbornia octodonta																			
Filices (Pteridaceae)																			
Acrostichum aureum					x		?	?	?	?	x	x	x	?	?	?		x	
Acrostichum speciosum																			
Palmae																			
Nypa fruticans							?	?	?	?	?	x	?	?	?	?	?		?
Phoenix spinosa																		x	
Phoenix paludosa																			
Phoenix reclinata					x							x							
Oncosperma filamentosa																			
Oncosperma tigillaria																			
Malvaceae																			
Hibiscus tiliaceus												x							
Thespesia populnea																			
Bignoniaceae																			
Dolichandrone spathacea																			
Pandanaceae																			
Pandanus candelabrum									x							x			
Pandanus tectorius																			
Pandanus kanehirae																			
Rubiaceae																			
Scyphiphora hydrophyllacea																			
Salvadoraceae																			
Salvadora persica																			
Salvadora oleides																			
Tiliaceae																			
Brownlowia lanceolata																			
Brownlowia riedelii																			
Brownlowia argentata																			
Verbenaceae																			
Clerodendrum inerme																			
Caesalpiniaceae																			
Cynometra remiflora var. bijuga																			
Intsia bijuga																			
Myristicaceae																			
Myristica hollrungei																			
Lepiniopsis trilocularis																			
Bombacaceae																			
Camptostemon schultzii																			
Theaceae																			
Pelliciera rhizophorae																			

Table 3. Biogeographical distribution zones of mangal. The location numbers refer to the localities listed in Table 2.

Distribution zones

Location No.	
01–06	East Africa
07–15	Red Sea / Persian Gulf
16–23	India
24–43	Malaysia / Guinea
44–63	Micro- and Macronesia
64–82	Australia / New Zealand
83	Hawaii
84–97	Pacific America
98–121	Atlantic America
122–139	West Africa

man, 1976, 1977; Kiener, 1973; MacNae, 1968), although this mangal is of high interest for biogeography and ecology of extreme environments (arid climate and high water salinity). Moreover the Persian Gulf is of historical interest for mangrove distribution studies, because the earliest descriptions of mangroves, documented in the available literature, refer to this region (those of Nearchus and Theophrast, 305-325 B.C., after MacNae, 1968), and mangroves exists there up to now (Basson et al., 1977; Dickson, 1955; Bolay and Schedler, 1978; and pers. observ.).

India

On the western coast of the Indian subcontinent one can distinguish three main areas of the mangal, the Kathiawar peninsula, the Bombay-Ratnagiri region, and the Kerala coast. On the eastern coast there are four main mangrove areas: the Tamil Nadu coast with the Cauvery delta, the Krishna and Godavari delta (Andhra Pradesh), the Mahanadi delta, and the Ganges and Brahmaputra deltas (Bengal). The latter region is better known as the Sunderbans.

The mangroves of India are well described by a number of authors, e.g. Banerjee (1964), Blasco (1975, 1977), Blasco et al. (1975), Chatterjee (1958), Joshi et al. (1975), Sidhu (1973, 1975). (For a more detailed quotation see Blasco, 1977.)

Malaysia/Guinea

This zone comprises all the mangal from Burma to Guinea and from the Philippines to South Japan. Throughout this region, except the northeastern part with the Riu-Kiu-Islands and South Japan, the largest number of mangrove species (up to 42) are to be found. The mangal of the northeastern region is characterized by a lesser amount of mangrove species with predominance of *Kandelia candel*. This species forms a one-species stand on Iriomote-Shima (24°20′N, South Japan), the northern limit of mangrove growth in this region.

Floristic compositions and ecological descriptions of the mangrove formations of Malaysia/ Guinea are given by a great number of authors, e.g. Backer and Bakhuizen van den Brink (1963, 1965, 1968), Diemont et al. (1975), Ding Hou (1958, 1960), Exell (1954), Frodin (1980), Hosokawa et al. (1977), Hsieh and Jang (1969), Percival and Womersley (1975), Vu Van Cuong (1954) and Watson (1928).

Australia/New Zealand

The shorelines of the mainland of Australia, the northern islands and the Great Barrier Reef together with New Zealand are included in this mangal zone. In Australia the latitudinal range of mangroves is between 11° and 38°S. After Wells (1980) the southernmost mangrove distribution of the world is that stand of *Avicennia marina* var. *australasica* at Corner Inlet, Victoria (38°45′S, 146°30′ E) (note, that in this contribution the varieties of *Avicennia marina* are classified under the species name, see Table 1).

The mangal of Australia is one of the best known of the world by e.g. Clarke and Hannon (1967), Chippendale (1972), Hopkins (1972), Jones (1971), MacNae (1966), Saenger and Hopkins (1975), Saenger et al. (1977), Stoddart (1980), Thom et al. (1975) and Walsh (1974).

Micronesia/Macronesia

The mangal flora of the Micronesian Islands has been described by Fosberg (1975) and Hosokawa et al. (1977, citing Hosokawa 1957). Both authors give species lists from the islands Palau, Yap, Truk, Ponape, Kusaie, Guam, Saipan, and the Marshal Isles. Fosberg present additional lists from the Eastern and Western Caroline Atolls, Rota, Tinian, the Northern Marianas, the Northern and Southern Gilbert Atolls, and Nauru (Tables 2 and 4). There are some discrepancies between Fosberg's and Hosokawa's species lists. In the lists of Hosokawa *Rhizophora mucronata* var. *stylosa*, *Dolichandrone spathacea*, *Xylocarpus moluccensis*, *Cerbera manghas*, *Pemphis acidula*, *Lepiniopsis trilocularis*, and *Hibiscus tiliaceus* do not occur. In Fosberg's lists some mangal associates, e.g. *Acrostichum speciosum*, *Pandanus Kanehirae*, *Cynometra bijuga* and others are absent. Even two species of the typical mangrove genera differ in the two publications: Hosokawa lists *Avicennia marina* Vierh. (not var. *alba*!) and *Sonneratia caseolaris* Engl. and Fosberg *A. alba* and *S. alba*.

In my text I refer mainly to Fosberg's presentation, but it is still not clear which of the two authors give the correct species composition.

Pacific America

In major parts of this region the mangal formations are composed of *Avicennia germinans*, *Rhizophora mangle*, *Laguncularia racemosa*, and *Conocarpus erecta* (Chapman, 1977a; West, 1977). Besides these dominant species of the western mangal at some places of the Atlantic coasts of Central and South America *Avicennia bicolor*, *A. tonduzii*, *Rhizophora harrisonii* and the endemic Theaceae *Pelliciera rhizophorae* are to be found. According to Cuatrecasas (cited after West, 1977) *R. mangle* does not occur on the Pacific coast of South America. The southern limit of mangrove growth is in this zone at only

3°40′S at Punta Malpelo, Tumbes (Peña, 1971).

Atlantic America

This zone includes the Atlantic coasts of the Americas from Florida to Brasil, the Caribbean Islands and the Bermuda Islands. This mangal is typified by the same four main mangrove species as on the Pacific coast. The floristic composition of this distribution group is completed by *Avicennia schaueriana*, found in the Lesser Antilles, the Guianas and Brazil, *Rhizophora racemosa* and *R. harrisonii*, found in Southeast Venezuela, the Guianas and at the mouth of the Amazon river, and *Xylocarpus guianensis*, found in the Guianas.

A lot of authors have described the mangal of Atlantic America, especially the northern part of the distribution area, e.g. Boyé (1962), Britton and Millspaugh (1962), Chapman (1939-1945), Cintrón et al. (1973), Davis (1940), Graham (1964), Jonker (1959), Lot-Helgueras et al. (1975), Pons and Pons (1975), McMillan (1971), Stoddart et al. (1973), Thom (1967) and West (1977).

West Africa

This group is characterized by the predominance or presence of three *Rhizophora* species *(R. mangle, R. harrisonii* and *R. racemosa)* and one *Avicennia* species *(A. germinans*, which is sometimes named as *A. africana)*. *Conocarpus erecta* and *Osbornia octodonta* are two other well-known mangrove species of this distribution zone.

In spite of the descriptions of Chapman (1977b), Giglioli and Thornton (1966), Grewe (1941), Jordan (1964), Keay (1953), Lawson (1966), Savory (1953) and others, there is a lack of information about this mangal zone.

Literature cited

Backer, C.A., Bakhuizen van den Brink, R.C., Jr. (1963-1968). Flora of Java, Vol. I-III, Wolters-Noordhoff N.V., Groningen (The Netherlands)

Backer, C.A., van Steenis, C.G.G.J. (1951). Sonneratiaceae in Flora Malesiana sér. I, vol. 4(3): 280-289

Basson, P.W., Burchard, J.E., Hardy, J.T., and Price, A.R.G.

(1977). Biotopes of the Western Arabian Gulf. Marine Life and Environments of Saudi Arabia. ARAMCO Dept. of Loss Prevention and Environmental Affairs, Dhahrar, Saudi Arabia, 284 pp.

Blasco, F. (1975). Mangroves of India. Inst. Fr. Pondichéry, Trav. Sect. Sci. Tech. 14: 180 pp.

Blasco, F. (1977). Outlines of ecology, botany and forestry of the mangals of the Indian subcontinent. In: Ecosystems of the

World Vol. 1: Wet Coastal Ecosystems (V.J. Chapman, ed.) Elsevier, Amsterdam–Oxford–New York, pp. 241-260

Blasco, F., Caratini, C., Chanda, S., Thanikaimoni, G. (1975). Main characteristics of Indian mangroves. In: Proc. Int. Symp. on Biology and Management of Mangroves (Walsh, G.E., Snedaker, S.C., Teas, H.J., eds.) Univ. Florida Press, Gainesville, Fla., pp. 71-87

Boyé, M. (1962). Les palétuviers du littoral de la Guyane française. Cahiers d'outre-mer 15: 271-290.

Boyé, M., Baltzer, F., Caratini, C., Hampartzoumian, A., Olivry, J.C., Plaziat, J.C., Villiers, J.F. (1975). Mangrove of the Wouri estuary, Cameroon. In: Proceedings of International Symposium on Biology and Management of Mangroves, 1974. Honolulu, Hawaii (Walsh, G.E., Snedaker, S.C. and H.J. Teas, eds.), Vol. II, Institute of Food and Agricultural Sciences, University of Florida, Gainesville, Fla., pp. 431-455

Breteler, F.J. (1977). America's Pacific species of *Rhizophora*. Acta Bot. Neerl. 26: 225-230

Britton, N., Millspaugh, C.F. (1962). The Bahama Flora, Hafner, New York–London

Carter, J. (1959). Mangrove succession and coastal change in South-West Malay. Inst. Brit. Geographers Transact. 26: 79-88

Chapman, V.J. (1939-1945). 1939 Cambridge University Expedition to Jamaica - Part 1. A study of the botanical processes concerned in the development of the Jamaican shore-line. J. Linn. Soc. London–Botany, 52: 407-446

Chapman, V.J. (1970). Mangrove phytosociology. Tropical Ecology 11: 1-10

Chapman, V.J. (1975). Mangrove biogeography. In: Proc. Int. Symp. Biology and Management of Mangroves (Walsh, G.E., Snedaker, S.C., Teas, H.J., eds.) Univ. Florida Press, Gainesville, Fla., pp. 3-22

Chapman, V.J. (1976). Mangrove Vegetation, J. Cramer, Vaduz, 427 pp.

Chapman, V.J. (1977a). Introduction. In: Ecosystems of the World, Vol. 1: Wet Coastal Ecosystems (V.J. Chapman, ed.), pp. 1-29, Elsevier, Amsterdam–Oxford–New York

Chapman, V.J. (1977b). Africa B. The remainder of Africa. In: Ecosystems of the World. Vol. 1: Wet Coastal Ecosystems (V.J. Chapman, ed.) Elsevier, Amsterdam–Oxford–New York, pp. 233-240

Chapman, V.J. (1977c). Wet Coastal formations of Indo-Malesia and Papua New Guinea. In: Ecosystems of the World. Vol. 1: Wet Coastal Ecosystems (V.J. Chapman, ed.), pp. 261-270, Elsevier, Amsterdam–Oxford–New York.

Chatterjee, D. (1958). Symposium on mangrove vegetation. Science & Culture 23: 329-335

Cintrón, G., Lugo, A.E., Pool, D.J., Morris, G. (1978). Mangroves of Arid Environments in Puerto Rico and Adjacent Islands. Biotropica 10 (2): 110-121

Clarke, L.D., Hannon, N.L. (1967). The mangrove swamp and salt marsh communities of the Sydney district. I. Vegetation, soils and climate. J. Ecol. 55: 753-771.

Daugherty, H.E. (1975). Human impact on the mangrove forests of El Salvador. In: Proc. Int. Symp. Biology and Management of Mangroves (Walsh, G.E., Snedaker, S.C., Teas, H.J., eds.), Univ. Florida Press, Gainesville, Fla., pp. 816-824.

Day, J.H. (1974). The ecology of the Morrumbene Estuary, Moçambique. Trans. Roy. Soc. South Africa 41: 43-97

Dickson, V. (1955). The Wildflowers of Kuwait and Bahrain. Allen and Unwin, London.

Diemont, W.H., van Wijngaarden, W. (1975). Sedimentation patterns, soils, mangrove vegetation and land use in the tidal areas of West-Malaysia. In: Proc. Int. Symp. Biology and Management of Mangroves (Walsh, G.E., Snedaker, S.C., Teas, H.J., eds.), Univ. Florida Press, Gainesville, Fla., pp. 513-528

Ding Hou (1958). Rhizophoraceae. In: Flora Malesiana sér. I, vol. 5: 429-472

Ding Hou (1960). A Review of the genus *Rhizophora*, with special reference to the Pacific species. Blumea 10 (2): 625-634

Ding Hou (1978). Addenda, corrigenda et emendanda. In: Flora Malesiana sér. I, vol. 8 (3): 550-551

Dwivedi, S.N., Parulekar, A.H., Goswami, S., Untawale, A.G. (1975). Ecology of mangrove swamps of the Mandovi estuary, Goa, India. In: Proc. Int. Symp. Biology and Management of Mangroves (Walsh, G.E., Snedaker, S.C., Teas, H.J., eds.), Univ. Florida Press, Gainesville, Fla., pp. 115-125

Evans, G., Murray, J.W., Biggs, H.E.J., Bate, R., Bush, P.R. (1973). The oceanography, ecology, sedimentology and geomorphology of parts of the Trucial Coast Barrier Island Complex, Persian Gulf. In: The Persian Gulf (B.H. Purser, ed.), pp. 233-277, Springer, Berlin–Heidelberg–New York

Exell, A.W. (1954). Combretaceae. In: Flora Malesiana sér. I, vol. 4: 533–589

Fosberg, F.R. (1975). Phytogeography of Micronesian mangroves. In: Proc. Int. Symp. on Biology and Management of Mangroves (Walsh, G.E., Snedaker, S.C., Teas, H.J., eds.), Univ. Florida Press, Gainesville, Fla., pp. 23-42

Frodin, D.G. (1980). Mangroves of New Guinea: some remarks on their distribution and taxonomy. Abstract of the 2nd International Symposia on Biology & Management of Mangroves and Tropical Shallow Water Communities, Papua New Guinea, July 20–August 2, 1980

Graham, Shirley A. (1964). The genera of *Rhizophoraceae* and *Combretaceae* in the Southeastern United States. J. Arnold Arboretum 45: 285-301

Grant, D.K.S. (1938). Mangrove woods of Tanganyika Territory, their silviculture and dependent industries. Tanzania Notes and Records 1938 (5): 5-16

Grewe, F. (1941). Afrikanische Mangrovelandschaften. Verbreitung und wirtschaftsgeographische Bedeutung. Wiss. Veröff. Deutsch. Museum f. Länderkde. (Leipzig) N.F., 9: 105-177

Halwagy, R., Macksad, A. (1972). A. contribution towards a flora of the state Kuwait and Neutral Zone. Bot. J. Linn. Soc. 65: 61-79

Hosokawa, T., Tagawa, H., Chapman, V.J. (1977). Mangals of Micronesia, Taiwan, Japan, The Philippines and Oceania. In: Ecosystems of the World, Vol. 1: Wet Coastal Ecosystems

(V.J. Chapman, ed.) pp. 271-291, Elsevier, Amsterdam–Oxford–New York

Hsieh, A-Tsai, Yang, Tsai-i (1969). Nomenclature of Plants in Taiwan

Hutchins, L.W., Scharff, Margaret (1947). Maximum and minimum monthly mean sea surface temperatures charted from the 'World Atlas of Sea Surface Temperatures'. J. Marine Res. 6 (3): 264-268

Jones, W.T. (1971). The field identification and distribution of mangroves in eastern Australia. Queensland Nat. 20: 35-51

Jonker, F.P. (1959). The genus *Rhizophora* in Surinam. Acta Bot. Neerl. 8: 58-60

Jordan, H.D. (1964). The relation of vegetation and soil to development of mangrove swamps for rice growing in Sierra Leone. J. Appl. Ecol. 1: 209-212

Joshi, G.V., Jamale, B.B., Bhosale, L.J. (1975). Ion regulation in mangroves. In: Proceedings of the First International Symposium on Biology and Management of Mangroves (Walsh, G.E., Snedaker, S.C., Teas, H.J., eds.), Univ. Florida Press, Gainesville, Fla., pp. 595-607

Keay, R.W.J. (1953). *Rhizophora* in West Africa. Kew Bull. 1953: 121-127

Kiener, A. (1972). Écologie, biologie et possibilités de mise en valeur des mangroves malgaches. Bulletin de Madagascar, 22: (308): 49-84

Lecompte, O. (1969). Combretaceae. In: Aubréville, A. (ed.) Flore du Cambodge, du Laos et du Vietnam, No. 10, Paris

Lötschert, W. (1960). Die Mangrove von El Salvador. Natur und Volk 90 (7): 213-224

Lot-Helgueras, A., Vázquez-Yanes, C., Menéndez, F.L. (1975). Physiognomic and floristic changes near the northern limit of mangroves in the gulf coast of Mexico. In: Proc. Int. Symp. on Biology and Management of Mangroves (Walsh, G.E., Snedaker, S.C., Teas, H.J., eds.) Univ. Florida Press, Gainesville, Fla., pp. 52-61

Lugo, A.E., Cintrón, G. (1975). The mangrove forests of Puerto Rico and their management. In: Proc. Int. Symp. Biology and Management of Mangroves (Walsh, G.E., Snedaker, S.C., Teas, H.J., eds.) Univ. Florida Press, Gainesville, Fla., pp. 825-846

MacDonald, K.G. (1977). Plant and animal communities of Pacific North American salt marshes. In: Ecosystems of the World. Vol. 1: Wet Coastal Ecosystems (V.J. Chapman, ed.), pp. 167-191, Elsevier, Amsterdam–Oxford–New York

MacNae, W. (1966). Mangroves in eastern and southern Australia. Aust. J. Bot. 14: 67-104

MacNae, W. (1968). A general account of the fauna and flora of mangrove swamps and forests in the Indo-West-Pacific region. Advances in Marine Biology 6: 73-270

MacNae, W., Kalk, M. (1962). The ecology of mangrove swamps at Inhaca Island, Mozambique. J. Ecol. 50: 19-35

Mc Millan, C. (1971). Environmental factors affecting seedling establishment of the black mangrove on the Central Texas coast. Ecology 52: 927-930.

Peña, G.M. (1971). Biocenosis de los manglares del Peru. An. Cient. 9: 38-45

Percival, Margaret, Womersley, J.S. (1975). Floristics and Ecology of the mangrove vegetation of Papua New Guinea. A companion volume to the Handbook Flora of Papua New Guinea. Papua New Guinea National Herbarium Department of Forests, Div. of Botany, Lae, Papua New Guinea Botany Bulletin No. 8

Pons, T.L., Pons L.J. (1975). Mangrove vegetation and soils along a more or less stationary part of the coast of Surinam, South America. In: Proc. Int. Symp. Biology and Management of Mangroves (Walsh, G.E., Snedaker, S.C., Teas, H.J., eds.) Univ. Florida Press, Gainesville, Fla., pp. 529-547

Richmond, T.A. de, Ackermann, J.M. (1975). Flora and fauna of mangrove formations in Viti Levu and Vanua Levu, Fiji. Proceedings of the International Symposium on Biology and Management of Mangroves (Walsh, G.E., Snedaker, S.C., Teas, H.J., eds.) October 8-11, 1974 Honolulu, Hawaii. Inst. Food and Agricult. Sci., Univ. Florida, Gainesville, Fla.

Ross, P. (1975). The mangrove of South Vietnam: The impact of military use of herbicides. In: Proc. Int. Symp. Biology and Management of Mangroves (Walsh, G.E., Snedaker, S.C., Teas, H.J., eds.), Univ. Florida Press, Gainesville, Fla., pp. 695-709

Saenger, P., Specht, Marion M., Specht, R.L., Chapman, V.J. (1977). Mangal and coastal salt-marsh communities in Australasia. In: Ecosystems of the World. Vol. 1. Wet Coastal Ecosystems (V.J. Chapman, ed.), pp. 293-345, Elsevier, Amsterdam–Oxford–New York

Saenger, P., Hopkins, M.S. (1975). Observations on the mangroves of the southeastern Gulf of Carpentaria, Australia. In: Proc. Int. Symp. Biology and Management of Mangroves 1974, Honolulu, Hawaii (Walsh, G.E., Snedaker, S.C., Teas, H.J., eds.) Inst. of Food and Agricultural Sciences, University of Florida, Gainesville, Fla., pp. 126-136

Savory, H.J. (1953). A note on the ecology of *Rhizophora* in Nigeria. Kew Bull.1953: 127-128

Sidhu, S.S. (1963). Studies on the mangroves of India. I. East Godavari region. Indian For.89 (5): 337-351

Sidhu, S.S. (1975). Culture and growth of some mangrove species. In: Proc. Int. Symp. Biology and Management of Mangroves (Walsh, G.E., Snedaker, S.C., Teas, H.J., eds.) Univ. Florida Press, Gainesville, Fla., pp. 394-401

Specht, R.L. (1958). The climate, geology, soil and plant ecology of the northern portion of Arnhem Land. In: Records of the American-Australian Scientific Expedition to Arnhem Land. Vol. 3: Botany and Plant Ecology (R.L. Specht and C.P. Mountford, eds.), Melbourne University Press, Melbourne, pp. 333-414

Stearn, W.T. (1958). A key to West Indian mangroves. Kew Bull.1958: 33-37

Steinke, T.D. (1972). Further observations on the distribution of mangroves in the Eastern Cape Province. Journal of South African Botany 3: 165-178

Stoddart, D.R. (1980). Mangroves as successional stages, inner reefs of the northern Great Barrier Reef. J. Biogeography 7: 269-284 (Oxford)

Stoddart, D.R., Bryan, G.W., Gibbs, P.E. (1973). Inland mangroves and water chemistry, Barbuda, West Indies. Journal of Natural History 7: 33-46

Taylor, F.J. (1980). The New Zealand mangrove association. Abstract of the 2nd International Symposia on Biology & Management of Mangroves and Tropical Shallow Water Communities. Papua New Guinea, July 20-August 2, 1980

Thom, B.G. (1967). Mangrove ecology and deltaic geomorphology: Tabasco, Mexico. J. Ecol. 55: 301-343

Thom, B.G., Wright, L.D., Coleman, J.M. (1975). Mangrove ecology and deltaic-estuarine geomorphology: Cambridge Gulf-Ord River, Western Australia. Journal of Ecology 63: 203-232

Tomlinson, P.B. (1978). *Rhizophora* in Australasia – some clarification of taxonomy and distribution. Journal of the Arnold Arboretum (Harvard University) 59: 156-169

Tomlinson, P.B., Womersley, J.S. (1976). A species of *Rhizophora* new to New Guinea and Queensland, with notes relevant to the genus. Contr. Herb. Austral. 19: 1-10

Van Steenis, C.G.G.J. (1958). Introduction to *Rhizophoraceae*. Flora Malesiana, sér. I, vol. 5: 429-444

Van Steenis, C.G.G.J. (1977). *Bignoniaceae*. In: Flora Malesiana, sér. I, vol. 8: 114-186

Vierhapper, F. (1907). Beiträge zur Kenntnis der Flora Südarabiens und der Inseln Sokótra, Sémha und Abd el Kûri. I. Theil. Gefäßpflanzen der Inseln Sokótra, Sémha und Abd el Kûri. Denkschriften der Kaiserlichen Akademie der Wissenschaften, math.-naturw. Klasse 71, 1. Halbband, Wien, S. 321-490 + 17 Tafeln

Vu Van Cuong, H. (1964). Nouveautés pour la Flore du Camboge, du Laos, et de Vietnam (*Rhizophoraceae, Sonneratiaceae, Myrtaceae*). Addisonia 4: 343-347

Walsh, G.E. (1974). Mangroves: A review. In: Reimold, R.J. and W.H. Queen (eds.) Ecology of Halophytes, pp. 51-174. Academic Press, New York–London

Walter, H., Steiner M. (1936/37). Die Ökologie der Ost-Afrikanischen Mangroven. Z. Bot. 30: 65-193

Wells, A.G. (1980). Distribution of mangrove species in Australia. Abstracts of the 2nd International Symposia on Biology & Management of Mangroves and Tropical Shallow Water Communities. Papua New Guinea, July 20–August 2, 1980

West, R.C. (1956). Mangrove swamps of the Pacific coast of Columbia. Ann. Assoc. Am. Geogr. 46: 98-121

West, R.C. (1977). Tidal salt-marsh and mangal formations of Middle and South America. In: Ecosystems of the World. Vol. 1: Wet Coastal Ecosystems (V.J. Chapman, ed.). Elsevier, Amsterdam–Oxford–New York, pp. 193-213

Wiggins, Ira L. (1980). Flora of Baja California. Stanford University Press, Stanford, California

Womersley, J.S. (1980). The nomenclature and taxonomy of the mangrove flora of Papua New Guinea and adjacent areas. Abstract of the 2nd International Symposia on Biology & Management of Mangroves and Tropical Shallow Water Communities. Papua New Guinea, July 20–August 2, 1980

Woolston, F.P. (1973). Ethnobotanical items from the Wellesley Island, Gulf of Carpentaria. Occ. Pap. Univ. Queensland Anthropol. Mus. No. 1: 95-103

Zahran, M.A. (1975). Biogeography of mangrove vegetation along the Red Sea coasts. In: Proc. Int. Symp. on Biology and Management of Mangroves (Walsh, G.E., Snedaker, S.C., Teas, H.J., eds.), Univ. Florida Press, Gainesville, Fla., pp. 43-51

Zahran, M.A. (1977). Africa. A. Wet formations of the African Red Sea Coast. In: Ecosystems of the World, Vol 1: Wet Coastal Ecosystems (V.J. Chapman, ed.), Elsevier, Amsterdam–Oxford–New York, pp. 215-231

CHAPTER 4

Survey and adaptive biology of halophytes in western Rajasthan, India

DAVID N. SEN, KISHAN S. RAJPUROHIT and F.W. WISSING*

*Departments of Botany and Geography, University of Jodhpur,
Jodhpur 342001, India*

1. Introduction

I. Physiography

Western Rajasthan (part of Indian arid zone) is located between the 24° 0′ to 30° 05′ N latitudes and 70° 0′ to 75° 42′ E longitudes, to the west of Aravalli mountains, extending up to the Thar desert of Pakistan. It covers nearly 233 100 sq km area of Rajasthan State. Like other hot deserts of the world, a considerable part of the Rajasthan desert is comprised of saline unproductive land, in the form of saline lakes, salt depressions, swamps and saline lands covered by desert wind blown sand. Such salt basins are the main characteristic feature of the hot deserts, where evapotranspiration exceeds rainfall. The saline areas of the Indian arid zone can be divided into two parts: (i) the salt lakes, such as the Sambhar, Didwana and Kuchaman, located in the eastern part of the Rajasthan; and (ii) the salt basins like the Pachpadra, Thob, Sanwarla, Bap (Bap-Malhar), Pokran, Lunkaransar, Lanela (Kharia, Kanodwala and Mitha Ranns), Sakhi and Khaju-walla located in the western half of Rajasthan (Fig. 1).

Actually these salt basins are low-lying areas, almost levelled with a 1% slope, which are located in the hilly and sand-dune tracts where local run-off collects. This collected water evaporates and these basins dry up. Upon dehydration, the surface layers become covered with a white crust of salt in the form of patches which consist of clay soils with evaporite deposits like sodium chloride, sodium sulphate, potassium chloride, magnesium sulphate, gypsum, nitrates and other salts.

Judging from the extensive aerial photographic study done by Ghose et al. (1966) and Singh et al. (1971), it appears that all the three salt basins under the present investigation are part of the Luni river system and similar to the saline lands of Western Australia. Ghose (1964) stated that Sanwarla, Pachpadra, Thob and other minor depressions in the northern part of the central Luni basin are the 'wet playas' (saline). A centripetal drainage system was not at all conspicuous and as such channels which had been disconnected from the head water branches after the arid conditions were set in.

The salinity in the salt basins generally increases from the periphery to the centre, while in some cases the gypsum deposition in the upper horizon of the profile or sand accumulation at the surface layers leads to lower salinity levels in the centre. In most cases salinity of inland areas is related to high aridity and a saline water table from rocks rich in sodium salts. There are low physiographic gradients so that water accumulates rather than draining away (Hayward, 1954).

II. Origin of saline areas

The origin of different salt basins of western Rajas-

---

* Abteilung Ökologie, Fakultät für Biologie, Universität Bielefeld, D-4800 Bielefeld, F.R. Germany.

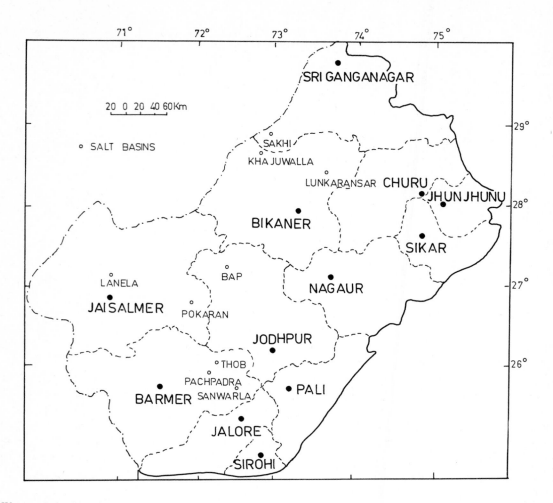

Fig. 1. Western Rajasthan, showing some important salt basins.

than is a much debated question and different theories have been put forward to explain it. Previously it was thought that the salts of this region were derived from various weathering products. This theory could not find support, however, due to the fact that such a huge amount of chloride – which is the main constituent of salt – could not be accumulated through weathering. Blanford (1876) and Hume (1924) suggested that the salt of these basins was a surface deposit left by the desiccation of the sea in the late tertiary period. Absence of calcium and magnesium salts in some salt lakes, however, ruled out this theory. It has also been suggested that these salt basins originated on the old rock-salt beds of marine formation, but no

evidences of marine depositions have been given.

Holland and Christie (1909) carried out experiment using an aspirator at the Pachpadra salt basin in western Rajasthan to collect airborne particles. From the analysis of these particles they advanced the view that the salts are carried by wind from the Rann of Kutch either as particles or sprays of sea water and deposited in the salt basins. The difference between the composition of salt of inland basins and sea salt and how wind can deposit salt at particular places, are two major objections to this wind-borne salt theory. Besides this, the composition of salt varies from one salt basin to another.

It is widely accepted that salts of these salt basins

have been formed by the atmospheric decomposition of felspathic rocks of this region, granite and gneiss of the Aravallis, which leached out by rain water and carried to the depressions without outlets. Having interpreted his aerial photographs, Ghose (1965) came to the conclusion that these basins were part of old river systems and that their salinity might also be due to the hillocks buried under the sand, doming the underground water flow.

Ghose (1964) gave his theory of geomorphological aspects of the formation of salt basins of Pachpadra, Sanwarla and Thob. According to him, before the arid cycle was set, the Luni river had a well organized drainage system. The mouth of its streams were blocked by their own silt deposits and other finer materials. After the onset of aridity, the scanty rainfall could not keep these streams alive. At present, water from catchment areas sinks down and flows subterraneously along these channels, with all the soluble salts washed down from the catchments. But, due to siltation of the channels, these streams are unable to reach the trunk streams. The result is subterraneous water logging, capillary rise of water, evaporation, and gradual increase in salt concentration. The above phases lead to development of the salt basins in western Rajasthan.

III. Salinity and climate

The most serious problem of arid areas is the scarcity of water. In hot deserts, like the Indian arid zone, most of the rainfall occurs during the summer, which leads to a high rate of evapotranspiration. During that time the effective utilization of water by plants is also high. In addition to the low precipitation, the wide variability in rainfall is a characteristic feature of the Indian arid zone. Here, rain falls mainly in thunderstorms so that precipitation is either meagre in amount, or is heavy in a brief period. This leads to rapid evaporation or to heavy storm run-off. In these regions the lack of moisture in certain years may be absolute or be caused by insufficient precipitation. Certain parts of arid regions are characterized by salinity.

It is well documented in the earlier climatic analyses of western Rajasthan by Bharucha and Meher-Homji (1965), Meher-Homji (1962a), Krishnan and Shankarnarayan (1964), Krishnan and Rakhecha (1965), Krishnan (1968), Gaussen et al. (1971), Harsh (1976), Charan (1978) and Charan et al. (1978a) that this region is characterized by a high rate of evaporation and low precipitation. These climatic conditions lead to the formation of salt basins in the low-lying areas of hot deserts (Raheja, 1966). They contain extra concentrations of soluble salts (Raheja, 1966; Rajpurohit and Sen, 1979).

It is a well documented phenomenon that in low-lying salt basins of arid regions, scanty and erratic rainfall which is unable to leach out the salts from rhizosphere regions, leads to the accumulation of an increasing amount of soluble salts in the upper horizon of the soil profile (USDA Handbook No. 60, 1954; Raheja, 1966; Waisel, 1972). The Pachpadra, Sanwarla and Thob salt basins have low rainfall which is distributed mainly over a short rainy season (July, August and September).

Seasonal variation in soil salinity in saline habitats is well documented and is directly influenced by the fluctuations in soil moisture levels. Decrease in soil moisture and intensity of rainfall lead to an increase in soil salinity, due to capillary movement of the salts in the soil (Rajpurohit and Sen, 1979). A considerable seasonal variation in the soil salinity was recorded at all the investigated spots of the three salt basins (Rajpurohit, 1980). Higher salinity levels in June have been observed due to the evaporative mechanism in summer.

The soil moisture percentage is mainly affected by rainfall (Rajpurohit and Sen, 1980; Rajpurohit et al., 1979). In salt basins of the Indian desert (Rajpurohit, 1980), the soil moisture percentage at the surface at a depth of 20 and 50 cm was high during months in the rainy season, dropping off sharply in February and June. At a depth of 20 and 50 cm the soil moisture percentage in all the investigated spots was comparatively higher and showed less difference seasonally than those recorded at the surface.

A considerable seasonal variation in the total soluble salts was recorded in the values of electrical conductivity as well as in chloride ion content due to the upward and downward movement of salts in the soil profile. Higher values of chloride ion content, as well as electrical conductance was recorded in the dry season (February, June and November) and it decreased in the rainy season (July, August and September) reaching a minimum in August.

Great fluctuations in electrical conductance as well as in chloride ion content were recorded throughout the course of study at the surface layers of all the spots investigated (Rajpurohit, 1980). Reduction in surface salinity due to premonsoonal showers in June was also accompanied by increased salinities in July at a depth of 20 cm, but extensive rains in July and August reduced the electrical conductance and chloride ion content to a certain extent at a depth of 20 and 50 cm. This leaching of salts to lower levels explains the seasonal decrease in salt content of surface layers in the rainy season.

When compared to those of the deeper layers, it appears that total soluble salts in the form of electrical conductance and chloride ion percentages remain higher at surface level.

Rajpurohit (1980) recorded a distinct variability in chloride ion content and electrical conductance among different spots of a locality.

2. Survey of saline vegetation

Natural vegetations of western Rajasthan have been studied by various workers since the 19th century. King (1879) studied systematically the floristic and vegetational associations which was followed by Blatter and Hallberg's (1918-1921) classification of Indian desert flora into 13 divisions. Champion (1936) gave an inadequate classification for the vegetation of the Indian subcontinent and later on, Chatterji (1939) divided the vegetation of India into eight phytogeographical regions on the basis of rainfall. The vegetation of different regions did not coincide with other important factors, e.g. edaphic and physiognomic factors. Mathur (1960) proposed five main forest types of Rajasthan and showed a demarcation line between western and eastern elements on the basis of vegetation types. Raheja and Sen (1964) divided the region into five phytogeographical units on the basis of climate and vegetation. During their in-

vestigations on the floristic elements of the semi-arid zones of India, Bharucha and Meher-Homji (1965) and Meher-Homji (1962b) gave the distribution of plant species based on evidence from modern bio-climatological studies. Bhandari (1978) gave percentages of the floristic elements for the Indian arid zone and studied the floristic composition. Similarly, Charan (1978) studied the vegetation of the extreme western part of this region from various ecological and phytogeographical aspects.

The vegetation of the saline areas of western Rajasthan has not been studied in its proper perspective although frequent references on the occurrence of saline plants are to be found in several floras and papers since the time of the Flora of British India. The saline areas present very interesting aspects for ecological, physiological and phytogeographical studies. Only certain physiologically specialized and ecologically adapted plants grow in this sensitive ecosystem and survive in the saline soil.

I. Classification

(i) On the basis of adaptability to saline soil
The vegetation of some Indian salt basins has been classified by Sen and Rajpurohit (1978) as follows:

(a) True halophytes: Plants growing in saline soil – *Suaeda fruticosa, Cressa cretica, Aeluropus lagopoides, Salsola baryosma, Haloxylon recurvum* and *Zygophyllum simplex.*

(b) Facultative halophytes: Plants which can grow in saline as well as in non-saline soils. (i) Mainly saline but can grow in non-saline areas – *Trianthema triquetra, Launaea nudicaulis, Tamarix dioica, Eragrostis ciliaris, E. pilosa, Salvadora persica* and *Cleome brachycarpa.* (ii) Mainly non-saline but can grow in saline areas – *Pulicaria wightiana, Euphorbia granulata* and *Indigofera cordifolia.*

(c) Transitional halophytes: Plants growing only at the transition of saline and non-saline areas – *Haloxylon salicornicum, Sporobolus helvolus* and *S. marginatus.*

(d) Marshy halophytes: Plants growing in water-logged areas of salt basins – *Scirpus roylei, S. tuberosus, Mariscus squarrosus* and *Tamarix troupii.*

(e) True glycophytes: Plants growing in non-saline soil–*Dactyloctenium sindicum, D. aegyptium, Cenchrus setigerus, C. biflorus, Citrullus colocynthis, Cucumis callosus* and *Solanum surattense.*

Later, Rajpurohit (1980) reported a total of 122 plant species (Table 1) including ten true halophytes, 48 facultative halophytes and 64 glycophytes and classified them into the following three groups:

(a) True halophytes: Plants which mainly attain optimal growth on the saline soil (above 0.5% NaCl level).

(b) Facultative halophytes: Plants which can grow and achieve optimal growth on saline soil (at 0.5% NaCl level) like halophytes, as well as on non-saline soil like glycophytes.

(c) Glycophytes: Plants of non-saline habitat which always grow and achieve optimal growth at non-saline niches of the salt basin.

In the above study, 64 plant species of saline areas have been included in the group of glycophytes, plants that are unable to achieve optimal growth in the habitats which have NaCl concentrations of more than 0.5%. These plant species are mainly ephemerals which appear only during the rainy season when NaCl salinity goes down below 0.5% or ephemerals and perennials which grow on the non-salty sand accumulated habitats.

(ii) On the basis of Raunkiaer's life-forms

Life-forms of vascular plants characterize the physiognomy of a plant community (Hanson and Churchill, 1961). The physiognomic aspect of the saline areas, which are parts of desert, is halophytic or xerophytic, with plants widely dispersed, leaving bare large areas. Despite the fact that different life-forms are common in various plant formations under various climates, only very few habitats have

a plant cover which consists of a single life-form only. The spectrum of life-forms, dominating under a certain climate represents some mode of adaptation of plants to that specific climate (Raunkiaer, 1934).

Such studies have been made for different parts of Rajasthan by Das and Sarup (1951), Meher-Homji (1964), Agarwal (1974) and Charan et al. (1978b). Classification of saline vegetation of Israel has been done by Waisel (1972) and of Egypt by Zahran (1981).

Being part of the arid zone, saline areas of western Rajasthan have a dominantly therophytic nature (Table 1) (Rajpurohit, 1980). The ephemerals and other herbs under the therophytes are more abundant and have the highest percentage (49.18%). Next to therophytes are nanophanerophytes, their percentage being 16.39. The other life-forms are geophytes, chamaephytes, mega-micro-phanerophytes, lianes, i.e. climbing phanerophytes and hemicryptophytes, their percentage being 12.30, 8.20, 6.56, 4.09 and 3.28, respectively (Table 2). Compared to the normal life-form spectrum of Raunkiaer (1934), the percentage of geophytes in the salt basins of western Rajasthan is three times higher. The epiphytes, helophytes, parasites and hydrophytes are absent in the life-form spectrum of these areas.

II. Comparison of vegetation structure with other saline areas

There are certain species of plants that are so fastidious regarding their habitat conditions that they can not grow in any other habitat except a particular one. This led ecologists to term them as plant 'indicators' of their specific habitat. From the mere presence of such indicators and other sociological data one can easily and effectively understand their habitat (Meher-Homji, 1962b).

The saline areas of the Indian arid zone have a mosaic of successive zones represented by characteristic taxa of limited distribution (Rajpurohit, 1980). These zones differ from one another in soil characteristics. Each zone supports a distinct plant grouping with its own characteristic species. Further, there is a distinct change in the taxa com-

Table 1. Classification and floral elements of different plant species in different localities.

Plant species	Family	Element	Classification		Localities		
			Adaptability to saline soil	Raunkiaer's life-forms	p	s	t
1	2	3	4	5	6	7	8
Acacia jaquemonti Benth. ex Hook	Mimosaceae	Indian	Gl.	N	+	−	−
A. nilotica (L.) Del.	Mimosaceae	Sudano-Rajasthanian	Gl.	M	−	+	−
Achyranthes aspera Linn.	Amaranthaceae	Pantropical	Gl.	N	+	−	+
Aeluropus lagopoides (L.) Trin. ex Thw.	Poaceae	Mediterranean	Hl.	H	+	+	+
Aerva persica (Burm. f.) Merrill.	Amaranthaceae	Pantropical	F.Hl.	N	+	+	−
A. pseudotomentosa Blatt. & Hallb.	Amaranthaceae	Indian desert	Gl.	N	+	−	−
Amaranthus hybridus Linn.	Amaranthaceae	Pluriregional	F.Hl.	Th	−	+	+
A. gracilis Desf.	Amaranthaceae	Pantropical	Gl.	Th	+	−	−
Anticharis senegalensis (Walp.) Bhand.	Scrophulariaceae	Saharo-Sindian	Gl.	Th	+	−	−
Aristida funiculata Trin. et Rupr.	Poaceae	Tropical African	Gl.	Th	+	+	−
Blepharis sindica T. Anders	Acanthaceae	Indian	Gl.	Ch	+	−	−
Boerhavia diffusa Linn.	Nyctaginaceae	Pantropical	F.Hl.	Th	+	−	−
B. repanda Roxb.	Nyctaginaceae	Indian	Gl.	Cp	+	+	+
Brachiaria ramosa (L.) Stapf.	Poaceae	Indian	F.Hl.	Th	+	−	+
Bulbostylis barbata (Rottb.) C.B. Clarke	Cyperaceae	Pantropical	F.Hl.	G	+	+	+
Calligonum polygonoides linn.	Polygonaceae	Mediterranean	Gl.	N	+	−	−
Calotropis procera (Ait.) R. Br.	Asclepiadaceae	Pantropical	Gl.	M	+	+	−
Cassia italica (Mill.) Lamk. ex Anders.	Caesalpiniaceae	Sudano-Rajasthanian	F.Hl.	N	+	−	−
Capparis decidua (Forsk.) Edgew.	Capparaceae	Sudano-Rajasthanian	Gl.	M	−	+	+
Cenchrus biflorus Roxb.	Poaceae	Sudano-Rajasthanian	Gl.	Th	+	+	+
C. ciliaris Linn.	Poaceae	Mediterranean	Gl.	Ch	+	−	−
C. prieurii (Kunth) Marei	Poaceae	Tropical African	Gl.	Th	+	−	−
C. setigerus Vahl.	Poaceae	Sudano-Rajasthanian	Gl.	Ch	+	+	+
Citrullus colocynthis (L.) Schrad.	Cucurbitaceae	Saharo-Sindian	Gl.	H	+	−	−
Chloris virgata Sw.	Poaceae	Warm countries	F.Hl.	Th	+	+	+
Cleome vahliana Farsen	Capparaceae	Sudano-Rajasthanian	F.Hl.	Th	+	−	+
Cleome viscosa Linn.	Capparaceae	Pluriregional	Gl.	Th	+	−	−
Coccinia grandis (L.) Voigt.	Cucurbitaceae	Pantropical	Gl.	Cp	+	−	−
Cocculus pendulus (Forst.) Diels.	Menispermaceae	Pantropical	Gl.	Cp	+	−	−
Convolvulus microphyllus Sieb. ex Spreng.	Convolvulaceae	Sudano-Rajasthanian	F.Hl.	H	+	+	−
Corchorus depressus (L.) Christensen	Tiliaceae	Sudano-Rajasthanian	Gl.	H	+	+	+
C. tridens Linn.	Tiliaceae	Pantropical	Gl.	Th	+	+	+
Cressa cretica Linn.	Convolvulaceae	Warm countries	Hl.	Ch	+	+	+
Crotalaria burhia Buch.-Ham.	Fabaceae	Indus Plain	Gl.	N	+	−	−
Cucumis callosus (Rottl.) Cogn.	Cucurbitaceae	Indo-Malayan	F.Hl.	Th	+	+	+
Cyperus arenarius Retz.	Cyperaceae	Tropical African	Gl.	G	+	−	−
C. bulbosus Vahl.	Cyperaceae	Pantropical	F.Hl.	G	+	+	+
C. compressus Linn.	Cyperaceae	Warm countries	F.Hl.	G	−	+	−
C. flavidus Retz.	Cyperaceae	Warm countries	F.Hl.	G	−	+	−
C. iria Linn.	Cyperaceae	Pantropical	F.Hl.	G	+	+	+
C. pengurei Rottb.	Cyperaceae	Warm countries	Gl.	G	−	+	−
C. rotundus Linn.	Cyperaceae	Warm countries	F.Hl.	G	+	+	+
C. triceps (Rottb.) Eng.	Cyperaceae	Indo-Malayan	F.Hl.	G	−	+	−
C. tuberosus Rottb.	Cyperaceae	Tropical African	F.Hl.	G	+	+	+
Dactyloctenium aegyptium (L.) P. Beauv.	Poaceae	Warm countries	F.Hl.	Th	+	+	+

Table 1 *(continued)*.

1	2	3	4	5	6	7	8
D. sindicum Boiss.	Poaceae	Sudano-Rajasthanian	Gl.	Ch	+	−	+
Dichanthium annulatum (Forsk.) Stapf.	Poaceae	Warm countries	F.Hl.	N	+	−	+
Digera muricata (L.) Mart.	Amaranthaceae	Saharo-Sindian	Gl.	Th	+	+	+
Digitaria ciliaris (Retz.) Koeler.	Poaceae	Indian	F.Hl.	Th	+	−	+
Echinochloa colonum (L.) Link.	Poaceae	Warm countries	F.Hl.	Th	+	−	+
Echinops echianatus Roxb.	Asteraceae	Indian	Gl.	Th	+	+	+
Eclipta prostrata Linn.	Asteraceae	Warm countries	F.Hl.	Th	+	+	+
Eleusine compressa (Forsk.) Asch. ex Schw.	Poaceae	Saharo-Sindian	Gl.	Ch	+	+	+
Enicostema hyssopifolia (Wild.) Verdoon	Gentianaceae	Pantropical	Gl.	Th	+	−	−
Ephedra foliata Boiss.	Gnetaceae	Mediterranean	Gl.	Cp	+	−	−
Eragrostis ciliaris (L.) R. Br.	Poaceae	Pantropical	F.Hl.	Th	+	+	+
E. pilosa (L.) P. Beauv.	Poaceae	Warm countries	F.Hl.	Th	+	+	+
E. tremula Hochst. ex Steud.	Poaceae	Tropical African	F.Hl.	Th	+	+	+
Euphorbia granulata Forsk.	Euphorbiaceae	Saharo-Sindian	Gl.	Th	+	−	+
E. jodhpurensis Blatt. & Hallb.	Euphorbiaceae	Indian desert	Gl.	Th	+	−	−
E. prostrata Ait.	Euphorbiaceae	Pantropical	Gl.	Th	+	−	−
Evolvulus alsinoides (L.) Linn.	Convolvulaceae	Indian	Gl.	Th	+	−	−
Fagonia cretica Linn.	Zygophyllaceae	Saharo-Sindian	F.Hl.	Th	+	+	+
Farsetia hamiltonii Royle	Brassicaceae	Saharo-Sindian	Gl.	Th	+	−	−
Fimbristylis miliacea (L.) Vahl.	Cyperaceae	Warm countries	F.Hl.	G	+	+	+
Gisekia pharnacioides Linn.	Molluginaceae	Pantropical	Gl.	Th	+	−	−
Haloxylon recurvum (Moq.) Bunge ex Boiss.	Chenopodiaceae	Indian	Hl.	N	+	−	−
H. salicornicum (Moq.) Bunge ex Boiss.	Chenopodiaceae	Indus plain	F.Hl.	N	+	−	−
Heliotropium marifolium Koem. ex Retz.	Boraginaceae	Indian	Gl.	Th	+	−	−
Heylandia latebrosa DC.	Fabaceae	Indian	F.Hl.	Th	+	+	−
Hibiscus obtusifolius Garcke	Malvaceae	Indian	F.Hl.	Th	+	−	−
Indigofera cordifolia Heyne ex Roth.	Fabaceae	Indo-Malayan	Gl.	Th	+	+	+
I. hochstetteri Baker	Fabaceae	Sudano-Rajasthanian	Gl.	Th	+	−	−
I. linifolia (L.) Retz.	Fabaceae	Pantropical	Gl.	Th	+	−	−
I. oblongifolia Forsk.	Fabaceae	Pantropical	Gl.	Th	+	−	−
Ipomoea pes-tigridis Linn.	Convolvulaceae	Indo-Malayan	Gl.	Th	+	−	−
Justicia simplex D. Don	Acanthaceae	Sudano-Rajasthanian	Gl.	Th	+	−	−
Launea chondrilloides (DC.) Hook.	Asteraceae	Saharo-Sindian	Gl.	Ch	+	−	−
L. nudicaulis Hook	Asteraceae	Indian	Gl.	Th	+	−	−
Leucas urticaefolia (Vahl.) R. Br.	Lamiaceae	Sudano-Rajasthanian	Gl.	Th	+	−	+
Lycium barbarum Linn.	Solanaceae	Saharo-Sindian	Gl.	M	+	−	−
Mariscus squarrosus (L.) C.B. Clarke	Cyperaceae	Tropical African	F.Hl.	G	+	+	+
Marsilea aegyptiaca Wild.	Marsileaceae	Saharo-Sindian	F.Hl.	G	−	−	+
Melanocenchrus jacquemontii Jaub. et Spach	Poaceae	Sudano-Rajasthanian	F.Hl.	Th	+	−	−
Oligochaeta ramosa (DC.) C. Koch	Asteraceae	Indian	Gl.	Th	+	−	−
Panicum antidotale (L.) Retz.	Poaceae	Pantropical	Gl.	N	+	−	+
P. turgidum Forsk.	Poaceae	Indus plain	Gl.	N	+	−	−
Phyla nodiflora (L.) Greene	Verbenaceae	Warm countries	Gl.	Ch	+	−	−
Phyllanthus fraternus Webster.	Euphorbiaceae	Pantropical	Gl.	Th	+	−	−
P. maderaspatensis Linn.	Euphorbiaceae	Pantropical	Gl.	Th	+	−	−
Pergularia daemia (Forsk.) Chiov.	Asclepiadaceae	Indian	Gl.	Cp	+	−	−
Polygala eriopetra DC.	Polygonaceae	Sudano-Rajasthanian	Gl.	Th	+	−	−
Portulaca oleracea Linn.	Portulacaceae	Warm countries	F.Hl.	Th	−	+	+
Prosopis juliflora (Swartz.) DC.	Mimosaceae	Tropical American	Gl.	M	+	+	+
Psoralea odorata Blatt. & Hallb.	Fabaceae	Indian desert	Gl.	N	+	−	−
Pulicaria crispa (Cass.) Benth. & Hook.	Asteraceae	Saharo-Sindian	F.Hl.	Th	+	+	+
P. rajputanae Blatt. & Hallb.	Asteraceae	Indian desert	F.Hl.	Th	+	−	−

Table 1 *(continued)*.

1	2	3	4	5	6	7	8
P. wightiana (DC.) C.B. Clarke	Asteraceae	Indian	F.Hl.	Th	+	+	+
Salsola baryosma (Roem. et Schult.) Dandy	Chenopodiaceae	Sudano-Rajasthanian	Hl.	N	+	+	+
Salvadora oleoides Decne.	Salvadoraceae	Sudano-Rajasthanian	F.Hl.	M	+	+	+
S. persica Linn.	Salvadoraceae	Sudano-Rajasthanian	F.Hl.	M	+	+	+
Schoenefeldia gracillis Kunth	Poaceae	Indian	Gl.	Th	+	—	—
Scirpus roylei (Nees) Parker	Cyperaceae	Tropical African	F.Hl.	G	+	—	—
S. tuberosus Desf.	Cyperaceae	Pantropical	F.Hl.	G	+	+	+
Sericostoma pauciflorum Stocks	Boraginaceae	Indian	Gl.	Ch	+	—	—
Sesbania bispinosa (Jacq.) W.F. Wight	Fabaceae	Pantropical	F.Hl.	Th	+	—	+
Sesuvium sesuvioides (Fenzl.) Verde.	Aizoaceae	Saharo-Sindian	F.Hl.	Th	+	+	+
Setaria verticillata (L.) P. Beauv.	Poaceae	Pantropical	F.Hl.	Th	—	—	+
Sporobolus helvolus (Trin.) Dur. et Schinz	Poaceae	Tropical African	Hl.	N	+	+	+
S. marginatus Hochst. et A. Rich.	Poaceae	Saharo-Sindian	F.Hl.	N	+	+	+
Suaeda fruticosa (L.) Forsk.	Chenopodiaceae	Warm countries	Hl.	N	+	+	+
Tamarix dioica Roxb.	Tamaricaceae	Indian	Hl.	M	+	—	+
T. ericoides Rottl.	Tamaricaceae	Indian	Hl.	N	+	—	—
T. troupii Hole	Tamaricaceae	Pluriregional	Hl.	N	+	+	+
Tephrosia purpurea (L.) Pers.	Fabaceae	Pantropical	Gl.	N	+	+	—
Tragus biflorus (Roxb.) Schult.	Poaceae	Warm countries	F.Hl.	Th	+	+	+
Trianthema portulacastrum (Linn.)	Aizoaceae	Pantropical	Gl.	Th	+	+	+
T. triquetra Willd. ex Rottl.	Aizoaceae	Tropical African	F.Hl.	Th	+	+	+
Tribulus terrestris Linn.	Zygophyllaceae	Warm countries	Gl.	Th	+	+	+
Vernonia cinerea (L.) Less.	Asteraceae	Pantropical	F.Hl.	Th	+	+	+
Zaleya redimita (Melville) Bhand.	Aizoaceae	Tropical African	Gl.	Ch	+	—	—
Zygophyllum simplex Linn.	Zygophyllaceae	Tropical African	Hl.	Th	+	+	+

Gl. = Glycophytes, Hl. = Halophytes, F.Hl. = Facultative halophytes, M = Mega-micro-phanerophytes, N = Nanophanerophytes, Ch = Chamaephytes, H = Hemicryptophytes, G = Geophytes, Th = Therophytes, Cp = Climbing phanerophytes, p = Pachpadra, s = Sanwarla, t = Thob, + = Present, — = Absent.

Table 2. Biological spectrum of some regions.

Region	Authority	Percentage distribution of the species among the life-forms										
		M	N	Cp	E	G	He	P	HH	H	Ch	Th
Salt basins of western Rajasthan	Rajpurohit (1980)	6.56	16.39	4.09	—	12.30	—	—	—	3.28	8.20	49.18
Halophytes of Israel	Waisel (1972)		17.00			—	—	—	—	34.50	29.10	19.40
Halophytes of Egypt	Zahran (1980)	5.25	13.15	—	—	28.98	10.52	—	—	—	42.10	—
Western Rajashtan	Charan (1978)	6.00	13.00	5.00	—	3.00	—	—	—	9.00	19.00	45.00
North India (semi-arid)	Meher-Homji (1964)	11.00	12.30	6.00	—	5.20	—	1.00	3.00	10.40	18.30	33.00
Indian desert (Rajasthan)	Das and Sarup (1951)	9.70	4.60	7.80	—	3.40	—	—	—	7.60	4.30	51.70
Normal	Raunkiaer (1934)	28.00	15.00	—	3.00	4.00	2.00	—	—	26.00	9.00	13.00

M = Mega-micro-phanerophytes, N = Nanophanerophytes, Cp = Climbing phanerophytes, E = Epiphytes, G = Geophytes, He = Helophytes, P = Parasites, HH = Hydrophytes, H = Hemicryptophytes, Ch = Chamaephytes, Th = Therophytes.

position from the periphery towards the centre of the salt basin. Some species that dominate at the periphery disappear slowly and gradually in the successive zones towards the centre. Similar vegetational zones, from the periphery of the salt basin towards the centre, were made by Satyanarayan (1964), Gupta and Saxena (1972), Saxena (1977) and Rajpurohit et al. (1979) in different salt basins of the Indian arid zone; Ananda Rao and Aggrawal (1964), Ananda Rao and Mukherjee (1972), Ananda Rao et al. (1974) did the same at different places of the Indian coast line.

The order of zonation is also controlled by the amount of salinity and water present in the soil as observed by Rajpurohit (1980), e.g. in transect 'A' at Pachpadra salt basin both salinity and water content were low at the periphery, high in between centre and periphery, and moderate at the centre, which shows a zone of desert xerophytes (*Eleusine compressa - Dactyloctenium sindicum - Cyperus arenarius*) at the periphery, a zone of halophytes (*Cressa cretica - Zygophyllum simplex - Suaeda fruticosa*) in the middle, and a zone of facultative halophytes (*Dichanthium annulatum - Chloris virgata - Dactyloctenium aegyptium*) in the centre. In transect 'B' less moist and less saline peripheral soil supported a *Dactyloctenium aegyptium - Sporobolus marginatus - Echinochloa colonum* association, while a central zone of high salinity and soil moisture supported a *Tamarix troupii* association. In transect 'C' of an abandoned salt pit at Pachpadra salt basin a peripheral zone of low salinity and low soil moisture content supported a zone of *Salvadora persica - Pulicaria rajputanae - Prosopis juliflora*; a next habitat of high salinity and medium water content supported a zone of *Suaeda fruticosa - Cressa cretica - Eclipta prostrata*; a habitat of medium salinity and high moisture content supported a zone of *Scirpus roylei - Mariscus squarrosus*, while an innermost zone of low salinity and high moisture content supported a zone of *Scirpus tuberosus* (Rajpurohit et al., 1979). Similarly, transects 'D' to 'F' located at Sanwarla and Thob salt basins supported different types of plant associations with variations in salinity and soil moisture. Gupta and Saxena (1972) found that *Suaeda fruticosa* and *Aeluropus lagopoides* are the most salt

tolerant halophytic plants, while in water-logged areas *Cyperus rotundus* and *Scirpus* spp. are dominant. Satyanarayan (1964) observed that as salinity reduces, the succession can proceed further to climax species i.e. *Salvadora persica* and other desert plants.

The distribution of saline plant communities depends upon the presence of soluble salts in the water or soil. The water of the habitat is the dominant ecological factor which determines the distribution of species (Rajpurohit et al., 1979). Similarly, Rajpurohit and Sen (1977) have also reported that salinity is a factor of prime importance in the growth and distribution of plants. Saxena (1977) observed that water-logged areas support *Cyperus* and *Scirpus* species, while a sand deposition at the periphery is largely dominated by *Prosopis juliflora*. Similarly, Zahran (1981) reported different vegetation associations in the differential saline habitats of Egypt in which *Typha australis* usually inhabited an area, where the soil was relatively less saline and the water was less shallow; *Phragmites australis* grew in swamps closer to the dryland often with a higher salt content; *Scirpus litoralis* prevailed in swamps associated with lakes; *Suaeda fruticosa* was common in the inland salt marshes, while *Halopeplis amplexicaulis* was very abundant in the seasonally flooded salt marshes associated with the sandy shore zone.

According to McMahon and Ungar (1978) populations of inland halophytes are subject to severe stress and Weaver (1918) cited the complete disappearance of *Suaeda depressa* and the population fluctuation of *Atriplex patula* var. *hastata* on Nebraska salt pan with precipitation. Similarly, in the salt basins of the Indian arid zone some halophytes are subject to severe stress due to ample salinity and lack of water. These conditions sometimes lead to the disappearance of vegetation. The amount of precipitation during the rainy season also influences fluctuations in the populations of various halophytes viz. *Zygophyllum simplex*, *Salsola baryosma*, etc. and facultative halophytes viz. *Cyperus* spp., *Scirpus tuberosus* and *S. roylei*.

3. Adaptations of saline plants

Salt basins of the Indian arid zone are of the inland types and differ greatly from coastal salines in vegetation make-up; they support a relatively small number of plant species, namely those capable of tolerating the high degree of salinity. These plants can easily survive in areas with extremely low soil water potentials, and have a high osmotic pressure in their cell sap (O'Leary, 1971).

Adaptation to a saline environment is an important property of plants which, at any time during their life cycle, must undergo the effect of salt. A saline environment induces morphological, anatomical, physiological and phenological adaptations in plants. In saline areas the effect of salinity on growth and salinity tolerance is temporary in glycophytes and permanent in halophytes; both try to adapt themselves accordingly. Plants which are well adapted to saline soils from various survival aspects are able to grow only in these habitats. Because the presence of salt in soil creates more favourable conditions for the growth of halophytes than glycophytes, the salt tolerance of halophytes increases both during their growth and development, and from one generation to another.

Resistance of plants to salinity leads to several adaptations. Most of them avoid salinity, some evade, and a few others tolerate it. Most of the plants avoid salinity by limiting reproduction, growth and germination during specific parts of the year, by limiting uptake of salt and by allowing roots to penetrate into non-saline soils. Evasion of salt has been achieved through the accumulation of salts into certain specific cells and trichomes, or, by secretion of excess salts through specially mechanized salt secreting glands.

As the habitats of halophytes in arid regions are mostly wet, there is always enough water in the soil. Dry and saline habitats are bare of vegetation.

The roots of the halophytes do not absorb the unchanged soil solution, but only a very diluted one. As the water is transpired, an accumulation of salts should take place in the leaves in the long run as is the case in *Suaeda fruticosa*.

Under physiological drought conditions, the leaves of saline plants play an important role and

develop certain xeromorphic characteristics, like succulence, reduction in surface area, thick cuticle or waxy layers on epidermis, hair cover, salt glands, etc.

Only a small number of the halophytes are able to excrete the salts. This salt excretion takes place through certain glandular cells. Salt excretory glands have been reported in some of the non-succulent halophytes of the Indian arid zone, viz. *Aeluropus lagopoides*, *Sporobolus helvolus*, *Chloris virgata*, *Cressa cretica*, *Tamarix dioica*, *T. ericoides*, and *T. troupii*.

The extreme halophytes lack the ability to excrete salt. They are highly succulent and thwart the rising of salt concentration by a permanent increase of their water content. They become more and more succulent during their development. This type of characteristic has also been found in Indian desert halophytes like, *Haloxylon recurvum*, *H. salicornicum*, *Portulaca oleracea*, *Salsola baryosma*, *Sesuvium sesuvioides*, *Suaeda fruticosa*, *Trianthema triquetra*, *Zygophyllum simplex*, etc. leading to thickening in leaves, elongation of cells, higher elasticity of cell walls and smaller relative surface areas, decrease in extensive growth, and high water content per unit of surface area. Leaves in some succulent halophytes like those in *S. fruticosa*, *S. baryosma* and *T. triquetra* are reduced in surface area, when exposed to a high salt content in the soil.

The third type of plant is known as the cumulative halophyte which lacks any regulatory mechanism. The salt concentration therefore rises during the growing season and, when a certain level is reached, the plant dies. In the Indian arid zone the main cumulative halophytes are: *Fagonia cretica*, *Vernonia cinerea*, *Eclipta prostrata*, *Scirpus* spp., *Cyperus* spp., etc.

Xerosucculents are characterized by a thick cuticle and a cover of waxy layers, e.g. *S. fruticosa*, *S. baryosma*, *H. recurvum* and *H. salicornicum*. Cuticle and waxy layers have also been reported on the leaf surfaces of *C. cretica*, *A. lagopoides*, *S. helvolus*, *S. marginatus*, *Chloris virgata*, etc.

Some halophytes, like *C. cretica*, *C. virgata*, *S. helvolus*, *S. marginatus*, *A. lagopoides*, etc show an additional mode of adaptation to their habitat. The leaves and stems of these plants remain covered with hair (trichomes) giving the plant a greyish appearance. Their effectiveness in reducing the loss of water is small (Waisel, 1972) but they are able to protect the leaf surface against dust.

Development of anthocyanins is a well known characteristic of plants exposed to drought, osmotic drought or physiological drought. Anthocyanins may develop in the leaves or stems of plants when a change in their metabolism results in a change in their ability to resist stress conditions. Some halophytes, like *S. fruticosa*, *S. baryosma*, *T. triquetra*, *Z. simplex*, etc., of this region exhibit the above characteristic under conditions of osmotic stress.

The effect of salinity as a specific and dominant factor in a saline environment determines to a great extent the ability of halophytes to reproduce and perpetuate. Information regarding germination behaviour of Indian halophytes is still scanty. Rajpurohit and Sen (1977) concluded that in the field conditions the highest germination percentage in *Cressa cretica*, *Haloxylon recurvum*, *Suaeda fruticosa*, and *Trianthema triquetra* can be achieved after rains which are heavy enough to leach out the salt from the close environment of the seeds.

It has been proved by several authors that the increase in salinity leads to dormancy of seeds in halophytes and glycophytes (Rajpurohit and Sen, 1977, 1981; Joshi and Iyengar, 1977). Rajpurohit (1980) evaluated the salt tolerance in seed germi-

nation of eight characteristic halophytes of the Indian desert. The results at the end of eight days are shown in Table 3.

It appears from Table 3 that seed germination is better in *Salsola baryosma* in low osmotic potentials, because seeds could germinate upto osmotica of −13 bars. Increase in germination was seen in *Salsola baryosma*, *Sesuvium sesuvioides* and *Trianthema triquetra* with a slight decrease in osmotic potentials upto −1 or −2 bars. Only *Cressa cretica*, *Salsola baryosma*, *Sporobolus helvolus* and *Trianthema triquetra* could germinate to a certain extent under osmotic stress of −7 bars, whereas *S. baryosma* showed germination even in −9 bars also.

4. Soil plant relationship

The saline plants of the Indian arid zone are totally dependent upon the availability of water in the rainy season (July–September), and the water deficiency is the most important factor controlling their growth and survival. The rainy season water (present in the soil) is depleted by the beginning of winter (November). In addition to tolerance to the lack of soil moisture, tolerance to salt is an important property for all plants which, at any time during their life cycle, are exposed to the effect of salt.

Table 3. Percentage seed germination in *Aeluropus lagopoides* (A.l), *Cressa cretica* (C.c), *Haloxylon recurvum* (H.r), *Salsola baryosma* (S.b), *Sesuvium sesuvioides* (S.s), *Sporobolus helvolus* (S.h), *Suaeda fruticosa* (S.f), and *Trianthema tiquetra* (T.t) (adapted after Rajpurohit, 1980).

Electrical conductance (mmho/cm) at 25°C	Osmotic potential (−bars)	Germination percentage after 8 days							
		A.l	C.c	H.r	S.b	S.s	S.h	S.f	T.t
0.0	0.0	100.0±0.0	93.3± 4.7	90.0± 0.0	70.0±0.0	96.6± 4.7	93.3± 9.4	70.0± 8.2	93.3±4.7
2.73	1.0	66.6±4.7	90.0± 8.2	30.0± 8.2	73.3±4.7	100.0± 0.0	86.6± 4.7	56.6± 9.4	96.5±4.7
5.51	2.0	56.6±4.7	86.6±12.5	20.0± 0.0	70.0±0.0	100.0± 0.0	73.3± 4.7	56.6±17.0	96.6±4.7
9.62	3.5	43.3±4.7	90.0± 8.2	13.3±10.0	70.0±8.2	73.3±22.1	46.6± 4.7	13.3± 4.7	86.6±9.4
13.92	5.0	13.3±4.7	13.3±12.5	0	50.0±8.2	26.6± 4.7	33.3±12.5	3.3± 4.7	46.6±9.4
19.43	7.0	3.3±4.7	10.0± 8.2	0	46.6±4.7	0	16.6± 4.7	0	6.6±4.7
25.00	9.0	0	0	0	40.0±8.2	0	0	0	0
30.55	11.0	0	0	0	23.3±4.7	0	0	0	0
36.20	13.0	0	0	0	13.3±4.7	0	0	0	0
41.75	15.0	0	0	0	0	0	0	0	0

I. Soil osmotic potential in relation to plant

It is necessary for plants growing under saline conditions to maintain a high concentration of osmotic substances to successfully overcome the water retaining forces of the surrounding soil. Halophytes growing in saline soils are able to change the osmotic potentials of their cell sap whenever it is required. Black (1960) found that mature leaves of *Atriplex vesicaria* maintained a gradient to the culture solution concentration of about —12 bars over the full concentration range of 0.1 M NaCl; Walter (1961) stated that halophytes which had moved into saline media showed osmotic adaptation of an average rate of 1 atm/day.

In the salt basins of the Indian arid zone Rajpurohit and Sen (1980) observed that the osmotic potential of plant leaves always remained lower than that of the soil of root zone (Figs. 2-3) and the results here were comparable to those of Beadle et al. (1957) and Black (1956, 1960) where the osmotic

potential of *Atriplex* leaves was always lower (—bars) than that of the soil solution. Von Faber (1923) calculated 148 and 163 atmospheres osmotic potential for *Rhizophora* and *Avicennia* leaves. The osmotic pressure in halophytes is higher (lower in —bars) in comparison to mesophytes and xerophytes (Migahid and El Shafei Ali, 1953). Harris et al. (1924) reported that the sap concentration may attain osmotic potential equivalent to 150 atm. in *Atriplex confertifolia*. In nature *A. polycarpa* exhibited plant water potentials as low as —58 bars (Moore et al., 1972b), whereas Harris and Lawrence (1917) reported osmotic potentials up to 50 atm. in plants growing on saline and dry substrata. Montasir (1938) studied the halophytic species of Lake Manzala in Egypt and found that their osmotic potentials ranged from 43.15 to 78.46 atm. Investigations carried out by Rajpurohit and Sen (1980) on the halophytes of the Indian arid zone show that minimum osmotic potentials were observed in *Salsola baryosma* (—69.11 bars) in Feb-

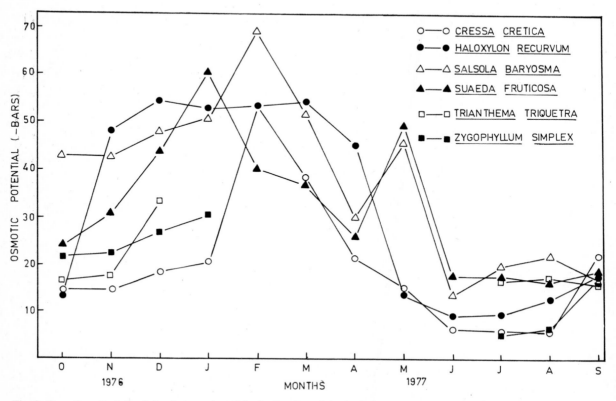

Fig. 2. Osmotic potentials of six plant species of the Pachpadra salt basin during 1976–1977.

ruary. *Suaeda fruticosa* (−60.22 bars) in January, *Haloxylon recurvum* (−54.24 bars) in March, *Cressa cretica* (−53.44 bars) in December and *Zygophyllum simplex* (−30.31 bars) in January.

The seasonal water potential pattern in *Suaeda monoica* reflected climatic changes closely. The fluctuations in water potential (between −22 and −48 bars) were observed in *S. monoica* by Waisel and Ovadia (1972), while osmotic potentials of leaf sap of the same plant range in summer between −63 and −73 bars (Ovadia, 1969). The osmotic pressure of *Z. simplex* plant sap also has a wide range varying from 13.879 to 27.045 atm. (Montasir and Abd El Rahman, 1951). Similarly, a wide range of the osmotic potential has been recorded from one season to the other for all the six species from the Indian arid zone studied by Rajpurohit and Sen (1980).

All of the six species from the Indian arid zone were also able to adjust themselves by rapidly changing their osmotic potentials, with a greater

range due to the change in the surrounding soil's osmotic potential. This is supported by Waisel (1972) who stated that it is probably true that the great majority of halophytic plants belong to the adjustable group and that their osmotic adjustment occurs rapidly. It emerges from Rajpurohit and Sen's study (1980) that during the rainy season higher soil moisture and the leaching down of the salts results in the increased osmotic potential of soil which in turn leads to the increased osmotic potential of plants. In *Atriplex* a sharp decrease in the osmotic potentials was apparently not related to precipitation and was not closely correlated with the gradually decreasing soil water potential, instead, it was closely associated with the development of rupture of vesicular epidermal hairs (Moore et al., 1972a). With the increase in salt concentration and decrease in soil moisture, plants try to adjust themselves to drought through the accumulation of salts (Rajpurohit, 1980). Field studies comparing the drought tolerance of *A.*

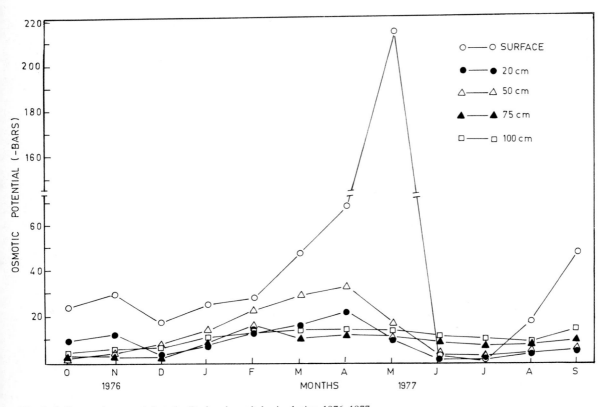

Fig. 3. Soil osmotic potential at the Pachpadra salt basin during 1976–1977.

vesicaria and *Maireana sedifolia* by Carrodus and Specht (1965) suggest that both of these species were able to reduce soil water to water potentials much lower than −15 bars. Improvement in water relations at high salinity levels (0 to −20 bars) were observed by De Jong (1978) for *Atriplex leucophylla* when compared with *Abronia maritima*. Thus, the accumulation of salts in plants decreased their osmotic potentials up to the level of highest stress, so that plants were able to take in a maximum of water during the hot summer; this in turn resulting in the gradual increase in their osmotic potentials. Some halophytic chenopods are able to survive when the water potential of the entire soil profile decreases to far below −15 bars (Jones, 1969; Sharma, 1976).

II. Soil chloride in relation to plant

Rainfall leaches salts down the soil profile , to as far down as the ground water, with a compensating upward movement as a result of capillary action (Jackson et al., 1956). On the basis of rainfall and soil moisture observations, Rajpurohit and Sen (1979) stated that decrease in soil moisture and intensity of rainfall lead to an increase in soil salinity (Table 4) due to the capillary movement of the salts in the soil.

Chloride ions (calculated as NaCl) have similar or even slightly more partial osmotic potential than the corresponding total osmotic potential of the leaves, i.e. at the end of July with *E. lanata* and at the end of June with *A. confertifolia* (Moore et al., 1972a). According to Hayward and Wadleigh (1949) salt accumulator plants are characterized by rapid salt uptake, regulation of internal salt concentration and tolerance of metabolic machinery to high electrolyte concentration. Not all the six plant species studied by Rajpurohit and Sen (1979) from the Indian arid zone, however, followed a similar pattern of Cl^- accumulation (Table 5). The measured values for Cl^- in leaves present a seasonal pattern of incoming and outgoing salts. The factors which might be responsible for inducing the changes are leaf age, growth rate, temperature, transpiration rate, electrolyte concentration of soil water, and probably matric potential of soil water (Sharma et al., 1972). The increase in the salinity of leaves may be due to a gradual accumulation of salts with increasing age (Wood, 1925). In all the species investigated, the ratio of chloride absorbed during a later period was much higher in young than old leaves (Jennings, 1968). Regarding the halophytes of the Indian arid zone, Wood's statement seems to be true only with respect to *C. cretica*. On the other hand, in *Z. simplex*, *H. recurvum* and *T. triquetra* newly appearing and younger leaves showed a Cl^- value more than that in the older leaves as was also reported earlier for *Atriplex* (Black, 1956; Sharma et al., 1972).

Reduction in Cl^- content in the leaves of *C. cretica*, *S. fruticosa* and *H. recurvum* during the

Table 4. Seasonal variations in soil Cl^- (in percentage) at the Pachpadra salt basin from October 1976 to September 1977.

Depth of soil sample (cm)	Months											
	Oct.	Nov.	Dec.	Jan.	Feb.	Mar.	Apr.	May	June	July	Aug.	Sept.
Surface	1.11	1.09	1.14	1.18	1.20	1.22	1.26	1.05	0.07	0.03	0.19	0.97
	±0.10	±0.14	±0.12	±0.14	±0.12	±0.11	±0.08	±0.09	±0.00	±0.00	±0.06	±0.10
20	0.85	0.89	0.89	0.85	0.80	0.85	0.89	0.76	0.15	0.25	0.74	0.76
	±0.05	±0.09	±0.06	±0.07	±0.09	±0.07	±0.08	±0.03	±0.00	±0.02	±0.06	±0.04
50	0.97	0.96	0.99	0.92	1.03	0.98	0.99	0.83	0.53	0.73	1.00	0.80
	±0.06	±0.03	±0.08	±0.06	±0.05	±0.07	±0.05	±0.06	±0.02	±0.05	±0.03	±0.04
75	1.32	1.35	1.44	1.43	1.53	1.78	2.56	2.60	1.11	1.15	1.22	1.27
	±0.10	±0.03	±0.04	±0.10	±0.04	±0.12	±0.09	±0.04	±0.06	±0.05	±0.05	±0.06
100	1.26	1.27	1.34	1.37	1.38	1.38	2.29	2.35	1.16	1.17	1.22	1.27
	±0.13	±0.08	±0.09	±0.07	±0.05	±0.06	±0.22	±0.15	±0.03	±0.05	±0.04	±0.04

period of increased growth in the months, June, July and August and in *T. triquetra* and *Z. simplex* in the months, September, October and November is supported by the findings of Greenway (1965). An increased growth rate resulted in a reduction of electrolyte concentration, as previously reported for *Atriplex* spp. (Sharma et al., 1972) while contradictory results were obtained for *S. baryosma* where there is an increase in Cl⁻ with increased growth rate.

In spite of higher temperatures during the summer months the salinity of plants did not rise when the soil water potential was kept high (Sharma et al., 1972). This is supported by the work of Rajpurohit and Sen (1979) on halophytes of the Indian arid zone. In *C. cretica* (April–June), *H. recurvum* (March–April), *S. baryosma* (March–April), and *S. fruticosa* (March–June) lower values of Cl⁻ percentage were observed. The salt concentration of *Atriplex* leaves increased with an increase in the electrolyte concentration of soil water, and with a decrease in the average osmotic potential of soil water from −2.4 to −7.4 bars, there was a corresponding increase in Cl⁻ concentration from 5.8 to 11.7% (Sharma, 1981); similarly, in *H. recurvum* (May) and *S. baryosma* (May–June) higher values of Cl⁻ percentage were observed. Greenway (1965) postulated that the build up of salinity in *Atriplex* leaves during the summer is due not only to an increased electrolyte concentration in growing medium and to high transpiration rates, but may also be associated with the drought tolerance of these plants. Wood (1925) suggested that the increase in salt content of leaves during the summer may be due to an increase in the salt content of the soil and also to a gradual accumulation of salt in the leaves, while Lachover and Tadmor (1965) implied that these increases were caused by decreases in soil water content.

Osmotic potential of the soil is one of the important factors controlling seasonal variability of salt concentration in *Atriplex* leaves (Sharma et al., 1972). The soils of the locality have lower osmotic potentials during February, March and April (Rajpurohit and Sen, 1980). While Cl⁻ values were lower only in three species, viz., *H. recurvum*, *S. baryosma* and *S. fruticosa*, these values were at a maximum in *C. cretica* during February and March and declined in April (Rajpurohit and Sen, 1979).

On the basis of the above discussion, it appears that seasonal variability of salt concentration in the Indian desert halophytic plant species is caused by a number of factors controlling plant chloride ion content. From the data presented, it is impossible to assess the exact contribution of individual factors. The problem needs detailed and intensive field study in different saline areas with different salinity levels.

Table 5. Seasonal variation in Cl⁻ (in percentage) on dry weight basis in leaves of six saline plants of the Pachpadra salt basin from October 1976 to September 1977.

Names of species	Months											
	Oct.	Nov.	Dec.	Jan.	Feb.	Mar.	Apr.	May	June	July	Aug.	Sept.
C. cretica	3.25	5.53	8.72	7.67	10.08	10.37	6.27	3.86	3.57	3.24	2.95	2.42
	±0.18	±0.61	±0.55	±0.81	±0.77	±0.79	±0.58	±0.34	±0.33	±0.23	±0.36	±0.41
H. recurvum	10.17	12.63	6.22	7.45	7.93	5.53	5.59	11.00	6.41	6.32	5.69	9.53
	±0.73	±0.81	±0.12	±0.33	±0.28	±0.09	±0.16	±0.72	±0.31	±0.08	±0.62	±0.66
S. baryosma	5.38	5.45	5.75	2.13	2.60	3.24	3.93	6.44	7.17	8.76	12.17	6.02
	±0.29	±0.57	±0.48	±0.13	±0.26	±0.26	±0.43	±0.58	±0.66	±0.61	±1.12	±0.36
S. fruticosa	19.58	22.40	22.47	27.14	17.83	16.07	14.24	16.76	13.33	16.13	16.19	15.64
	±1.40	±0.76	±1.38	±1.09	±0.88	±0.75	±0.63	±0.87	±0.80	±0.71	±0.53	±0.70
T. triquetra	5.76	4.46	6.89	*	*	*	*	*	*	8.87	7.95	6.55
	±0.37	±0.26	±0.41							±1.12	±0.49	±0.41
Z. simplex	22.12	16.30	20.95	13.33	*	*	*	*	*	36.16	34.17	18.20
	±0.81	±0.74	±1.04	±0.60						±1.88	±1.29	±0.69

* Plants not seen during these months.

Acknowledgements. Financial assistance from the CSIR to one of us (KSR) in the form of a Senior Research Fellowship, is duly acknowledged. Financial assistance from DAAD (German Academic Exchange Service) for a one-year-stay in Jodhpur (FWW) is also acknowledged.

Literature cited

Agarwal, S.K. 1974. The biological spectrum of the flora of Gogunda and Prasad (Udaipur, Rajasthan). *J. Biol. Sci.* 17: 67–71.

Ananda Rao, T. and Aggrawal, K.R. 1964. Ecological studies of Saurashtra coast and neighbouring islands. III. Okhamandal point to Diu Coastal areas. UNESCO, *Proc. Symp. Prob. Indian Arid Zone*, Jodhpur, pp. 31–42.

Ananda Rao, T. and Mukherjee, A.K. 1972. Ecological aspects along the shores of the Burabalanga tidal estuary, Balasore district, Orissa State. *Proc. Indian Acad. Sci.* 76: 201–206.

Ananda Rao, T., Shanware, P.G. and Mukherjee, A.K. 1974. Ecological studies on the coastal sand dunes and slacks in the vicinity of Digha, Midnapore district, West Bengal. *Indian For.* 100: 101–107.

Beadle, N.C.W., Whalley, R.D.B. and Gibson, J.B. 1957. Studies in halophytes. II. Analytical data on the mineral constituents on three species of *Atriplex* and their accompanying soils in Australia. *Ecology* 38: 340–344.

Bhandari, M.M. 1978. *Flora of Indian Desert.* Scientific Publisher, Jodhpur.

Bharucha, F.R. and Meher-Homji, V.M. 1965. On the floral elements of the semi-arid zones of India and their ecological significance. *New Phytol.* 64: 330–342.

Black, R.F. 1956. Effect of NaCl in water culture on the ion uptake and growth of *Atriplex hastata* L. *Aust. J. Biol. Sci.* 9: 67–80.

Black, R.F. 1960. Effect of NaCl on the ion uptake and growth of *Atriplex vesicaria* Heward. *Aust. J. Biol. Sci.* 13: 249–266.

Blanford, W.T. 1876. Physical geography of the great Indian desert. *Jour. Asiatic Soc. Bengal* 45: 86–103.

Blatter, E. and Hallberg, F. 1918–1921. The flora of Indian desert. *J. Bombay nat. Hist. Soc.* 26: 218–246; 525–551; 811–818; 968–987. 27: 40–47; 270–279; 506–519.

Carrodus, B.B. and Specht, R.J. 1965. Factors affecting the relative distribution of *Atriplex vesicaria* and *Kochia sedifolia* (Chenopodiaceae) in the arid zone of South Australia. *Aust. J. Bot.* 13: 419–433.

Champion, H.G. 1936. A preliminary survey of forest types of India and Burma. *Indian For. Rec.* 1: 1–286.

Charan, A.K. 1978. Phytogeography of western Rajasthan. Ph.D. Thesis, Jodhpur Univ., Jodhpur.

Charan, A.K., Sen, D.N. and Rajpurohit, K.S. 1978a. Ecoclimatic indices and evapotranspiration divisions of western Rajasthan. *Trans. Isdt. & Ucds.* 3: 85–87.

Charan, A.K., Sen, D.N. and Rajpurohit, K.S. 1978b. Biological spectrum of the flora of western Rajasthan, India. *Indian J. For.* 1: 226–228.

Chatterji, D. 1939. Studies on the endemic flora of India and Burma. *Jour. Roy. Asiatic Soc. Bengal Sci.* 5: 19–67.

Das, R.B. and Sarup, S. 1951. The biological spectrum of the Indian desert. *Univ. Raj. Stud. Biol. Sci.* 1: 36–42.

De Jong, T.M. 1978. Comparative gas exchange and growth responses of C_3 and C_4 beach species grown at different salinities. *J. Ecol.* (in press).

Gaussen, H., Legris, P., Gupta, R.K. and Meher-Homji, V.M. et al. 1971. International map of vegetation and environmental conditions. Sheet Rajputana ICAR, New Dehli. Inst. Fr. Pondicherry. *Trav. Sect. Sci. Tech. Inst. Francais de Pondicherry Hars. Ser.* No. 17.

Ghose, B. 1964. Geomorphological aspects of the formation of salt basins in western Rajasthan. UNESCO, *Proc. Symp. Prob. Indian Arid Zone*, Jodhpur, pp. 79–83.

Ghose, B. 1965. Genesis of desert plains in central Luni basin of western Rajasthan. *J. Ind. Soc. Soil. Sci.* 13: 123–126.

Ghose, B., Pandey, S., Singh, S. and Lal, G. 1966. Geomorphology of the central Luni basin, western Rajasthan. *Ann. Arid Zone* 5: 10–25.

Greenway, H. 1965. Plant responses to saline substrates. IV. Chloride uptake by *Hordeum vulgare* as affected by inhibitors, transpiration and nutrients in the medium. *Aust. J. Biol. Sci.* 18: 249–268.

Gupta, R.K. and Saxena, S.K. 1972. Potential grassland types and their ecological succession in Rajasthan desert. *Ann. Arid Zone* 11: 198–218.

Hanson, H.C. and Churchill, E.D. 1961. *The Plant Community.* Reinhold Publ. Co., New York.

Harsh, L.N. 1976. Distribution of biomass and nutrients in the ephemerals of desert ecosystem. Ph.D. Thesis, Jodhpur Univ., Jodhpur.

Harris, J.A., Gortner, R.A., Lawrence, W.F. and Valentine, A.T. 1924. The osmotic concentration, specific electrical conductivity and chloride content of tissue fluids of the indicator plants of Tooele Valley. *Utah J. Agri. Res.* 27: 893–924.

Harris, J.A. and Lawrence, W.F. 1917. The osmotic concentrations of the sap of the leaves of the mangrove trees. *Biol. Bull.* 32: 202–211.

Hayward, N.E. 1954. Plant growth under saline conditions. *Reviews of Research on Problems of Utilization of Saline Water.* UNESCO, Paris. pp. 37–72.

Hayward, H.D. and Wadleigh, C.H. 1949. Plant growth on saline and alkali soils. *Adv. Agron.* 1: 1–38.

Holland, T.H. and Christie, W.A.K. 1909. The origin of the salt depositions in Rajasthan. *Rec. Geol. Surv. India* 38: 154–186.

Hume, W.F. 1924. *Geology of Egypt*, Vols. 2. Cairo.

Jackson, E.A., Blackburn, G. and Clarke, A.R.P. 1956. Seasonal changes in soil salinity at Tintinara, South Australia. *Aust. J. Agric. Res.* 7: 20–44.

Jennings, D.H. 1968. Halophytes, succulence and sodium in plants – A unified theory. *New Phytol.* 67: 899–911.

Jones, R.M. 1969. Soil moisture and salinity under bladder salt bush (*Atriplex vesicaria*) pastures in the New South Wales riverine plain. *Aust. J. Exp. Agr. Anim. Husb.* 9: 603–609.

Joshi, A.J. and Iyengar, E.R.R. 1977. Germination of *Suaeda nudiflora* Moq. *Geobios* 4: 267–268.

King, G. 1879. Sketches of the flora of Rajasthan. *Indian For.* 4: 226–236.

Krishnan, A. 1968. Delineation of different climatic zones in Rajasthan and their variability. *Ind. J. Geog.* 3: 33–40.

Krishnan, A. and Rakhecha, P.R. 1965. Potential evapotranspiration by Thornthwaite and Leeper's method. *Ann. Arid Zone* 4: 32–35.

Krishnan, A. and Shankarnarayan, K.A. 1964. Criteria for the delimitation of the arid zone of Rajasthan. UNESCO, *Proc. Symp. Prob. Indian Arid Zone*, Jodhpur. pp. 380–387.

Lachover, D. and Tadmor, N. 1965. Quantitative studies of *Atriplex halimus* growing as plant fodder in the semi arid conditions of Israel. I – Seasonal variations in chloride content, in essential minerals and the presence of oxalates in the different parts of the plants. *Agron. Trop. Paris* 20: 309–322.

Mathur, C.M. 1960. Forest types of Rajasthan. *Indian For.* 86: 734–739.

McMahon, K.A. and Ungar, I.A. 1978. Phenology, distribution and survival of *Atriplex triangularis* Willd. in an Ohio salt pan. *Amer. Midl. Nat.* 100: 1–14.

Meher-Homji, V.M. 1962a. The bioclimates of India in relation to the vegetational criteria. *Bull. Bot. Surv. India* 4: 105–112.

Meher-Homji, V.M. 1962b. Phytogeographical studies of the semi-arid regions of India. Ph.D. Thesis, Univ. of Bombay, Bombay.

Meher-Homji, V.M. 1964. Life-forms and biological spectra as epharmonic criteria of aridity and humidity in the tropics. *J. Indian bot. Soc.* 43: 424–430.

Migahid, A.M. and El-Shafei Ali, M. 1953. Osmotic pressure of plants of different ecological types. *Bull. de l'Inst. du Desert*, Cairo 3: 9–17.

Montasir, A.H. 1938. Egyptian soil structure in relation to plants. *Bull. Fac. Sci. Cairo Univ.* 15: 47–56.

Montasir, A.H. and Abd El Rahman, A. 1951. Studies on the autecology of *Zygophyllum simplex* L. *Bull. de Inst. du Desert*, Cairo 1: 35–54.

Moore, R.T., Breckle, S.W. and Caldwell, M.M. 1972a. Mineral ion composition and osmotic relations of *Atriplex confertifolia* and *Eurotia lanata*. *Oecologia* 11: 67–78.

Moore, R.T., White, R.S. and Caldwell, M.M. 1972b. Transpiration of *Atriplex confertifolia* and *Eurotia lanata* in relation to soil, plant and atmospheric moisture stress. *Can. J. Bot.* 50: 1411–1418.

O'Leary, J.W. 1971. Physiological basis for plant growth inhibition due to salinity. In *Food, Fiber and the Arid Lands*, eds. W.G. McGinnies, B.J. Goldman and P. Paylore, pp. 332–336, The University of Arizona Press Tucson, Arizona.

Ovadia, S. 1969. Autectology of *Suaeda monoica* Forssk. ex Gmel. M.Sc. Thesis, Tel Aviv University, Tel Aviv.

Raheja, P.C. 1966. Salinity and aridity, new approaches to old problems. In *Aridity and Salinity – A Survey of Soils and Land Use*, ed. H. Boyko, pp. 43–127, Dr. W. Junk Publ., The Hague.

Raheja, P.C. and Sen, A.K. 1964. Resources in perspective. *Sov. Vol. CAZRI*, Jodhpur, pp. 1–28.

Rajpurohit, K.S. 1980. Soil salinity and its role on phytogeography of western Rajasthan. Ph.D. Thesis, Jodhpur Univ., Jodhpur.

Rajpurohit, K.S., Charan, A.K. and Sen, D.N. 1979. Micro distribution of plants in an abandoned salt pit at Pachpadra salt basin. *Ann. Arid Zone* 18: 122–126.

Rajpurohit, K.S. and Sen, D.N. 1977. Soil salinity and seed germination under water stress. *Trans. Isdt. Ucds.* 2: 106–110.

Rajpurohit, K.S. and Sen, D.N. 1979. Seasonal variation in chloride ion percentage of plants and soils of Pachpadra salt basin in Indian desert. *Indian J. Bot.* 2: 17–23.

Rajpurohit, K.S. and Sen, D.N. 1980. Osmotic potentials of plants and soils of Pachpadra salt basin in Thar desert. In *Arid Zone Research and Development*, ed. H.S. Mann, pp. 191–198. Scientific Publ., Jodhpur.

Rajpurohit, K.S. and Sen, D.N. 1981. Effect of salt stress on seed germination of some desert plants. *Ind. J. Bot.* (in press).

Raunkiaer, C. 1934. *The Life-forms of the Plants and Statistical Plant Geography*. Clarendon Press, Oxford.

Satyanarayan, Y. 1964. Habitats and plant communities of the Indian desert. UNESCO, *Proc. Symp. Prob. Indian Arid Zone*, Jodhpur. pp. 59–67.

Saxena, S.K. 1977. Vegetation and its succession in the Indian desert. In *Desertification and its Control*, ed. P.L. Jaiswal, pp. 176–192, ICAR, New Dehli.

Sen, D.N. and Rajpurohit, K.S. 1978. Plant distribution in relation to salinity in Indian desert. Abst. *Second International Congress of Ecology*, Jerusalem, p. 340.

Singh, S., Pandey, S. and Ghose, B. 1971. Geomorphology of the middle Luni basin of western Rajasthan, India. *Ann. Arid Zone* 10: 1–14.

Sharma, M.L. 1976. Soil water regimes and water extraction patterns under two semi-arid shrub (*Atriplex* spp.) communities. *Aust. J. Ecol.* 1: 249–258.

Sharma, M.L. 1982. Aspects of salinity and water relations of Australian chenopods. In *Contributions to the Ecology of Halophytes*, eds. D.N. Sen and K.S. Rajpurohit, pp. 155–172. Dr W. Junk Publ., The Hague.

Sharma, M.L. Tunny, J. and Tongway, D.J. 1972. Seasonal changes in sodium and chloride concentration of saltbush (*Atriplex* spp.) leaves as related to soil and plant water potential. *Aust. J. Agric. Res.* 23: 1007–1019.

USDA Handbook No. 60. 1954. Diagnosis of Improvement of Saline and Alkali Soils.

von Faber, F.C. 1923. Zur Physiologie der Mangroven. *Ber. Deut. Bot. Ges.* 2: 334–342.

Waisel, Y. 1972. *Biology of Halophytes*. Academic Press, New York.

Waisel, Y. and Ovadia, S. 1972. Biological flora of Israel. 3. *Suaeda monoica* Forssk. ex J.F. Gmel. *Israel J. Bot.* 21: 42–52.

Walter, H. 1961. The adaptation of plants to saline soils. UNESCO, *Arid Zone Res. Salinity Problems in the Arid Zone Proc. Tehran Symp.* 14: 129–134.

Weaver, J.E. 1918. The quadrat method in teaching ecology.

Plant World. 21: 267–283.

Wood, J.G. 1925. The selective absorption of chlorine ions and absorption of water by the leaves of the genus *Atriplex. Aust. J. Exp. Biol. Med. Sci.* 2: 45–56.

Zahran, M.A. 1982. Ecology of the halophytic vegetation of Egypt. In *Contributions to the Ecology of Halophytes*, eds. D.N. Sen and K.S. Rajpurohit, pp. 3–20. Dr W. Junk Publ., The Hague.

CHAPTER 5

Biology of *Atriplex*

DAVID B. KELLEY*, J.R. GOODIN and DON R. MILLER**

Department of Biological Sciences, Texas Tech University,
Lubbock, Texas 79409, USA

1. Introduction

At the close of a symposium on *Atriplex* biology held in Deniliquin, New South Wales, Australia, in October, 1969 (Jones, 1970), F.L. Milthorpe attempted to summarize, in grand terms, the proceedings of the symposium in a lecture entitled '*Atriplex* – Atlantean or Atrabilarian?', in which he questioned whether the genus was '...an Atlas ...on the shoulders of which the Australian interior rests...' or a '...sickly hypochondriac...contributing nothing to a well-managed economic ecosystem...'. In terms less quaint, he went on to examine the peculiarities of the genus (physiologial and economic), and to pick out some areas of *Atriplex* biology that might fall under closer scrutiny in the future.

We briefly discuss here some of the contributions made to a biological understanding of this genus since that 1969 conference. Although we hesitate to liken *Atriplex* to Atlas, and are reluctant to label the plant atrabilious, we feel that Milthorpe would agree that the genus occupies important niches in many (and, particularly, arid) ecosystems.

2. Distribution and ecology

Atriplex species are botanically interesting, particularly so because of their adaptations to many diverse – and generally harsh – environments. Besides the botanist's intrinsic interest in the physiology of this diverse group, it has become apparent that many of the habitats of the species of the genus (rangeland, desert, salt marsh, etc.) are becoming increasingly important, i.e., economically valuable, natural resources. Historically, man's encroachment on any such habitat has infrequently been beneficial to the native flora. It is in a strangely converse sense that man's use of these heretofore marginal use areas may depend on his ability to exploit, wisely, the plants which are best adapted to those areas. *Atriplex*, as a genus of manifold adaptations to stressful environments, is therefore of potential value in many of these resource 'frontiers' and is so examined in this review.

Biological features of Atriplex, family Chenopodiaceae

The genus comprises about 200 species worldwide. Most of the species are found in arid and saline habitats, although several species occur in salt marshes and some in fresh water marshes.

The halophytic species of the genus are usually considered facultative halophytes (Waisel, 1972) rather than obligate halophytes. These species transport accumulated salts to specially developed

---

* Department of Land, Air and Water Resources, University of California, Davis, CA 95616, USA.
** Biology Division, Oak Ridge National Laboratories, Oak Ridge, Tennessee 37830, USA.

Tasks for vegetation science, Vol. 2 ed. by D.N. Sen and K.S. Rajpurohit

trichomes on the leaf surfaces, and eventually to the surface of the leaf when the vesiculated hairs burst (Mozafar and Goodin, 1970; Kelley, 1974). It is possible that these halophytes have been restricted to saline areas because of their inability to compete with glycophytes under non-saline conditions (Chapman, 1975).

Many *Atriplex* species assimilate carbon via the C_4 pathway of photosynthesis and have corresponding anatomical characteristics (Laetsch, 1974). The genus is one of several which comprise both C_3 and C_4 species – a fact which raises some interesting taxonomic questions (Björkman and Berry, 1973; Laetsch, 1974). The implications of this photosynthetic anomaly are unclear; it is possible that potential productivity of the C_4 species is greater than that of the C_3 species, but this has not been verified (Gifford, 1974).

These features of *Atriplex* biology – the salt extrusion via the hairs and the C_4 carboxylation scheme – will be more fully discussed later.

Of particular significance to the utility of *Atriplex* species is their shrubby habit. In some cases, *Atriplex* dominates the landscape (Jones, 1970), and provides high protein forage for wildlife and domestic livestock (McKell, 1975). Goodin and McKell (1970) calculated potential yearly yields of 16,000 kg of dry forage per hectare per year on arid rangelands. Of further significance is the average protein of that forage, as much as 15% crude protein having been measured in leaf tissue. The ability of the plant to produce palatable forage with high concentrations of protein makes it a desirable rangeland shrub for areas of marginal land use.

This description of some of the biological peculiarities should give some idea of the potential utilization of members of the genus. Two excellent reviews of wildland shrub biology and utilization have appeared in recent years (McKell et al., 1972; McKell, 1975). In both cases, *Atriplex* species have been extensively explored as rangeland resources. A detailed listing of these potential uses is beyond the scope of this chapter but the uniqueness of *Atriplex*'s biological utility merits some attention.

As mentioned earlier, *Atriplex* is often a palatable, nutritious shrub (not in every case, however – McKell, 1975). Rangeland seeding of *Atriplex* is one approach to utilization (Goodin and McKell, 1970). Management of existing *Atriplex* populations for better productivity, and restricting misuse (overgrazing) promises to be an even better approach. The forage potential of *Atriplex* is large – a better biological understanding of the genus (potential productivity, response to grazing pressure, nutritive quality of forage) is essential.

Atriplex can be used in soil stabilization – particularly in arid lands (McKell, 1975). Nord and Countryman (1972) described several fire resistant *Atriplex* species – a potential use in areas like Southern California where chronic fire problems occur. Goodin and Mozafar (1972) speculated on the possibility of reclaiming salty soils by growing and harvesting *Atriplex* on those soils. Stark (1966) listed *Atriplex* as a possible roadside plant for Nevada highways, citing its ecological suitability and low-cost maintenance. Many other uses might be postulated; the adaptability of *Atriplex* species to stressful environments is a key to their utilization.

The importance of *Atriplex* as a component of wildlife habitats should not be overlooked. The genus provides browse and cover for deer and smaller mammals in the arid southwestern United States, and is an important food source for birds and rodents (McKell, 1975).

Recently perennial *Atriplex* spp. have been considered as a possible biomass crop plant for energy production. With rather large biomass productivity (particularly in C_4 species) in arid and semi-arid regions, this presents an interesting possibility for utilization within the framework of the relatively undisturbed ecosystem.

Cook (1972) discussed the influence of growing site on the nutritive qualities of shrubs. This may be extremely important in the case of *Atriplex* because of the ability of the plant to modify its growing site. Not only does the plant have an effect on the inorganic salt content of the soil beneath it (through the cycling of salt through the hairs and back to the soil via rainwater or leaf fall), it has been shown that the shrubs create islands of fertility in some desert situations through the accumulation of organic matter under the plant (McKell, 1975). This cycling function of *Atriplex*, particularly in nutrient-poor arid lands, is a key ecosystem process.

The biology of *Atriplex* species in large measure determines the utilization of the plants. The interdependence of *Atriplex* and its ecosystem can be demonstrated, and the exploitation of the plant under managed conditions appears to be hinged on the appropriateness of our use of the plant. Perhaps most importantly, exploitation of the various harsh habitats of *Atriplex* – arid and/or salty lands – can be aided by knowledgeable use of the species. The material presented here is not intended to be a complete examination of the genus, but rather a case study of interesting features of *Atriplex* biology.

3. The physiology of salt tolerance

The mechanisms of salt tolerance in *Atriplex* have been investigated fairly extensively, but some controversies remain in the interpretation of physiological and structural observations made. One of the most complete early collections of such data is the proceedings of the symposium mentioned earlier (Jones, 1970). Since then, numerous papers on *Atriplex* biology have appeared.

The physiology of salt tolerance in *Atriplex halimus* L. was studied by Mozafar (1969), who investigated the ability of the plant to grow in saline media with osmotic potentials of −0.36 to −10.1 atm. He found that low salinity levels of NaCl and KCl stimulated the growth of the plant to an osmotic potential of approximately −4.5 atm and past that point, increased salinity began to have a negative effect on the growth of the plant. Blumenthal-Goldschmidt and Poljakoff-Mayber (1968) showed also that low levels of NaCl stimulated the growth of *A. halimus* but an excess of NaCl in the medium, while tolerated to −12 to −16 atm, caused a decrease in growth. Mozafar (1969) and Mozafar and Goodin (1970) related the salt tolerance of the plant to the presence of vesiculated hairs on the upper and lower epidermis of the leaves by showing that the concentration of salt in the bladder cell of the hair was much higher than that of the leaf sap and xylem exudate – evidence that the hairs served as salt reservoirs to which excess salts were pumped for storage and later released to the exterior of the leaf when the bladder cell burst. Other

workers have assigned various functions to the hairs, including water absorption from the atmosphere and water storage (Wood, 1925; Black, 1954).

The hairs consist of two cells: a small dense stalk cell which contains chloroplasts, and a large, highly vacuolate bladder cell which is thought to be the primary salt repository cell in the leaf (Black, 1954; Osmond et al., 1969; Mozafar and Goodin, 1970). The transport of chloride ions into the bladder cell was shown to be dependent on a light-mediated transport mechanism (Lüttge and Osmond, 1970; Lüttge, 1971). Similar types of transport and related plant structures (glands, hydathodes, etc.) have been reviewed by Lüttge (1971). Thomson and his coworkers (Campbell et al., 1974; Thomson, 1975) have presented convincing ultrastructural evidence for an additional apoplastic transport of salts to the bladder cells. Smaoui (1971) described two types of hairs on *A. halimus* leaves: 'vesicular' trichomes which are highly vacuolate and function as salt repositories, and 'en massue' trichomes which have only small vacuoles and secrete polysaccharides. The vesicular trichomes are more numerous, especially on the older leaves, and the 'en massue' trichomes are found primarily on young leaves.

Waisel (1972) reviewed many aspects of salinity tolerance, including the functional anatomy of halophytes. He classified *Atriplex halimus* as a semihalophyte, rather than a halophyte, a distinction which indicates that *A. halimus* can grow in non-saline soils as well as soils that are highly saline. However, most research to date suggests that *A. halimus* shows a positive response to saline media (Blumenthal-Goldschmidt and Poljakoff-Mayber, 1968; Mozafar, 1969; Kelley, 1974; Miller, 1973) and the plant will be considered a true halophyte in this report.

In addition to remarkable salt tolerance, several *Atriplex* species possess a method of carbon fixation which is much more efficient than the normal Calvin cycle (Laetsch, 1968, 1974; Hatch et al., 1971; Black, 1973). This carbon fixation pathway, commonly called the C_4 pathway, was first demonstrated in tropical grasses – particularly sugar cane (Kortschak et al., 1965; Hatch and Slack, 1966) and later in several species of monocots and dicots (Hatch et al., 1967; Tregunna and

Downton, 1967; Johnson and Hatch, 1968; Laetsch, 1968, 1974; Downton et al., 1969; Osmond et al., 1969).

Species with the C_4 pathway are so called because they contain enzymes that synthesize C_4 dicarboxylic acids (oxaloacetate, malate, and aspartate) as initial photosynthetic products (Hatch et al., 1971). Species utilizing the Calvin cycle first synthesize a 3-carbon phosphorylated compound, 3-phosphoglyceric acid (hence the designation 'C_3 pathway') in the initial steps of carbon fixation (Downton, 1971a; Hatch et al., 1971). There have been several excellent reviews of the C_3 and C_4 pathways and their associated physiological processes (Hatch and Slack, 1970; Hatch et al., 1971; Zelitch, 1971; Marx, 1973; Björkman and Berry, 1973; Black, 1973; Laetsch, 1974).

Welkie and Caldwell (1970) listed the primary physiological characteristics of C_4 plants as a lack of photorespiration (evolution of CO_2 from respiratory processes in the light), low carbon dioxide compensation points, low carbonic anhydrase activity, high temperature optimum of net photosynthesis, high light saturation intensities, and greater efficiency of net photosynthesis when compared to plants that fix carbon by the normal C_3 pathway (hence the designation of C_4 plants as 'efficient' plants and of C_3 plants as 'non-efficient' plants).

4. Growth habit

Investigations of the growth habit of *A. halimus* L. and generation of tissue for the microscopy studies were made under the following conditions: Seeds of *A. halimus* L. were germinated in full-strength Hoagland nutrient solution (Hoagland and Arnon, 1950) under constant light (Grow-Lux fluorescent bulbs) with an intensity of approximately 200 foot candles. The germination trays consisted of a wire gauze frame covered with cheese cloth and placed in shallow trays. Nutrient solution was added to the trays to bring the level of the solution to the base of the wire frame; the solution was changed weekly, and the seedlings were allowed to grow until they were four weeks old at which time they were 6–8 cm tall.

At this stage they were transferred to 60 ml foil-covered test tubes. The hypocotyl of each seedling was wrapped with non-absorbent cotton to form a plug for the neck of the tube and to hold the seedling in place.

To the test tubes a full-strength Hoagland solution was added and the plants were allowed to grow for two more weeks in this solution. Control plants were grown in full-strength Hoagland solutions for the duration of the experiment. Salinization of the test plants was begun when they were approximately six weeks old. The base solution was full-strength Hoagland solution to which appropriate amounts of NaCl were added. The addition of NaCl was gradual with small increments (0.05 to 0.1 M NaCl) made each week after the sixth week until a final concentration of 0.345 M NaCl was reached when the plants were ten weeks old. Solutions in all test tubes were changed weekly to maintain known concentrations of NaCl and essential nutrients. A final concentration of 0.345 M NaCl was reached at ten weeks and the plants were allowed to grow for two more weeks at this concentration before leaves were selected for microscopic examination.

In a separate portion of the experiment, 60 seedlings 6–8 cm tall were transferred to foil-covered flasks of 250 ml capacity and placed in the greenhouse. Of these, 20 were grown in full-strength Hoagland solution for the duration of the experiment, 20 were salinized to a final NaCl concentration of 0.174 M NaCl in Hoagland solution, and 20 were salinized to a final NaCl concentration of 0.345 M NaCl in Hoagland solution. Plants were observed at weekly intervals for any signs of stress response. Salinization was completed in four weeks, and the plants were allowed to grow four more weeks before they were harvested and fresh and dry shoot weights were determined.

Osmotic potentials of the solutions used were as follows: full-strength Hoagland solution, −0.6 atm; 0.174 M NaCl in full-strength Hoagland solution, −7.2 atm; and 0.345 M NaCl in full-strength Hoagland solution, −14.4 atm. These osmotic potentials were measured on an Advanced Instruments osmometer by freezing point depression. The values given do not represent calculated values, but are instead actual values as measured.

The growth habits of plants under control and salinized conditions were noticeably dissimilar. Plants grown under control conditions had an erect habit with little or no lateral branching. In the twelve weeks of growth monitored, the plants attained a height of from 0.3 m to 0.5 m. There was no indication of leaf wilt during periods of high light intensity and maximum greenhouse temperatures. The plants were erect and the leaves were turgid during those periods.

Plants salinized to 0.174 M NaCl were characterized by a relatively erect habit with some lateral branching from nodes near the base of the plant. Some recumbency was noted, particularly during periods of high light intensity in the greenhouse. The leaves remained turgid at all times. These plants were no taller than the control plants (0.3 to 0.5 m) and possessed as many as three lateral branches from nodes near the base of the plant.

Those plants grown in high salt concentrations (0.345 M NaCl) had a noticeable recumbent posture with as many as six lateral branches from nodes near the base of the plant. Again, there was no noticeable leaf wilt during periods of high light intensity and maximum temperature in the greenhouse. The leaves from the lowermost two or three nodes of the main shoot were shed between the tenth and twelfth weeks of the experiment. This leaf abscission did not occur in the other treatments. The leaves on the upper part of the main shoot and on the lateral branches did not abscise. Preliminary indications of abscission were yellowing of the leaf tissue and some necrosis of the edges of the leaf.

Lateral bud activation of *A. halimus* plants grown in saline media has been reported (Gale and Poljakoff-Mayber, 1970). These workers suggested that the sprouting of the lateral buds was due to a breakdown of hormone-induced apical dominance, a phenomenon implicated previously in salinity damage of plants (Itai et al., 1968; Waisel, 1972). This hormonal imbalance could have occurred in two ways: (1) an actual ionic effect on hormones of the root and apical regions of the plant; i.e. a specific ion effect brought about by the salt in direct contact with the hormone producing tissue; or (2) a mechanical effect due to the periods of subtle physiological drought (not manifested by leaf wilt) that allowed the plants to assume a recumbent posture, thus affecting apical dominance and leading to lateral bud activation. Other workers have reported that plants grown in saline media exhibit signs of toxicity; e.g. decreased protein synthesis, due to inhibition of cytokinin activity and transport in the roots (Itai, 1967). Cytokinins produced in the roots have been shown to affect growth of the shoot and thus are indirectly responsible for some synthesis of auxins (Hall, 1973; Torrey, 1976). It is unlikely that the salinity effects described here were the result of one particular process alone, but were the result of two or more osmotic or specific ionic processes. It may be significant that the recumbency and lateral bud sprouting that occurred in both the low-salt and high-salt treatments did not lead to a reduction in biomass accumulation (discussed in the next section), an indication that toxic ion concentrations were not reached in either treatment.

Shoot growth in the plants approximated closely that reported by earlier workers (Blumenthal-Goldschmidt and Poljakoff-Mayber, 1968; Mozafar, 1969) in control and salinized plants. The growth was better in the presence of sodium chloride than in its absence, but growth of the plants in higher concentrations of sodium chloride was not optimal (Table 1).

Table 1. Fresh and dry weights of shoots in control and salinized plants (Values given are means ± standard error).

Treatment	Osmotic Potential (atm)	Shoot fresh Wt (g)	Shoot dry Wt (g)
Control	−0.6	17.5 ± 2.2	3.1 ± 0.36
* 0.174 M NaCl	−7.2	24.4 ± 2.6	4.3 ± 0.28
* 0.345 M NaCl	−14.4	20.4 ± 2.0	3.6 ± 0.40

* In full-strength Hoagland solution

Response of non-halophytic species to salinized media usually is characterized by an osmotic stress (manifested by leaf wilt or similar response) that leads to reduction in growth (Waisel, 1972). At low concentrations of salt, many plants exhibit an osmotic adjustment mechanism involving an increase in internal solute concentration and a con- comitant increase in growth (Oertli, 1966; Waisel, 1972). Halophytes (including *A. halimus*) are prob- ably capable of faster osmotic adjustment than non- halophytes (e.g. spinach) and can preserve or en- hance their metabolic abilities (e.g. photosynthesis, hormone production, protein synthesis) at the hig- her levels of internal solute concentration nec- essarily involved in adjustment to increased osmotic pressures in the growing medium (Waisel, 1972).

Brownell (1965) ascribed the increase in growth rate under salt stress to an increase in turgidity in the plant cells that takes place after osmotic adjust- ment has occurred. This is an osmotic effect that is not necessarily a specific ion effect.

A definite nutritional requirement for sodium and for chloride has been shown in a number of plants, including some *Atriplex* species (Broyer et al., 1954; Brownell, 1965; Waisel, 1972). Sodium chloride has been used to increase the yield of many crop plants including semi-halophytic sugar beets (Ulrich and Ohki, 1956) and glycophytic tomatoes, cotton, and garden beets (Waisel, 1972).

Halophyte response to low concentrations of sodium salts in the growing media is decidedly favorable, but the physiological roles of the salts in the various metabolic processes of the plant are not well understood (Jennings, 1968, 1976; Waisel, 1972). One theory of sodium ion enhancement of growth in halophytes is based on reports of sodium- activated ATPases in plant cells that are associated with the sodium-potassium pump at the plasmal- emma (Jennings, 1968, 1976). Jennings' theory – that there is net ATP synthesis as sodium enters the cell – remains unproven. In a more recent article (Jennings, 1976) he discussed potassium-sparing effects of sodium, the development of succulence, hormonal changes, and other effects of sodium on the growth of halophytes. Brownell (1965) de- termined that halophytes require large amounts of sodium as an essential nutrient, but did not offer an explanation of the physiological role of sodium ions except to say that phosphorylation of energy-rich compounds was increased under conditions of high sodium concentration in the external medium. Lae- tsch (1974) provides a good argument for the importance of sodium in the C_4 pathway of photo- synthesis (see the following sections of this chapter for discussion of the C_4 syndrome).

The observed decrease in growth of plants sub- jected to the 0.345 M NaCl treatment (Table 1) has several possible explanations. It should be noted that at this NaCl concentration, growth was better than in the control plants, but showed a decrease from the optimal growth noted in the 0.174 M NaCl treatments. The effects of salinity on the ultrastruc- ture of the leaf cell organelles appear to be primarily of an osmotic nature; viz. swelling of chloroplasts and mitochondria and some disruption of organel- lar structure. It is significant that the decrease in optimal growth conditions (as reflected by net growth) occurred concomitantly with the swelling of these organelles. The salinity effects on the organelles will be more fully discussed in a later section.

Salinity at the higher concentrations may have also affected the turgidity of the leaf cells, the assimilation of certain essential nutrients, e.g. nit- rogen, and may have disturbed the hormonal bal- ance of the plants so severely that growth was impaired. These effects have been noted in *A. halimus* (Blumenthal-Goldschmidt and Poljakoff- Mayber, 1968; Waisel, 1972) and several other species of halophytes (Brownell, 1965; Waisel, 1972; Jennings, 1976; Flowers et al., 1977).

Although no leaf wilt was observed in the plants in this investigation, it is probable that the relative turgidity of the cells was not high enough in the salt stressed plants, even during the night, to allow optimal cell expansion and thus maintain the growth rate at a high level. The fact that some growth did take place, and at a rate higher than the control plants, is perhaps an indication of the ability of *A. halimus* to undergo rapid osmotic adjustment. Mozafar (1969) and Blumenthal-Gold- schmidt and Poljakoff-Mayber (1968) reported that the growth rate dropped steadily as an osmotic potential in the medium of -10 to -16 atm was reached. At lower osmotic potentials (-10 to -22 atm) the growth rate was greatly inhibited, drop- ping below that of the control plants (Blumenthal- Goldschmidt and Poljakoff-Mayber, 1968). Ultra- structural damage at these lower osmotic potentials was excessive.

It has been reported that high concentrations of chloride in plant tissues can affect nitrogen metabolism as well as assimilation of other essential nutrients (Waisel, 1972). Organic acid metabolism (particularly oxalate) is greatly affected by high salinity substrates (Mozafar, 1969; Waisel, 1972). It is thought that oxalate is synthesized by the leaf cells to balance the ionic effects of sodium and potassium (Osmond, 1967). Evidence for oxalate accumulation under highly saline conditions was presented by Mozafar (1969). Ultrastructural evidence for effects of salinity on nitrogen metabolism will be discussed later. It is likely that nitrogen availability, or at least assimilation, is affected by highly saline media, with the net result a decrease in growth.

The process of photosynthesis may be affected by highly saline conditions. The C_4 pathway described earlier occurs in many species of halophytes (as well as closely related glycophytes) including several species of *Atriplex* (Welkie and Caldwell, 1970; Hatch et al., 1971; Waisel, 1972). This fact may be significant in considering the ecological conditions under which many halophytes, including *A. halimus*, are found, i.e. high light intensities and variable water potentials. The ability to photosynthesize under these conditions is critical to the survival of these species. Laetsch (1974) discussed this aspect of halophyte physiology in detail.

As mentioned earlier, Itai (1967) and Itai et al. (1968) reported hormonal imbalances due to high salinity in their studies on tobacco leaves. Hall (1973) and Torrey (1976) have discussed these effects in halophytes and glycophytes. It is very likely that the hormonal balance of *A. halimus* in high salt concentrations is severely disturbed by either a specific ion effect on the hormones, e.g. cytokinins and auxins, or hormone-producing tissues, e.g. roots and shoot apices, or an osmotic effect on the transport of these hormones. A hormonal imbalance in the plants as a result of high salinity could lead to a reduction in growth rate.

The effects of salinity on enzyme activity in the plant may cause a shift in certain metabolic patterns (Waisel, 1972). Workers in our lab have shown a salt effect on catalase, an enzyme involved in organic acid synthesis and part of the respiratory pathway in *A. halimus* (R. D. B. Whalley and J. R. Goodin, personal communication). The possible effects on ATPase activity have been noted earlier, and it is possible that these and other enzymes are affected by high salinity. The fact that growth of the plants in 0.345 M NaCl decreased, but did not cease, is an indication that most of the enzymes of photosynthesis and other metabolic pathways were still functional although the degree of activation may have been changed.

Sodium chloride plays a significant role in the growth of *A. halimus*. The mechanisms of salinity effects are complex and at this time largely obscure. The interactions of hormones, enzymes, and environmental conditions seem to be mediated by salts in the growing medium to some significant degree, but the effects on growth cannot be ascribed to any one process. This can be said of most halophytes. The list of reviews of salt tolerance in halophytes and non-halophytes appearing in recent years grows larger annually. Among those reviews that should be consulted for further information on the above are Black (1973); Caldwell (1974); Laetsch (1974); Poljakoff-Mayber and Gale (1975); Hellebust (1976); Jennings (1976); Flowers et al. (1977); and Greenway and Munns (1980).

5. Transpiration and water relations

Transpiration plays a relatively minor role in physiological processes under normal conditions (Oertli, 1971). Such a small amount of water is actually utilized by plants that this amount is almost always nonlimiting, regardless of the rate of transpiration. Only under severe water stress would the drying effects of transpiration interfere with normal metabolic processes. Transpirational cooling is quite effective in lowering leaf temperatures (Gates, 1968), but seldom does overheating produce any damage in the first place (Kramer, 1969). It has also been suggested that high transpiration rates are important in the absorption and transfer of nutrients (Curtis, 1926; Clements, 1934), but this has been disputed (Broyer and Hoagland, 1943; Russell and Barber, 1960) and remains unconfirmed. An adaptive advantage of transpiration is yet to be found.

Transpirational resistance can be broken up into several components (Kramer, 1969). Air boundary layer resistance is due to transient elevated humidities adjacent to leaf surfaces interfering with free atmospheric diffusion of the evaporating water. Such boundary layers in nature are almost non-existent because of their extreme sensitivity to anything but static conditions. Cuticular resistance is so high compared to that of the alternative pathway through the stomata that cuticular evaporation can be disregarded under normal conditions. Resistance to the flow of water vapor in this latter pathway consists of the intercellular space resistance, which is a function of mesophyll geometry, and the pore resistance. The very existence of a mesophyll resistance, or at least its importance, is still unresolved at this time. Theoretically, it limits evaporation from the moist mesophyll cells and is inversely proportional to their water vapor pressure; it can be considered a type of evaporative resistance.

Other factors affecting transpiration rates come into existence in saline soils. Saline solutions, as well as those of any other type, necessarily have lower vapor pressures than those of the pure liquids. Resultant evaporation is thereby decreased, leading to elevated solution temperatures (Waisel, 1972). Under such conditions in halophytes, however, no heat buildup has ever been detected; it is possible that opposing forces counter these buildups by other positive effects on transpiration. Although changes in salinity do not affect transpiration rates in halophytes as much as in glycophytes, an increasing concentration of NaCl in particular almost always decreases transpiration to some extent. Specific ion effects may be involved in response to NaCl salinity, but are probably of lesser importance than the ionic concentrations. It can only be said that transpiration rates in halophytes, while lower in general than those of glycophytes under similar conditions, are quite variable; no further generalizations can be made.

The presence of water in plants can be considered a type of steady-state system, the primary resistances to the flow of water in it occurring in the roots and in the leaves (Kramer, 1969). If the cortical symplast theory of Crafts and Broyer (1938) is correct, such resistances in the roots would be determined by diffusion rates across cortical plasma membranes into the symplast, at least until the endodermis is reached. The driving forces involved in this case are water potential differences between the root and the surrounding soil solution; in saline soils, favorable gradients are obtained by lowering the osmotic components of internal water potentials.

A further drop in water potential occurs with entrance into the vessel elements. Although root pressure (as evidenced by guttation in certain plants) has been considered as an important process in the ascent of xylem sap (White, 1938; White et al., 1958; Rufelt, 1956), its magnitude is usually too small to have any effect under less than water-saturated conditions. An opposing view, the cohesion theory, emphasizing the electrostatic attraction of water molecules for each other, has also met with opposition due to the well-known phenomenon of cavitation in extended, vertical, capillary columns of water. Oertli (1971) has explained away this opposition, however, with the observation that the xylem sap is a very pure liquid, containing only water and solutes which have been transported across a plasma membrane of the symplast, and having been derived from the equally pure cytoplasm of the mother cell. A liquid of this purity would be devoid of dust particles or other foreign matter, thereby having no nuclei around which air bubbles could form. If this is indeed the case, the ascent of xylem sap under normal conditions would be due to transpirational pull, creating a favorable potential gradient for absorption into the xylem by decreasing the pressure component of its water potential.

The driving forces for transpiration are vapor pressure differences between the leaf interiors and the adjacent atmosphere; the steeper this gradient, the more water that transpires, other things being equal. It is possible that the dense, vesiculated hair covering of *Atriplex* leaves acts as a baffle to boundary layer air currents, thereby increasing surface humidity (and vapor pressure) which would in turn decrease transpiration rates, but this has never been demonstrated. Water evaporation from the vesicles themselves would seem to be prevented due to the waxy, hydrophobic nature of the surrounding sheaths (Mozafar and Goodin, 1970). Although the presence of salt crystals, remnants from previously ruptured vesicles, would bring about some absorption of atmospheric water, the amounts involved would probably be negligible compared to the amount of transpiration occurring (Waisel, 1960).

Experiments conducted by Miller (1973) on *Atriplex halimus* showed conclusively that salinization with NaCl reduces rates of transpiration. These results are consistent with those of other workers (Kaplan and Gale, 1972; Moore et al., 1972; Gates, 1972). Miller's experiments were conducted over a 65-day period, and represented a gradual salinization up to 0.2 M NaCl. Plants grown in 0.1 M NaCl exhibited more rapid growth than either 1X Hoagland controls or 0.2 M NaCl (all treatments contained the 1X Hoagland solution). Succulence also increased in response to salinity, as reported by others (Black, 1958; Jennings, 1968; Gates, 1972).

Miller also grew *A. halimus* plants in Carbowax 6000 to compare the stress imposed by a solution which does not allow for osmotic adjustment. The rapid wilting pointed to the irreversible nature of ion flow. With no incoming source of ions, all the internal ions were secreted into the vesicles within a matter of a few days. They could not have been simply diluted by growth because so little growth occurred in that time span. It is important for halophytes such as *Atriplex* to take in vast quantities of excess ions since they are constantly being lost back to the environment via the vesicles. The existence of this mechanism provides halophytes with an adaptive advantage in saline soils.

6. Salt effects on ultrastructure

Plants that have the C_4 pathway are characterized by a particular leaf anatomy distinguished by a readily-apparent sheath of cells surrounding the vascular bundle (the bundle sheath) and containing large chloroplasts with large amounts of starch (Laetsch, 1968; Welkie and Caldwell, 1970; Downton, 1971a). The mesophyll cell layer surrounding the bundle sheath and separated from it by thick cell walls typically contains less starch and smaller chloroplasts (Laetsch, 1968, 1971; Welkie and Caldwell, 1970; Downton, 1971a).

A major cytological difference between bundle sheath and mesophyll cells of many C_4 plants has been discussed in the literature at length. Rhoades and Carvalho (1944) noted that there is a distinct dimorphism in the chloroplasts of bundle sheath and mesophyll cells of maize plants. Hodge et al. (1955) published the first electron micrographs of dimorphic chloroplasts in *Zea mays*. Since that time, many workers have noted distinct differences in the chloroplasts of the cells (Johnson, 1964; Laetsch et al., 1965; Laetsch 1968, 1969, 1971, 1974; Downton et al., 1969; Osmond et al., 1969b; Black and Mollenhauer, 1971; Johnson and Brown, 1973). The differences in chloroplast structure involve the presence or absence of grana, presence or absence of starch grains, size of the chloroplasts, presence or absence of a peripheral reticulum on the chloroplasts, and number and concentration of chloroplasts in the cells (Goodchild, 1971; Black and Mollenhauer, 1971). Other differences in the ultrastructure of the C_4 plants involve other cell organelles, including mitochondria, ribosomes, endoplasmic reticulum, and peroxisomes (Laetsch, 1968, 1969, 1971, 1974; Black and Mollenhauer, 1971; Downton, 1971b; Goodchild, 1971; Gracen et al., 1972). Laetsch (1974) and Goodchild (1971) suggest that the cell walls and plasmodesmata between mesophyll and bundle sheath cells might be a key to transport of photosynthetic intermediates between the cells. The presence of a suberized layer around the bundle sheath may be an important factor in such transport (Laetsch, 1974). The arrangement of chloroplasts and mitochondria in the bundle sheath cells may have a relation to C_4 photosynthesis with specialization of the mitochondria for transamination and decarboxylation reactions (Downton, 1971b).

Black and Mollenhauer (1971) suggest that the most reliable anatomical criteria for determining the photosynthetic capacity of a plant are based on the number and concentration of chloroplasts, mitochondria, and peroxisomes in the bundle sheath cells.

Particular attention has been paid to the occurrence of peroxisomes in C_4 plants (Black and Mollenhauer, 1971; Frederick and Newcomb, 1971; Newcomb and Frederick, 1971; Tolbert, 1971; Smaoui, 1972b). Peroxisomes (leaf microbodies) are quite numerous in C_3 plants and less numerous in C_4 plants (Newcomb and Frederick 1971). The same workers showed that in the C_4 plants the bundle sheath cells have more peroxisomes than the mesophyll cells and the mesophyll cell peroxisomes are smaller than the bundle sheath cell peroxisomes. Peroxisomes are intimately tied to photorespiration, a process which depends on the oxidation of glycolate in the peroxisomes (Tolbert, 1971; Beevers, 1979). It has been postulated that the presence of peroxisomes in the C_4 plant bundle sheath cells is an indication that C_3 carbon metabolism takes place in these cells and that the mesophyll cells are responsible for C_4 metabolism (Newcomb and Frederick, 1971; Tolbert, 1971). Laetsch (1974) points out that this may not be the case and offers a convincing argument for his position. His examination of the C_4 syndrome in higher plants remains the most lucid of those made up to this time.

The effects of salinity on the ultrastructure of plant cells have been investigated by several workers on various halophytic species (Ziegler and Lüttge, 1966; Blumenthal-Goldschmidt and Poljakoff-Mayber, 1968; Osmond et al., 1969a; Shimoney et al., 1973). The work to date has not been conclusive and warrants further investigation (Waisel, 1972). The book edited by Poljakoff-Mayber and Gale (1975) offers several chapters (notably those by Poljakoff-Mayber and by Thomson) that establish the status of these investigations.

We report here on our own work on leaf cell structure of *Atriplex halimus* as it is affected by saline media. The general leaf anatomy of several halophytes, including two species of *Atriplex*, was discussed by Fahn (1974). Observations on ultrastructure of *Atriplex* tissue have been made as noted above.

Tissue preparation for microscopy

Small, fully-developed leaves at the third node from the apex were selected for microscopical observation from the plants grown in test tubes. Care was taken to assure that the leaves selected were leaves that developed *after* the salinization process was complete. The entire plant was taken to the laboratory and the leaves cut from the plant and placed immediately into the fixative. Proper fixation of the leaves was made more difficult by the tendency of the vesiculated hairs on the surface to burst when fixed using normal fixation materials and methods. The determination was made that the hairs were bursting primarily during the first step of fixation when they initially encountered the glutaraldehyde fixation medium, so the glutaraldehyde was osmotically adjusted by the addition of NaCl to a concentration of 5 M NaCl (which was approaching saturation). The presumed osmotic potential of the hairs, according to Mozafar (1969), ranges from -235 atm to -787 atm. The presumed osmotic potential of the hairs (-235 atm) is much lower than the osmotic potential of 5 M NaCl in 6.5% glutaraldehyde (-150 to -175 atm), so the use of an isosmotic solution was not possible. The most effective solution for fixation was found to be 6.5% glutaraldehyde in sodium cacodylate buffer (as described below), which preserved the smaller hairs, but not the larger ones.

The preparation of the leaf tissue for light and electron microscopy was the same. Morphologically similar leaves, selected as described above, were fixed in 6.5% glutaraldehyde in 0.1 M sodium cacodylate buffer (pH 6.8). In addition, the glutaraldehyde and sodium cacodylate buffer solution was used as a fixative to which NaCl was added to lower the osmotic potential. Solutions containing 1 M NaCl and 5 M NaCl were used to fix similar leaves. The leaves were placed in fixative on plastic trays and sections approximately 1 mm^2 were made. These sections were then placed in vials of fixative and kept at 2–4° C for 8 hours. The sections were rinsed in cold 0.1 M Na-cacodylate buffer (pH 6.8) three times, and postfixed in cold 1% osmium tetroxide in sodium phosphate buffer for 4 hours. The tissue was rinsed with distilled water, taken through a graded ethanol-water dehydration series, and embedded in a low viscosity plastic embedding medium (Spurr, 1969).

Sections 1 μm thick for light microscopy were made with a glass knife on an LKB Ultratome microtome and mounted on a glass slide. Sections were stained for light microscopy with Mallory's stain (Richardson et al., 1960), a mixture of Azure II and methylene blue. These sections were examined and photographed on a Zeiss microscope.

Thin sections (500–700 Å) were made with a diamond knife on an LKB Ultratome and mounted on 200 or 300 mesh copper grids. The sections were stained in uranyl acetate for 5 minutes and lead citrate for 5 minutes (Reynolds, 1963) and examined with an Hitachi HS–8 electron microscope.

The leaves of *Atriplex halimus* have a typical 'Kranz' type anatomy (Johnson and Brown, 1973) in which the vascular bundles are surrounded by a layer of cells (the bundle sheath cells) which are morphologically distinct from the surrounding mesophyll cells. Fahn (1974) reported a definite layer of palisade (mesophyll) cells around the bundle sheath, but in the plants we examined this second layer was not distinct (Fig. 1). Other workers,

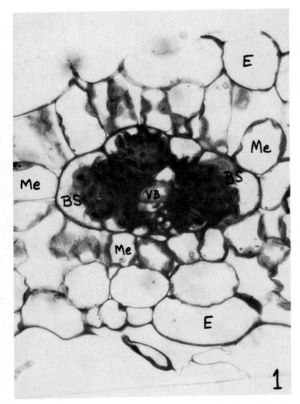

Fig. 1. Cross section of leaf from a control plant showing vascular bundle (VB), bundle sheath cells (BS), mesophyll cells (Me), and large epidermal cells (E). 480×.

notably Björkman and Berry (1973), reported this second layer in other species of *Atriplex*. The mesophyll cells of the plants we examined could be considered 'radially arranged' (Johnson and Brown, 1973) around the bundle sheath but they do not form a second concentric sheath per se. There are definite morphological differences in the organelles of the bundle sheath and mesophyll cells, as well as differences in organelle arrangement and concentration. The bundle sheath is partially open abaxially in most cases (Fig. 1).

Other characteristics of *A. halimus* leaves observable at the light microscope level include: numerous vesiculated hairs on both surfaces of the leaf (Fig. 2), a high concentration of organelles in the bundle sheath cells and a low concentration of organelles in the mesophyll and epidermal cells, a great deal of intercellular space in the mesophyll region, and large, vacuolate epidermal cells (Fig. 1). We could determine no structural differences in control and salinized plants at the light microscope level.

The ultrastructure of leaf cells in control plants (those grown in straight Hoagland solution) and the low salinity group (those salinized to 0.174 M NaCl in Hoagland solution) were similar. There were no differences due to salinity at the lower concentration that could be observed with the electron microscope, and this study is accordingly limited to a comparison of plants in the high salt group (0.345 M NaCl in Hoagland solution) with control plants.

Of primary interest in this study was the ultrastructure of the chloroplasts of the various types of leaf cells. Blumenthal-Goldschmidt and Poljakoff-Mayber (1968) reported that salinity could affect the structure of chloroplasts and mitochondria in leaf cells. The dimorphism in chloroplasts from bundle sheath cells and mesophyll cells has been well documented in *A. halimus* L. (Fahn, 1974; Blumenthal-Goldschmidt and Poljakoff-Mayber, 1968) and several other species of *Atriplex* and other C_4 plants (Laetsch, 1968, 1971, 1974; Downton et al., 1969; Black and Mollenhauer, 1971).

The morphology of hair, bundle sheath, and mesophyll cells from control plants is shown in Figs 3 through 9. The bundle sheath cells are dense, largely avacuolate and contain many large chlorop-

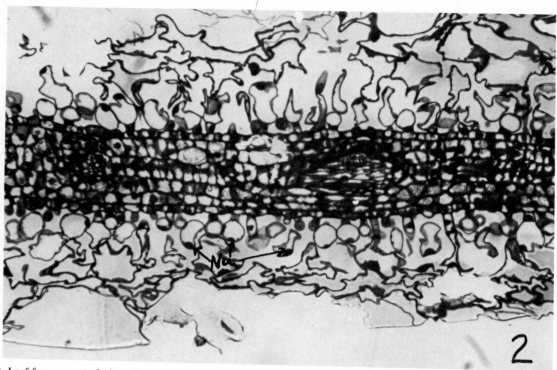

Fig. 2. Leaf from a control plant showing numerous two-celled hairs. Note nuclei of bladder cells (Nu). 250×.

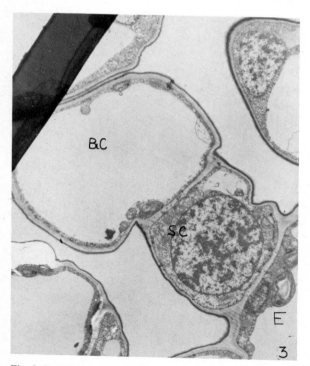

Fig. 3. Hair from a control plant. Note stalk cell (SC), bladder cells (BlC), and epidermal cell (E). 1600×.

lasts and mitochondria (Figs. 4 and 6). Mesophyll cells have a large vacuole, fewer and smaller organelles and have a peripheral cytoplasm that is not as dense as that of bundle sheath cells (Figs. 5,8, and 9).

The chloroplasts of the two types of cells are dissimilar in structure. Bundle sheath chloroplasts are long and comparatively thick (Figs. 4 and 6) whereas mesophyll chloroplasts are generally shorter and thinner (Fig. 5). The bundle sheath chloroplasts have more grana and the stroma lamellae are shorter (Fig. 4). Mesophyll chloroplasts have fewer grana and longer lamellae between grana (Fig. 5). The differences in the two types of chloroplasts discussed are not striking, but may be significant in assigning them the physiological functions of C_4 plants (Laetsch, 1968, 1974). The differences, however, become more evident with salinity treatments. Mesophyll cell chloroplasts of several C_4 plants and a few C_3 plants have been shown to have a peripheral reticulum, i.e., a modification of the inner membrane of the chloroplast composed of a series of anastomosing tubules

(Laetsch, 1968, 1971, 1974; Black and Mollenhauer, 1971; Gracen et al., 1972). Bundle sheath cells of *A. halimus* do not have a peripheral reticulum (Figs. 10–13). The chloroplast of mesophyll cells from salinized plants have a peripheral reticulum in most cases (Figs. 14 and 17), but we did not find a peripheral reticulum on the chloroplasts of control plants. The presence of the peripheral reticulum in other C_4 plants has not been related to salinity treatments (Laetsch, 1971; Black and Mollenhauer, 1971; Gracen et al., 1972), but has been routinely shown in unsalinized plants. It is possible that the salinization of the plants caused a swelling of the tubules to the extent that the peripheral reticulum became visible under the electron microscope. This conjecture appears more likely when a more obvious effect of salinity on chloroplast ultrastructure is considered; in all cases, the effects of salinity on granal and lamellar spacing of the chloroplasts are evidenced by a swelling, i.e. separation, of the membrane structures. This can be clearly seen by comparing bundle sheath cell chloroplasts from control plants (Fig. 4, 6, and 7) with bundle sheath cell chloroplasts from salt-treated plants (Figs. 10–13) and mesophyll cell chloroplasts from salt-treated plants (Figs. 14–17). The swelling is more pronounced in the bundle sheath chloroplasts than in the mesophyll chloroplasts. This swelling of granal and lamellar compartments was also evident in chloroplasts in other cells of the leaf, most notably the epidermal cells (Figs. 21–22) and the hair cells (Figs. 18–20). Blumenthal-Goldschmidt and Poljakoff-Mayber (1968) reported a similar swelling of chloroplast lamellae and mitochondrial cristae under highly saline conditions; in extreme cases the swelling was so pronounced that the granal structure became unrecognizable.

There are other observable effects of salinity on the chloroplasts. Extensive lipid accumulation occurred in some of the chloroplasts of salt-treated plants (Figs. 12, 14, and 16). The chloroplasts of the bladder and stalk cells of the vesiculated hairs also show extensive lipid accumulation in control as well as salinized plants (Figs. 18–19). This could be expected in the control plant hair cells because of the high concentration of salt reported there (Mozafar, 1969; Mozafar and Goodin, 1970).

An interesting similarity can be noted in chloroplasts from high-salt treatments and chloroplasts from mesophyll cells of maize plants grown in nitrogen-deficient media. Hall et al. (1972) published micrographs of maize mesophyll cells from various mineral-deficient solutions. The chloroplasts of nitrogen-deficient plants in that study showed lipid accumulation that is very similar to that noted in our salt-treated plants. The possibility exists that in addition to the osmotic effects of highly saline media (which lead to swelling of the organelles), the growth of the plants may be affected by specific ion interference with the metabolism or assimilation of various minerals. We cannot offer a mechanism by which this might take place except to say that it is known that salinity affects the transport of various plant substances, e.g. hormones (Itai, 1967; Itai et al., 1968), and there may be some similar relationship with cell constituents whose syntheses depend on extracellular nitrogen taken up by the roots of the plant.

Thomson (1974) observed that lipid bodies ('plastoglobuli') occur in many types of chloroplasts in a wide range of plants. He did not indicate that their occurrence was a stress response, but did mention studies of 'aging' chloroplasts which showed greater accumulation of plastoglobuli. He assigned carbon storage functions to them but chose not to discuss them as reservoirs of degradation products following stress. The evidence for any of these theories remains, at best, controversial.

Mitochondria also show effects of high salinity treatments. In the mesophyll cells, the mitochondria are small and round (Figs. 5 and 9), and the bundle sheath cells have numerous mitochondria that are much larger than the mesophyll mitochondria (Figs. 4,6,10, and 11). Salt-treated plants have mitochondria that show no cristal structure, probably due to osmotic disruption and swelling (Figs. 10, 11, and 13). Again, the effects of salinity on the ultrastructure of these organelles is much more pronounced in the bundle sheath cells than in the mesophyll cells. It is interesting to note that in the stalk cell of the hairs the chloroplasts show typical effects of salt stress, i.e. lipid accumulation and swelling of membrane compartments, whereas the mitochondria appear to have nearly normal cristal

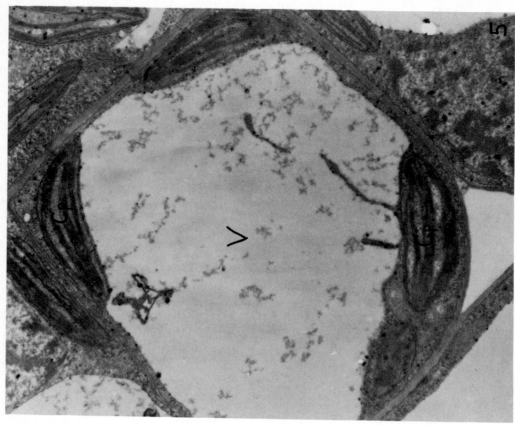

Fig. 5. Mesophyll cell from control plant. Note large vacuole (V), chloroplasts (Cp), and mitochondria (M). 4380×.

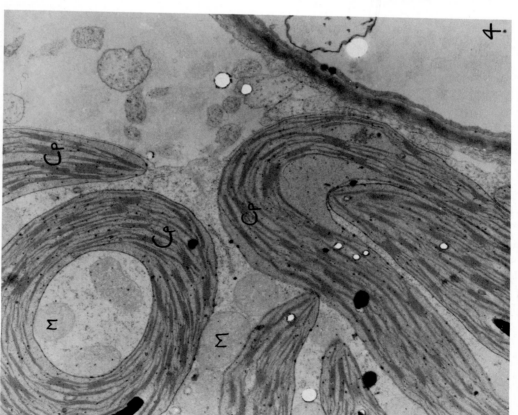

Fig. 4. Portion of bundle sheath cell from control plant showing dense concentration of organelles, including mitochondria (M) and chloroplasts (Cp). 4380×.

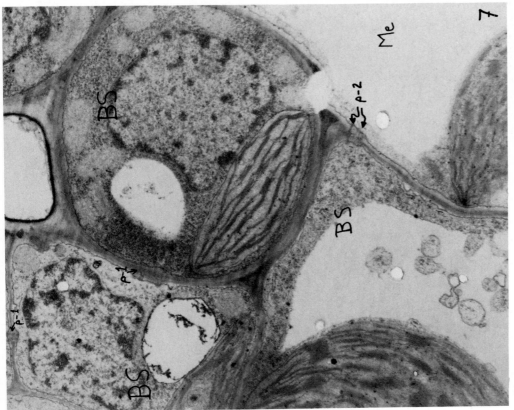

Fig. 7. Bundle sheath cell and adjacent mesophyll cell from control plant. Note plasmodesmata between adjacent bundle sheath cells (p-1) and between a bundle sheath and a mesophyll cell (p-2). 4380×.

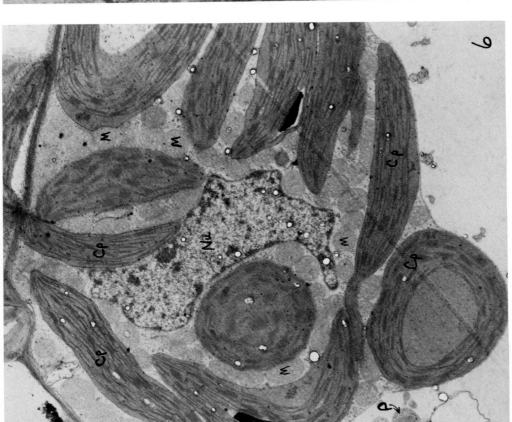

Fig. 6. Bundle sheath cell from control plant showing large, misshaped nucleus (Nu), chloroplasts (Cp), and peroxisome with definite crystal structure (P). 2400×.

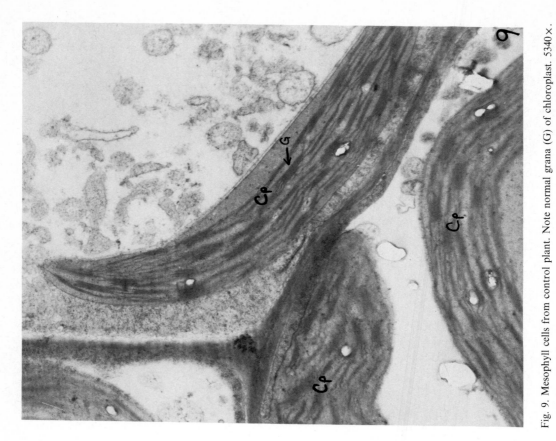

Fig. 9. Mesophyll cells from control plant. Note normal grana (G) of chloroplast. 5340×.

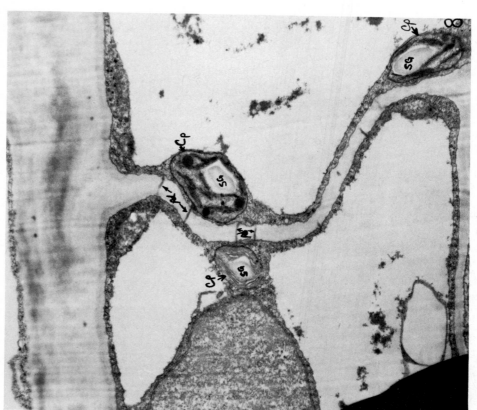

Fig. 8. Portions of adjacent mesophyll cells from a control plant. Note chloroplasts (Cp) with starch grains (SG) and several plasmodesmata (Pd). 12,000×.

94

structure with only slight swelling (Figs. 19–20). This may reflect a relative sensitivity to salt, but the chloroplasts appear functional as evidenced by the presence of large starch grains within the chloroplasts (Figs. 3 and 19).

All of the various cells studied have prominent nuclei. The bundle sheath cells have large, somewhat misshaped nuclei often situated in the center of the cell (Figs. 6,7, and 12). The nuclei of the mesophyll cells are located peripherally (Figs. 5,8, and 17) and are somewhat smaller than the bundle sheath nuclei in relation to cell size. The nucleus of the stalk cell is especially large in relation to cell size, seemingly filling up the central portion of the cell (Figs. 3 and 20). The bladder cells have prominent nuclei in the thin layer of peripheral cytoplasm (Figs. 2–3). The nuclei did not show any significant response t salinity treatment. Blumenthal-Goldschmidt and Poljakoff-Mayber (1968) reported some swelling of the nuclear envelope at extremely high osmotic pressures, but we could discern no differences in the nuclei of control and salinized plants.

Peroxisomes (leaf microbodies) are abundant and morphologically variable in the leaf cells. The distribution of peroxisomes in the leaves of C_4 plants, including *Atriplex halimus*, has been studied by several workers (Newcomb and Frederick, 1971; Smaoui, 1972a, 1972b; Tolbert, 1971). The bundle sheath cells have numerous peroxisomes which are generally spherical, dense, and may contain a crystalline core (Figs. 6 and 13). The mesophyll peroxisomes are somewhat smaller, less abundant, and may also contain a crystalline nucleoid structure (Figs. 14 and 16). Other microbody-like organelles can be seen in the stalk cell of the hair (Fig. 19), but the crystalline structures observed may not be correctly identified as peroxisomes. They are morphologically similar to other microbodies reported (Newcomb and Frederick, 1971).

High salinity treatments do not appear to affect the morphology of the peroxisomes. The various types described above are present in their respective cell types regardless of treatment. The occurrence of numerous peroxisomes in the bundle sheath cells is compatible with the theory that photorespiration (an indicaor of C_3 carbon fixation) takes place main-ly in bundle sheath cells of C_4 plants (Newcomb and Frederick, 1971; Tolbert, 1971). The enzymatic functions of the various morphological types of peroxisomes cannot be determined by morphology alone. It is probable that the peroxisomes of bundle sheath, mesophyll, and hair cells have different enzymatic functions.

Bundle sheath cells in *A. halimus* have small vacuoles with various inclusions, notably occasional myelinated figures (terminology as in Blumenthal-Goldschmidt and Poljakoff-Mayber, 1968) (Fig. 10) and plasma droplets or strands or cytoplasm (Fig. 4). The organelles of the bundle sheath cells tend to be clumped centripetally to the vascular tissue (Fig. 1). Mesophyll cells have much larger vacuoles with some myelinated figures and plasma strands or droplets (Figs. 5, 8, and 15). The mesophyll cytoplasm is usually spread peripherally around the cell (Fig. 5). The hair cells differ greatly in vacuolar characteristics: the stalk cell has a few very small vacuoles (Fig. 20) and the bulk of the cell is filled with cytoplasm, whereas the bladder cell has an extremely large vacuole with a very small amount of cytoplasm in relation to total cell volume (Fig. 3). There are no apparent differences due to salinity in vacuolar arrangement or inclusions.

Endoplasmic reticulum and ribosomes are abundant in bundle sheath and mesophyll cells, and in the stalk cells (Figs. 12, 16, and 19). Cell walls of bundle sheath cells are thick (Figs. 4 and 7) but not necessarily thicker than mesophyll cell walls (Figs. 5 and 16). Numerous plasmodesmata traverse cell walls between bundle sheath and bundle sheath cells (Fig. 7), bundle sheath and mesophyll cells (Fig. 7), and mesophyll and mesophyll cells (Fig. 8). The presence of a suberized layer between bundle sheath and mesophyll cells, as reported by Laetsch (1971), could not be confirmed. These features – endoplasmic reticulum, ribosomes, cell walls, and plasmodesmata – do not appear to vary in morphology with salt treatment.

Special mention must be made of the hair cells and their importance in salinity tolerance in the plant. The hairs are unique structures; the stalk cell is the only pathway through which salts can pass in order to enter the bladder cell (and are therefore specialized for ion transport as well as salinity

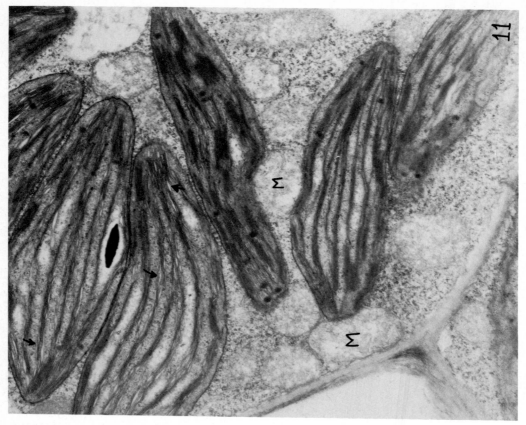

Fig. 11. Bundle sheath cell from salt-treated plant. Note swollen chloroplast lamellae (arrows) and swollen mitochondria (M). 6600×.

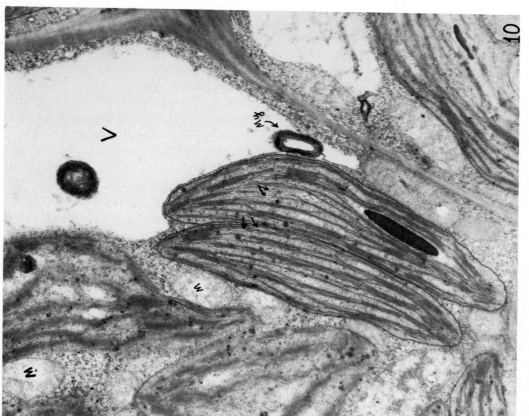

Fig. 10. Bundle sheath cells from salt-treated plant. Note myelinated figure (My) in vacuole (V) and swollen chloroplast lamellae (arrows). 5340×.

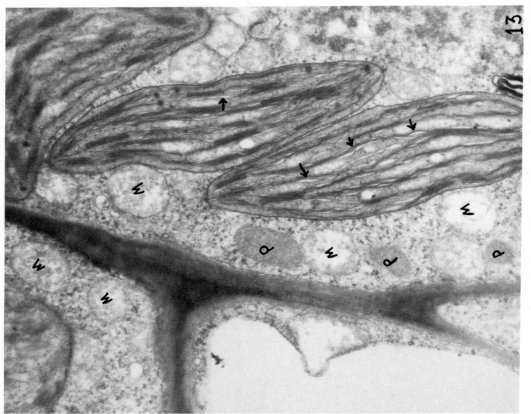

Fig. 13. Bundle sheath cell from salt-treated plant. Note the large peroxisomes (P), swollen chloroplast lamellae (arrows), and disrupted mitochondria (M). 6600×.

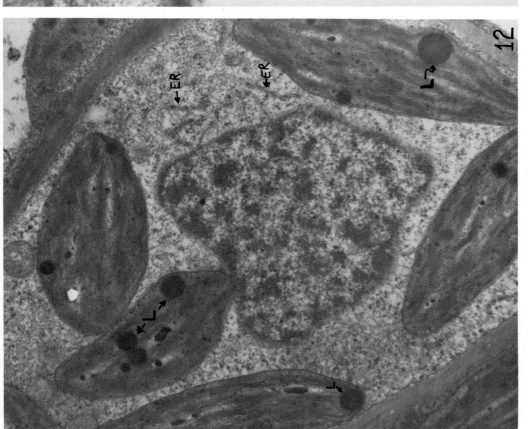

Fig. 12. Bundle sheath cell from salt-treated plant showing lipid accumulation (L) in chloroplasts, extensive endoplasmic reticulum (ER), and dense cytoplasm. 5340×.

97

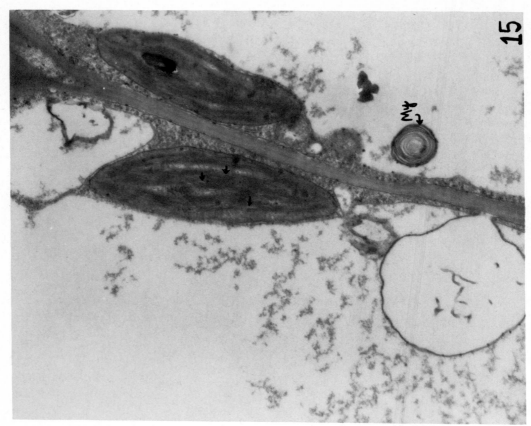

Fig. 15. Portions of mesophyll cells from salt-treated plant. Note myelinated figure (My) in vacuole and swollen chloroplast lamellae (arrows). 9000×.

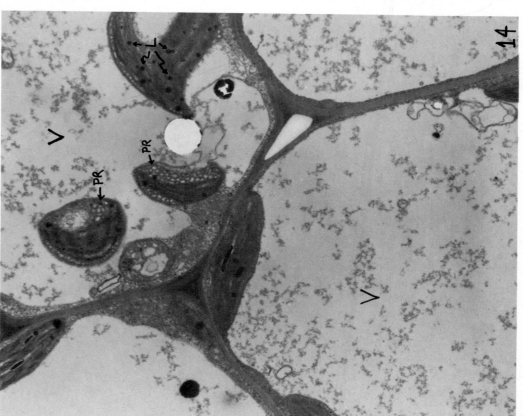

Fig. 14. Mesophyll cells from salt-treated plant. Note large vacuoles (V), peripheral reticulum of chloroplasts (PR), and extensive lipid accumulation in chloroplasts (L). 2400×.

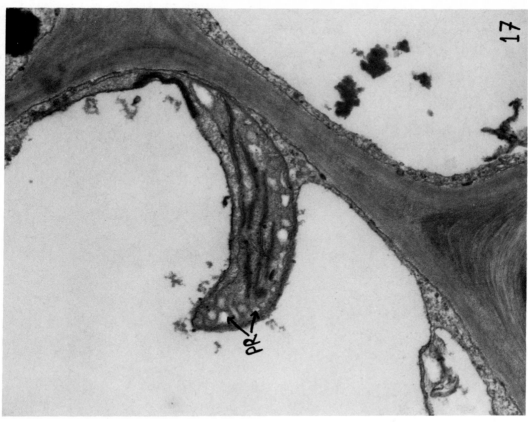

Fig. 17. Mesophyll cell from salt-treated plant. Note peripheral reticulum (PR) on chloroplast. 9000×.

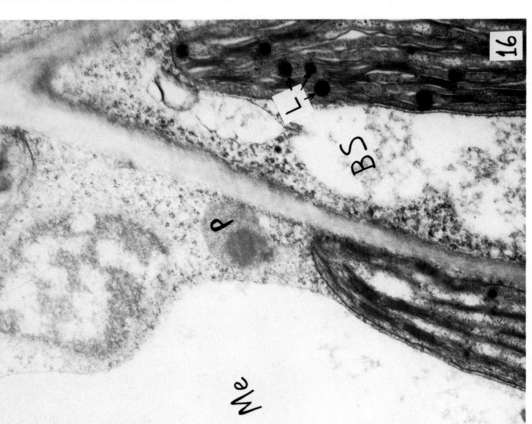

Fig. 16. Adjacent mesophyll cell and bundle sheath cell. Note peroxisome (P) in mesophyll cell and lipid droplets (L) of bundle sheath chloroplasts 9000×.

99

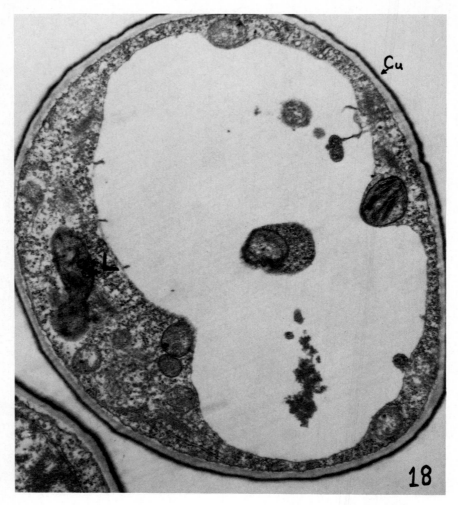

Fig. 18. Portion of bladder cell of hair from control plant. Note thick cuticle (Cu) and chloroplasts with lipid droplets (L) 5340×.

tolerance), and the bladder cell where the salts are concentrated (to a point that surpasses the saturation point of NaCl) must be unusually salt tolerant.

The stalk cell (Figs. 3, 19, and 20) contains many mitochondria, several functional chloroplasts, and a dense cytoplasm. A feature noted in conjunction with the hair cells is the presence of numerous microtubules in the epidermal cell immediately adjacent to the stalk cell, in the stalk cell itself, and in the cytoplasm of the bladder cell immediately adjacent to the stalk cell (Figs. 3 and 20). The function of non-nuclear microtubules in plant cells, although reported in many cells (Ledbetter and

Porter, 1963; Newcomb, 1969), is not well known. They have been implicated as cytoskeletal elements in plant cells and as functioning organelles in movement of cytoplasmic materials (Newcomb, 1969). In addition, Newcomb (1969) described their supposed function as repositories of protein subunits which can be available for use in other cell activities.

The function of microtubules in the hair cells may be as cytoskeletal elements which maintain the structure of the cell. Their presence around the peripheries of the stalk cell cytoplasm tend to support this view (Fig. 20). Cytoplasmic streaming has been noted in the hair cells (Black, 1954, 1956)

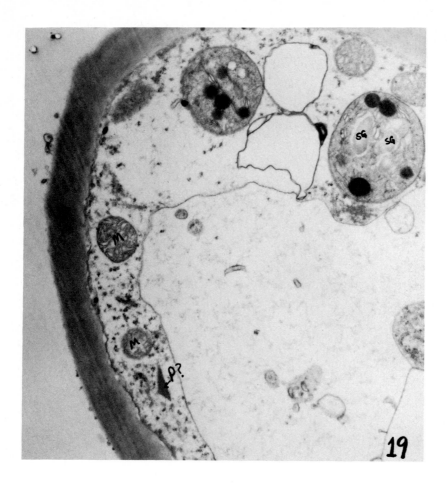

Fig. 19. Portion of stalk cell (in cross section) from control plant. Note chloroplasts with large starch grains (SG). Mitochondria (M) have distinct cristae. Note wedge-shaped crystalline structure (peroxisome?) (P?) 5340 × .

and the microtubules may have some function in this process. If so, a case can be made for their importance in salt transport through the stalk cell, a process which would most likely depend on ATP-mediated ion transport into the cytoplasm of the stalk cell and then out of the stalk cell into the bladder cell. The presence of mitochondria and chloroplasts in epidermal, stalk, and bladder cells would support this view because of the importance of these organelles in ATP production. Lüttge (1971) suggests that the energy for the active transport of ions (a so-called salt pump) may come not from the organelles of the hair cell but from mesophyll (and perhaps epidermal) cells adjacent to

the hair apparatus. In any event, the presence of the microtubules in the hair cells is significant and may be related to the transport of salts through the cells.

Also of interest in the hair cells is the presence of a thick, multi-layered cuticle which is continuous with the cuticle of the epidermis (Figs. 18–21). The presence of this cuticle supports Mozafar's (1969) conclusion that the cells can neither acquire nor lose water through the intact hair cells. This cuticular layer may be especially important in the transport of ions from the leaf to the hairs. If, as Thomson (1975) suggested, the movement of ions is apoplastic as well as symplastic, the cuticle cannot be continuous between the stalk and bladder cells.

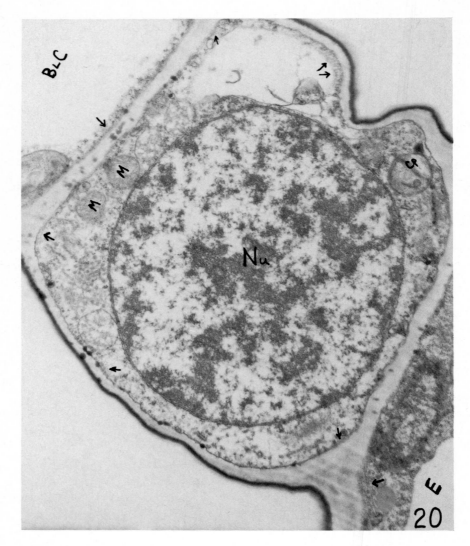

Fig. 20. Stalk cell from hair of control plant. Note large nucleus (Nu), chloroplast (Cp), numerous mitochondria (M), and microtubules (arrows) in stalk cell and adjacent bladder and epidermal cells. 5340×.

This interpretation must, therefore, place extreme reliance on the capabilities of the bladder cell and its plasmalemma to actively transport salt into the cell from the apoplastic wall where the salts accumulate by hydrostatic pressure or other means. Lüttge (1975) discussed the energy requirements and structure/function relationships of salt glands (including *Atriplex* leaf hairs). While the current interpretations of salt gland function, as noted above, are not entirely compatible, the arguments for both apoplastic and symplastic ion transport appear to be equally valid and deserve more work.

The vesiculated hairs of *Atriplex halimus* L. function as unique mechanisms of salt tolerance; an entire study could be devoted to their structure, physiology, and development.

7. Conclusions

With limitations on energy resources and serious consideration being given to energy inputs versus technological advancements in productivity, greater emphasis is being given to those plants which have a

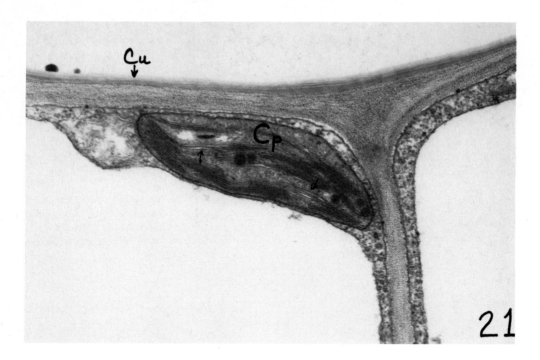

Fig. 21. Epidermal cell from salt-treated plant. Note cuticle (Cu), and chloroplast (Cp) with swollen lamellae (arrows). 11250×.

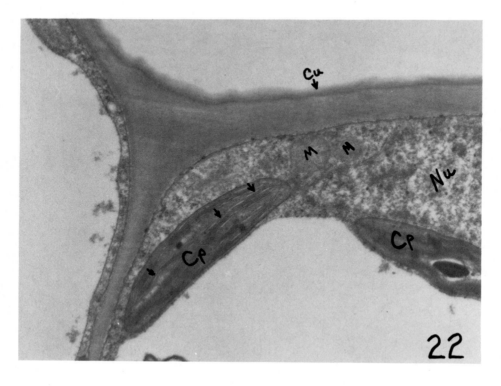

Fig. 22. Epidermal cell from salt-treated plant. Note cuticle (Cu), nucleus (Nu), mitochondria (M), and chloroplasts (Cp) with swollen lamellae (arrows). 6675×.

103

potential for productivity with minimal inputs of water, fertilizers, mechanical harvesting and processing technology. With those limitations, one looks to the arid and semi-arid regions of the world for species with genetic adaptations which have assured their success over hundreds or thousands of years. Many of the desirable traits of arid species have been known for centuries, but because of ignorance or simply oversophistication in the developed countries, we have ignored many excellent possibilities.

The cosmopolitan genus *Atriplex* with approximately 200 species represents a group of plants with a tremendous genetic diversity and special features for adaptation to stress situations, including hal-

ophytism (Goodin and Mozafar, 1972; Goodin, 1977, 1979). The demands imposed on natural resources by population and energy pressures have caused a reassessment of our unused or underused lands. Regions heretofore considered too dry or too salty have suddenly become an integral part of the resource. With proper management, halophytes like *Atriplex* can make a significant contribution to mankind and to world productivity.

Acknowledgements. We are indebted to Dr. J.D. Berlin for technical advice and assistance with the microscopy portion of this study, and to the Institute for the Study of Physiological Stresses in Plants at Texas Tech University for financial assistance with various phases of the work.

Literature cited

Beevers, H. 1979. Microbodies in higher plants. *Ann. Rev. Plant Physiol.* 30: 159–193.

Bernstein, L. and Hayward H. E., 1958. Physiology and salt tolerance. *Ann. Rev. Plant Physiol.* 9: 25–46.

Björkman, O. and Berry, J. 1973. High efficiency photosynthesis. *Scientific American*, Oct., 1973, 80–93.

Black, C. C. 1973. Photosynthetic carbon fixation in relation to net CO_2 uptake. *Ann. Rev. Plant Physiol.* 24: 253–286.

Black, C. C. an Mollenhauer, H. H. 1971. Structure and distribution of chloroplasts and other organelles in leaves with various rates of photosynthesis. *Plant Physiol.* 47: 15–23.

Black, R. F. 1954. The leaf anatomy of Australian members of the genus *Atriplex*. I. *Atriplex vesicaria* Heward and *A. nummularia* Lindl. *Aust. J. Bot.* 2: 269–286.

Black, R. F. 1956. Effects of NaCl in water culture on the ion uptake and growth of *Atriplex hastata* L. *Aust. J. Biol. Sci.* 9: 67–80.

Black, R. F. 1958. Effect of sodium chloride on leaf succulence and area of *Atriplex hastata*. *Aust. J. Bot.* 6: 306–321.

Blumenthal-Goldschmidt S. and Poljakoff-Mayber, A. 1968. Effect of substrate salinity on growth and submicroscopic structure of leaf cells of *A. halimus* L. *Aust. J. Bot.* 16: 469–478.

Brownell, P. F. 1965. Sodium as an essential micronutrient for a higher plant *(Atriplex vesicaria)*. *Plant Physiol.* 40: 460–468.

Broyer, T. C. and Hoagland, D. R. 1943. Metabolic activity of roots. *Amer. J. Bot.* 30: 261–273.

Broyer, T. C., Carlton, A. B., Johnson, C. M. and Stout, P. R. 1954. Chlorine as a micronutrient. *Plant Physiol.* 29: 526–532.

Caldwell, M. M. 1974. Physiology of desert halophytes. In *Ecology of Halophytes*, eds. R. Reimold and W. H. Queen, pp. 355–378. Academic Press, New York.

Campbell, N., Thomson, W. W. and Platt, K. 1974. The apoplastic pathway of transport to salt glands. *J. Exp. Bot.* 25: 61–69.

Chapman, V. J. 1975. The salinity problem in general, its importance, and distribution with special reference to natural halophytes. In *Plants in Saline Environments*, eds. A. Poljakof-Mayber and J. Gale, pp. 7–24. Springer-Verlag, New York.

Clements, H. F. 1934. Significance of transpiration. *Plant Physiol.* 9: 165–172.

Cook, C. W. 1972. Comparative nutritive values of forbs, grasses, and shrubs. In *Wildland Shrubs – Their Biology and Utilization*, eds. C. M. McKell, J. P. Blaisdell, and J. R. Goodin, pp. 303–310. USDA For. Serv. Tech. Rept. INT–1.

Crafts, A. S. and Broyer, T. C. 1938. Migration of salts and water into xylem of the roots of higher plants. *Amer. J. Bot.* 25: 529–535.

Curtis, O. F. 1926. What is the significance of transpiration? *Science* 63: 267–271.

Downton, W. J. S. 1971a. Adaptive and evolutionary aspects of C_4 photosynthesis. In *Photosynthesis and Photorespiration*, eds. M. D. Hatch, C. B. Osmond, and R. O. Slatyer, pp. 3–17. Wiley-Interscience, John Wiley and Sons, New York.

Downton, W. J. S. 1971b. The chloroplasts and mitochondria of bundle sheath cells in relation to C_4 photosynthesis. In *Photosynthesis and Photorespiration*, eds. M. D. Hatch, C. B. Osmond, and R. O. Slatyer, pp. 419–425. Wiley-Interscience, John Wiley and Sons, New York.

Downton, W. J. S., Bisalputra, T. and Tregunna, E. B. 1969. The distribution and ultrastructure of chloroplasts in leaves differing in photosynthetic carbon metabolism. II. *Atriplex rosea* and *Atriplex hastata* (Chenopodiaceae). *Can. J. Bot.* 47: 915–919.

Fahn, A. 1974. *Plant Anatomy*, Second Ed. Pergamon Press, Oxford.

Flowers, T. J., Trobe, P. F. and Yeo, A. R. 1977. The mechanism of salt tolerance in halophytes. *Ann. Rev. Plant Physiol.* 28:

89–121.

Frederick, S. E. and Newcomb, E. H. 1971. Ultrastructure and distribution of microbodies in leaves of grasses with and without CO_2-photorespiration. *Planta* 96: 152–174.

Gale, J. and Poljakoff-Mayber, A. 1970. Interrelations between growth and photosynthesis of salt bush (*Atriplex halimus* L.) grown in saline media. *Aust. J. Biol. Sci.* 23: 937–945.

Gates, C. T. 1972. Ecological responses of the Australian native species *Acacia harpophylla* and *Atriplex nummularia* to soil salinity: effects on water content, leaf area, and transpiration rate. *Aust. J. Bot.* 20: 261–272.

Gates, D. M. 1968. Transpiration and leaf temperature. *Ann. Rev. Plant Physiol.* 19: 211–238.

Gifford R. M. 1974. A comparison of potential photosynthesis, productivity, and yield of plant species with differing photosynthetic metabolism. *Aust. J. Plant Physiol.* 1: 107–117.

Goodchild, D. J. 1971. Chloroplast structure: assessment. In *Photosynthesis and Photorespiration*, eds. M. D. Hatch, C. B. Osmond, and R. O. Slatyer, pp. 426–427. Wiley-Interscience, John Wiley and Sons, New York.

Goodin, J. R. 1977. Salinity effects on range plants. In: *Rangeland Plant Physiology, Range Science Series No. 4*, ed. R. E. Sosebee, pp. 141–153. Society for Range Management, Denver, Colorado.

Goodin, J. R. 1979. *Atriplex* as a forage crop for arid lands. In *New Agricultural Crops*, ed. G. A. Ritchie, pp. 133–148. AAAS Symp. 38, Amer. Assoc. Adv. Sci., Washington.

Goodin, J. R. and Mozafar, A. 1972. Physiology of salinity stress. In: *Wildland Shrubs--Their Biology and Utilization*, eds. C. M. McKell, J. P. Blaisdell, and J. R. Goodin, pp. 255–259. U.S. Dept. Agr. Forest Serv. Tech. Rep. INT-1.

Goodin, J. R. and McKell, C. M. 1970. *Atriplex* spp. as a potential forage crop in marginal agricultural areas. *Int. Grassland Cong., Proc., Univ. Queensland Press, Brisbane* 11: 158–162.

Gracen, V. E., Hilliard, J. H. and West, S. H. 1972. Presence of peripheral reticulum in chloroplasts of Calvin cycle cells. *J. Ultrastruct. Res.* 38: 262–264.

Greenway, H. and Munns, R. 1980. Mechanisms of salt tolerance in nonhalophytes. *Ann. Rev. Plant Physiol.* 31: 149–190.

Hall, J. D., Barr, R., Al-Abbas, A. H. and Crane, F. L. 1972. The ultrastructure of chloroplasts in mineral-deficient maize leaves. *Plant Physiol.* 50: 404–409.

Hall, R. H. 1973. Cytokinins as a probe of developmental processes. *Ann. Rev. Plant Physiol.* 24: 415–444.

Hatch, M. D., Osmond, C. B. and Slatyer, R. O. eds. 1971. *Photosynthesis and Photorespiration*. Wiley-Interscience, John Wiley and Sons, New York.

Hatch, M. D. and Slack, C. R. 1966. Photosynthesis by sugar cane leaves. A new carboxylation reaction and pathway of sugar formation. *Biochem. J.* 101: 103–111.

Hatch, M. D. and Slack, C. R. 1970. Photosynthetic CO_2 fixation pathways. *Ann. Rev. Plant Physiol.* 21: 141–162.

Hatch, M. D., Slack, C. R. and Johnson, H. S. 1967. Further studies on a new pathway of photosynthetic carbon dioxide fixation in sugar cane and its occurrence in other plant species.

Biochem. J. 102: 417–422.

Hellebust, J. A. 1976. Osmoregulation. *Ann. Rev. Plant Physiol.* 27: 485–507.

Hoagland, D. R. and Arnon D. I. 1950. The water-culture method for growing plants without soil. Bull.No. 347. Calif. Agri. Exp. Sta., Berkeley.

Hodge, A. J., McLean, J. D. and Mercer, F. V. 1955. Ultrastructure of the lamellae and grana in the chloroplast of *Zea mays* L. *Biophys. Biochem. Cytol.* 1: 605–614.

Itai, C. 1967. Shoot and root interaction under different water regimes. Ph.D. Thesis, Hebrew University, Jerusalem.

Itai, C., Richmond, A. and Vaadia, Y. 1968. The role of cytokinins during water and salinity stress. *Israel J. Bot.* 17: 187–195.

Jennings, D. H. 1968. Halophytes, succulence, and sodium in plants – a unified theory. *New Phytol.* 67: 899–911.

Jennings, D. H. 1976. The effects of sodium chloride on higher plants. *Biol. Rev.* 51: 453–486.

Johnson, Sr. C. 1964. An electron microscope study of the photosynthetic apparatus in plants, with special reference to the Gramineae. Ph.D. Diss., Univ. Texas, Austin.

Johnson, Sr. C. and Brown, W. V. 1973. Grass leaf ultrastructural variations. *Amer. J. Bot.* 60: 727–735.

Johnson, H. S. and Hatch, M. D. 1968. Distribution of the C_4-dicarboxylic acid pathway of photosynthesis and its occurrence in dicotyledonous plants. *Phytochemistry* 7: 375–380.

Jones, R. 1970. *The Biology of Atriplex*. CSIRO, Canberra.

Kaplan, A. and Gale, J. 1972. Effect of sodium chloride salinity on the water balance of *Atriplex halimus*. *Aust. J. Biol. Sci.* 25: 895–903.

Kelley, D. B. 1974. Salinity effects on growth and fine structure of *Atriplex halimus* L. M. S. Thesis, Texas Tech Univ., Lubbock, Texas.

Kortschak, H. P., Hartt, C. E. and Burr, G. O. 1965. Carbon dioxide fixation in sugar cane leaves. *Plant Physiol.* 40: 209–213.

Kramer, P. J. 1969. *Plant and Soil Water Relationships*. Mc-Graw-Hill, New York.

Laetsch, W. M. 1968. Chloroplast specialization in dicotyledons possessing the C_4-dicarboxylic acid pathway of photosynthetic CO_2 fixation. *Amer. J. Bot.* 55: 875–883.

Laetsch, W. M. 1969. Relationship between chloroplast structure and photosynthetic carbon-fixation pathways. *Sci. Progr.* 57: 323–351.

Laetsch, W. M. 1971. Chloroplast structural relationships in leaves of C_4 plants. In *Photosynthesis and Photorespiration*, eds. M. D. Hatch, C. B. Osmond, and R. O. Slatyer, pp 323–349. Wiley-Interscience, John Wiley and Sons, New York.

Laetsch, W. M. 1974. The C_4 syndrome: a structural analysis. *Ann. Rev. Plant Physiol.* 25: 27–52.

Laetsch, W. M., Stetler, D. A. and Vlitros, A. J. 1965. The ultrastructure of sugar cane chloroplasts. *Z. Pflanzenphysiol.* 54: 472–474.

Ledbetter, M. C. and Porter, K. R. 1963. A 'microtubule' in plant cell fine structure. *J. Cell Biol.* 19: 239–250.

Lüttge, U. 1971. Structure and function of plant glands. *Ann. Rev. Plant Physiol.* 22: 23–44.

Lüttge, U. 1975. Salt glands. In *Ion Transport in Plant Cells and Tissues*, eds. D. A. Baker and J. L. Hall, pp. 126–158. North Holland Publ. Co., Amsterdam.

Lüttge, U. and Osmond, C. B. 1970. Ion absorption in *Atriplex* leaf tissue. III. Site of metabolic control of light dependent chloride secretion to epidermal bladders. *Aust. J. Biol. Sci.* 23: 17–25.

Marx, Jean L. 1973. Photorespiration: Key to increasing plant productivity? *Science* 179: 365–367.

McKell, C. M. 1975. Shrubs – a neglected resource of arid lands. *Science* 187: 803–809.

McKell, C. M., Blaisdell, J. P. and Goodin, J. R., eds. 1972. *Wildland Shrubs – Their Biology and Utilization*. General Technical Report INT-1, U. S. Forest Service, Washington, D.C.

Miller, D. R. 1973. Ionic balance and transpiration rates in *Atriplex halimus* L. M. S. Thesis, Texas Tech University, Lubbock.

Moore, R. T., White, R. S. and Caldwell, M. M. 1972. Transpiration of *Atriplex confertifolia* and *Eurotia lanata* in relation to soil, plant, and atmospheric moisture stresses. *Can. J. Bot.* 50: 2411–2418.

Mozafar, A. 1969. Physiology of salt tolerance in *Atriplex halimus* L.: Ion uptake and distribution, oxalic acid content, and catalase activity. Ph.D. Diss., Univ. of Calif., Riverside.

Mozafar, A. and Goodin, J. R. 1970. Vesiculated hairs: a mechanism for salt tolerance in *Atriplex halimus* L. *Plant Physiol.* 45: 62–65.

Newcomb, E. H. 1969. Plant microtubules. *Ann. Rev. Plant Physiol.* 20: 253–288.

Newcomb, E. H. and Frederick, S. E. 1971. Distribution and structure of plant microbodies (peroxisomes). In *Photosynthesis and Photorespiration*, eds. M. D. Hatch, C. B. Osmond, and R. O. Slatyer, pp. 442–457. Wiley-Interscience, John Wiley and Sons, New York.

Nord, E. C. and Countryman, C. M. 1972. Fire relations. In *Wildland Shrubs – Their Biology and Utilization*, eds. C. M. McKell, J. P. Blaisdell and J. R. Goodin, pp. 88–97. USDA For. Serv. Gen. Tech. Rept. INT-1.

Oertli, J. J. 1966. Effects of external salt concentrations on water relations in plants. III. Concentration dependence of the osmotic differential between xylem and external medium. *Soil Sci.* 104: 56–62.

Oertli, J. J. 1971. The stability of water under tension in the xylem. *Z. Pflanzenphysiol.* 65: 195–209.

Osmond, C. B. 1967. Acid metabolism in *Atriplex*. I. Regulation of oxalate synthesis by the apparent excess cation absorption in leaf tissue. *Aust. J. Biol. Sci.* 20: 575–587.

Osmond, C. B., Lüttge, U., West, K. R., Pallaghy, C. K. and Schacher-Hill, B. 1969a. Ion absorption in *Atriplex* leaf tissue. II. Secretion of ions to epidermal bladders. *Aust. J. Biol. Sci.* 22: 797–814.

Osmond, C. B., Troughton, J. H. and Goodchild, D. J. 1969b. Physiological, biochemical, and structural studies of photo-synthesis and photorespiration in two species of *Atriplex*. *Z. Pflanzenphysiol.* 61: 218–237.

Poljakoff-Mayber, A. and Gale, J., eds. 1975. *Plants in Saline Environments*. Springer-Verlag, Berlin.

Reynolds, E. S. 1963. The use of lead citrate at high pH as an electron-opaque stain in electron microscopy. *J. Cell Biol.* 17: 208–212.

Rhoades, M. M. and Carvalho, A. 1944. The function and structure of the parenchyma sheath plastids of the maize leaf. *Bull. Torrey Bot. Club* 71: 335–345.

Richardson, K. C., Jarrett, L. and Finke, E. H. 1960. Embedding in epoxy for ultrathin sectioning in electron microscopy. *Stain Tech.* 35: 313–325.

Rufelt, H. 1956. Influence of the root pressure on the transpiration of wheat plants. *Physiol. Plant.* 9: 154–164.

Russell, R. S. and Barber, D. A. 1960. The relationship between salt uptake and the absorption of water by intact plants. *Ann. Rev. Plant Physiol.* 11: 127–140.

Shimony, C., Fahn, A. and Reinhold, L. 1973. Ultrastructure and ion gradients in the salt glands of *Avicennia marina* (Forssk.) Vierh. *New Phytol.* 72: 27–36.

Smaoui, A. 1971. Differenciation des trichomes chez *Atriplex halimus* L. *C. R. Acad. Sci. Ser.* D 273 (15): 1268–1271.

Smaoui, A. 1972a. Structures protiques dans la cellule pedicel-laire des trichomes de *Chenopodium album* L. *C. R. Acad Sci. Ser.* D. 275(3): 373–377.

Smaoui, A. 1972b. Distribution des peroxysomes dans les feuilles de deux Chenopodiacees differant par leur structure et leur metabolisme photosynthetique: *Atriplex halimus* L. et *Chenopodium album* L. *C. R. Acad. Sci. Ser.* D. 275(10): 1031–1034.

Spurr, A. R. 1969. A low-viscosity epoxy resin embedding medium for electron microscopy. *J. Ultrastruc. Res.* 26: 31–43.

Stark, N. 1966. Review of highway planting information appropriate to Nevada. Desert Res. Inst., Univ. Nevada Coll. Agric. Bull. B–7.

Thomson, W. W. 1974. Ultrastructure of mature chloroplasts. In *Dynamic Aspects of Plant Ultrastructure*, ed. A. W. Robards, pp. 138–177. McGraw-Hill, United Kingdom.

Thomson, W. W. 1975. The structure and function of salt glands. In *Plants in Saline Environments*, eds. A. Poljakoff-Mayber and J. Gale, pp. 118–146. Springer-Verlag, Berlin.

Tolbert, N. E. 1971. Microbodies-peroxisomes and glyoxysomes. *Ann. Rev. Plant Physiol.* 22: 45–74.

Torrey, J. G. 1976. Root hormones and plant growth. *Ann. Rev. Plant Physiol.* 27: 435–459.

Tregunna, E. B. and Downton, J. 1967. CO_2 compensation in members of the Amaranthaceae and some related families. *Can. J. Bot.* 45: 2385–2387.

Ulrich, A. and Ohki, K. 1956. Chloride, bromine, and sodium as nutrients for sugar beet plants. *Plant Physiol.* 31: 171–181.

Waisel, Y. 1960. Ecological studies on *Tamarix aphylla* (L.) Karst. II. The water economy. *Phyton* 15: 17–27.

Waisel, Y. 1972. *Biology of Halophytes*. Academic Press, New York.

Welkie, G. W. and Caldwell, M. M. 1970. Leaf anatomy of species in some dicotyledon families as related to the C_3 and

C$_4$ pathways of carbon fixation. *Can. J. Bot.* 48: 2135–2146.

White, P. R. 1938. Root-pressure--an unappreciated force in sap movement. *Amer. J. Bot.* 25: 223–227.

White, P. R., Schuker, E., Kern, J. R. and Fuller, F. H. 1958. Root pressure in gymnosperms. *Science* 128: 308–309.

Wood, J. G. 1925. The selective absorption of chlorine ions and the absorption of water in the genus *Atriplex. Aust. J. Exp. Biol. Med. Sci.* 2: 45–56.

Zelitch, I. 1971. *Photosynthesis, Photorespiration and Plant Productivity.* Academic Press, New York.

Ziegler, H. and Lüttge, U. 1966. Die Salzdrüsen von *Limonium vulgare.* II. Mitteilung: Die Feinstructur. *Planta* 70: 193–206.

Reference added in proof. A significant and timely review of *Atriplex* biology appeared in 1980 after this chapter had gone to press. This book, 'Physiological Processes in Plant Ecology. Toward a Synthesis with *Atriplex*' by C.B. Osmond, O. Björkman and D.J. Anderson and published by Springer-Verlag, Berlin, updates information on the work done on the genus since the 1969 symposium in Australia (Jones, 1969).

PART TWO

Ecological and ecophysiological problems

DAVID N. SEN and KISHAN S. RAJPUROHIT

Introduction

It appears difficult to estimate the actual total area of salt-affected soils throughout the world. A recent survey indicated that the irrigated area of 103 countries totaled 203 million hectares (Anonymous, 1970). In general it can be concluded that salinity problems are found in all countries having areas where arid and semi-arid climates exist. The great majority of these areas appear to have been affected by excess soluble salts for thousands of years, but there is a very large area of land which has become saline within the period of recorded history. Secondary salinity is a great environmental problem faced by man today.

The excess salts, especially in alkali soils, cause physical and chemical changes in soil structure, such as swelling of soil particles which interfere with drainage and aeration. Plants can adapt to some extent in the low osmotic potentials prevailing in saline soils by absorbing ions. A major portion of research work on salinity has been done keeping in view NaCl as the most common salt present. Sodium as well as chloride ions are readily absorbed by the root system of both glycophytes and halophytes. This absorbed sodium after transportation to leaves is either excreted through salt glands or contained in the shoot till the concentration is too high. In glycophytes only a very small amount of sodium reaches the leaves as most of it is retained in the roots or stem itself (Jacoby, 1964, 1965; La Haye and Epstein, 1969, 1970). Salinity affects metabolic processes of plants and induces changes in their morphology and anatomy.

It has been shown that salinity effects the time and rate of germination, the size of plant, branching, leaf size, and species zonation (Rajpurohit et al., 1978; Rajpurohit, 1980). Succulence is one of the most common features of halophytes, which is often considered to be an adaptation to reduce the internal salt concentration. The various affects of salinity on plants, occurring in response to salinity which are typical for halophytes have been reviewed by Waisel (1972), and Poljakoff-Mayber and Gale (1975). Strogonov (1962) and Waisel (1972) have enlisted numerous structural changes ascribed to salinity. These are: (a) increase of succulence; (b) changes in number and size of stomata; (c) thickening of the cuticle; (d) inhibition of differentiation; (e) extensive development of tyloses; (f) earlier occurrence of lignification; (g) changes in diameter and number of xylem vessels, etc. Stunting of growth is caused in halophytes which is not necessarily always a disadvantage. A big vegetative body with a large leaf area, commonly found in arid and semi-arid regions, although beneficial for photosynthesis, leads to exposure of a large transpiring area. Low yields (fresh and dry weights) although disadvantageous from an agricultural point of view, may ecologically-wise be a very useful strategy (Goodman, 1973).

The reduced growth of plants under conditions of salinity was ascribed to physiological drought, which indicated shortage of water within the plant even when growing under moist but saline soil conditions (see

Schimper, 1898; quoted by Strogonov, 1964). The lowered osmotic potential of the soil water, resulting from high concentration of soluble salts, was thought to prevent uptake of water by the plant. A corollary to the physiological drought hypothesis was that equi-osmolar concentrations of different salts would have the same effect on plant growth. This hypothesis prevailed for a long time despite the early work of Osterhout (1906) who showed that diluted sea water (a mixed salt solution) was less damaging to plant growth than the equi-osmolar concentrations of single salts. Uptake of water is determined by the gradient of total water potential (w) between the root and soil, and the resistance to liquid flow in the soil-root system. Loss of water from the plants is determined by the gradient of water potential between leaf and atmosphere. Under saline conditions, the driving forces for water uptake is usually, if not always, maintained. This is brought about by osmotic adjustment by plant tissues. The overall rate of water turnover (uptake and transpiration) is generally reduced.

Literature cited

Anonymous. 1970. *Irrigation Statistics of the World*. ICID Bul., International commision on Irrigation and Drainage 48, Naya Marg. Chanakyapuri, New Delhi 21, India, pp. 76–78.

Goodman, P.J. Physiological and ecotypic adaptation of plants to salt desert conditions in Utah. *J. Ecol*. 61: 473–494.

Jacoby, B. 1964. Function of bean roots in sodium retention. *Plant Physiol*. 39: 445–449.

Jacoby, B. 1965. Sodium retention in excised bean stem. *Physiol. Plantarum* 18: 730–739.

La Haye, P.A. and Epstein, E. 1969. Salt toleration by plants: enhancement with calcium. *Science* 166: 395–396.

La Haye, P.A. and Epstein, E. 1970. Salt toleration by plants: enhancement with calcium. *Science* 167: 1388.

Osterhout, W.J. v. 1906. On the importance of physiologically balanced solutions for plants. *Bot. Gaz*. 42: 127–134.

Poljakoff-Mayber, A. and Gale, J. eds. 1975. *Plants in Saline Environments*. Springer Verlag, Berlin.

Rajpurohit, K.S. 1980. Soil salinity and its role on phytogeography of western Rajasthan. Ph.D. Thesis Univ. of Jodhpur, Jodhpur.

Rajpurohit, K.S., Sen, D.N. and Charan, A.K. 1978. Ecology of halophytes of Indian desert. I. *Aeluropus lagopoides* (Linn.) Trin ex Thw. *Geobios* 5: 188–190.

Schimper, A.F.W. 1898. Pflanzengeographie auf physiologischer Grundlage. 1st ed. Gustav Fischer, Jena.

Strogonov, B.P. 1962. *Fisiologithcheskie Osnovy Soleustoitchivosti Rastenii* (Physiological basis of salt tolerance in plants). Akademia Nauk SSSR, Moskva.

Strogonov, B.P. 1964. *Physiological Basis of Salt Tolerance of Plants*. I.P.S.T., Jerusalem.

Waisel, Y. 1972. *Biology of Halohytes*. Academic Press, New York.

CHAPTER 1

Mangrove species zonation: why?

SAMUEL C. SNEDAKER

Division of Biology and Living Resources, Rosenstiel School of Marine and Atmospheric Science, 4600 Rickenbacker Causeway, Miami, FL 33149, USA

1. Introduction

Mangroves are a taxonomically-diverse group of woody spermatophytes which possess a common ability to survive and perpetuate themselves along sheltered tropical coastlines in saline environments under tidal influence. A halophytic existence is made possible through a wide range of morphological, anatomical and physiological adaptations which has elicited much scientific interest (Walsh, 1974; Chapman, 1976). Although mangroves share a common ability to exist as halophytes in a common environment, they frequently appear in rather predictable mono-specific zones parallel to shorelines, tidal channels, and the banks of rivers and streams influenced by the sea. Thus, mangrove species zonation has been a dominant theme in a voluminous literature on mangroves which exceeds 7,000 titles (B. Rollet, pers. comm.[1]). However, when one searches through this literature for a coherent scientific explanation as to why mangroves so frequently appear in zones, at least four bodies of knowledge and opinion can be identified which, in part reflect the scientific interests and professional biases of the authors[2]: plant succes-

sion, geomorphology, physiological ecology and population dynamics. The purpose of this chapter is to critically review the basic proposition in each of these schools of thought and integrate the pertinent parts into a unified perspective explaining why mangroves are zoned. As will be seen, the sequence of presentation (i.e. plant succession, geomorphology, physiological ecology and population dynamics) partially recapitulates the chronological development of knowledge on mangrove species zonation.

Prior to a discussion of the possible reasons underlying the zonation of mangroves, it is necessary to point out that the 'classical' pattern of zonation may represent just one of several patterns of spatial organization. For example, the works of Thom (1967) and Thom et al. (1975) in Mexico and Australia, respectively, describe spatial patterning atypical of classical zonation. West (1956) acknowledged his inability to identify zonation patterns in the mangrove forests along the Pacific coast of Columbia. Throughout the Indus delta area (249,500 ha) of Pakistan, species zonation is less apparent than zones of differing structure in the same species (Khan, 1966). In the floristically-richer and larger delta of the Ganges River (407,000 ha), species zonation appears on a microscale along streams, creeks and rivers (khals, bharanis and gangs) and around depressions (bils) (M. Ismail, pers. comm.). However, a pattern of zonation also occurs at a macroscale across the entire delta, known as the Sundarbans forest, in which there is

---

[1] B. Rollet has prepared an annotated bibliography of some 5,000 titles which is expected to be published by UNESCO's Division of Marine Sciences.

[2] Two other factors also contribute to the lack of a consensus on mangrove zonation. Anecdotal reports and purely descriptive studies have led to conclusions lacking rigorous scientific and evidentiary support. Also, mangroves and the mangrove environments are so variable that locally developed conclusions cannot always be validly extrapolated.

Tasks for vegetation science, Vol. 2 ed. by D.N. Sen and K.S. Rajpurohit

a zonation of species and forest structure (density and height) along a northeast-southwest salinity gradient (Curtis, 1933; Choudhury, A.M., 1968). Along the east coast of Africa, the coastal mangrove forests of Tanzania are reported to be zoned (Walter, 1971) whereas in Mozambique they are not (Macnae and Kalk, 1962). In our own studies, the classical zonation first described by Davis (1940) was found to be so rare in southwest Florida (Ten Thousand Islands), that another spatial classification scheme had to be devised as a basis for research (Snedaker and Lugo, 1973; Lugo and Snedaker, 1974). In the extensive dwarf mangrove forest of southeast Florida, each of the mangrove species common to southern Florida occurs randomly intermingled without any statistical evidence of local clustering (Snedaker and Stanford, 1976).

This qualification of the zonation argument does not imply that mangrove species in zoned versus unzoned environments have differing environmental requirements or tolerances. In fact, that is one of the questions addressed in this chapter.

2. Zonation as an expression of plant succession

The interpretation of mangrove species zonation as being a spatial expression of plant succession seems to have its origin in the western hemisphere where much of the early work focused on the land building role of mangroves. In this regard, Curtiss (1888) appears to have been among the first to describe how the *Rhizophora mangle* L. create land and '... in the course of time build up a foundation for other species'. Curtiss (1888) was followed by many other investigators who further described the land building role of the *R. mangle* (e.g. Sargent, 1893; Phillips, 1903; Pollard, 1903; Vaughan, 1909; Harshberger, 1914; Harper, 1917; Gifford, 1934; Davis, 1938, 1940).[3] It is interesting to note that Davis (1940) cited several of these early works and

stated: 'Their descriptions were nearly all written without either a thorough study of the general ecology of mangrove vegetation or enough field and experimental work to verify their conclusions about the land-building role of mangroves.' Davis (1940) can probably be credited with the formal interpretation of mangrove species zonation being an expression of plant succession as his paper is considered a classic on the ecology and ecologic role of mangroves. At the time of his investigations, plant succession was being generally defined as the successive occupation by different plant communities on the same site (Weaver and Clements, 1938). To Davis (1938, 1940), the existence of zones, with the forward zone advancing into the sea, made it seem reasonable to argue that mangrove vegetation was neither a 'climax forest' nor an 'association', but rather '... a number of seral communities arranged in fairly definite zones which form a prisere...'. From this viewpoint Davis (1940) was able to outline the successional sequence of plant communities from the 'poineer *Rhizophora* family' to the 'tropical forest climax association' and describe the successional interrelationships of the seres which he equated with definable communities. The whole basis of the successional argument seems to hinge on the ability of the *R. mangle* to build and colonize new land (i.e. primary succession) by trapping water-borne debris within its arching prop root system into which fall the viviparous projectile-shaped propagules that colonize the new land (Curtiss, 1888; Davis, 1940; Richards, 1964). In this successional sequence, the continual 'accretion of soils' (Davis, 1940) allows the *Avicennia germinans* (L.) Stearn to invade and eventually replace *R. mangle* as the dominant, thus forming a new inner zone, etc., until 'Finally in the competition with the saw-grass vegetation, the mangrove is worsted and gradually thins out and disappears' (Harshberger, 1914).

The plant succession basis for mangrove zonation is logical and scientifically appealing, and thus it is understandable why so many of the mangrove studies throughout the world have interpreted zonation in the successional context. The most definitive syntheses of successional zonation have been prepared by Chapman (1970, 1976) in which se-

---

[3] The purported 'land building' role of mangroves continues to be a popular theme in the literature, e.g. Stephens (1962), Savage (1972), and Carlton (1974). Savage (1972) states that *A. germinans* '... is at least as important' as *R. mangle* in both land building and stabilization, and recommends its use for those purposes.

quences of halophytic community replacement are outlined for major areas of the world (Walsh, 1974). Chapman (1976, 1977), however, does acknowledge that there are numerous variations which can be attributed to differing local environmental factors and man's use of the environment. Nevertheless, mangrove species zonation as an expression of plant sucession is by far the most widely accepted paradigm to explain zonation. Although widely accepted, the paradigm has not been without its critics. Egler (1950) presented strong evidence that each of the mangrove zones behaves differently, both in terms of development and control (Watson, 1928). Egler (1950) also considered the land building role attributed to mangroves '...to be part of arm-chair musings of air-crammed minds.' Chapman (1976) attempted to resolve this difference in scientific opinion by proposing a 'succession of successions' in which hurricanes and fire (in south Florida) re-initiate successional sequences. But even again, Chapman (1977) qualifies the argument of successional zonation by stating that '...insufficient information is available to show whether zonation is always successional.'

Whereas there seemed to be a compatibility between the concepts of mangrove zonation and plant succession in the 1930's and '40's, the theory of plant succession that has subsequently evolved is no longer entirely suitable as a frame of interpretative reference; the earlier phytosociological-oriented definitions have given way to a more holistic viewpoint. It appears that it would be difficult if not impossible to empirically relate the characteristics of successional zonation with the general pattern for ecosystem development outlined by Odum, E.P. (1969). Only the development in situ of the pioneer *Rhizophora* community, from colonization through maturation, appears to follow some of the trends associated with that concept of change in developing (i.e. successional) ecosystems (Odum, E.P., 1969). Budowski (1963) prepared a list of some 20 empirically-derived characteristics of secondary succession in tropical forests which distinguish between early and later seral stages. This author (SCS) has compared the successional-zonation sequence of Davis (1940) to these 20 characteristics (Table 1); the comparison exhibits a low

degree of correspondence. Specifically, only seven of the characters appear to have analogies in the pattern of mangrove species zonation, nine are not valid and four are inconclusive. It suggests that insufficient data and information are available for mangroves which would allow for a sound appraisal of their successional status. Quite probably, this identifies an ecological anomaly worthy of vigorous scientific attention.

3. Zonation as a response to geomorphic change

Although Davis (1940) attributed a geologic role to mangroves, he placed emphasis on mangroves as a causal agent influencing geomorphic change in the coastal environment. In fact, up until the time of Egler (1950), the body of opinion held that mangroves, specifically *R. mangle*, were such a dominating geologic force that in recent times they had created significant new land and island areas in south Florida; Davis (1940) suggests that in an increment of 30 to 40 years, some 1500 acres (ca 600 ha) were created in Biscayne and Florida bays. In contrast, in the Old World, particularly in the Sundarbans Forest in the Ganges delta, and to a lesser extent in the deltas of such rivers as the Indus and Irrawaddy, it was unquestionably accepted that sediment deposition builds land (Curtis, 1933; Khan, 1966; Choudhury, A.M., 1968) faster than can be colonized by mangroves[4] due to the extraordinary magnitudes of the delta building processes (Coleman, 1969; Curray and Moore, 1971). Of course, this could reasonably be considered an extreme example. Nevertheless, these two extremes define the limits of the question (of zonation) to which coastal geomorphologists have also made a contribution.

Both West (1956) and Vann (1959), working on the alluvial coasts of Colombia, emphasized the control of geomorphic processes and change on

---

[4] Between 1947 and 1978, some 436,500 ha of new accretions have formed at the mouth of the combined Ganges-Brahmaputra-Meghna rivers in the Bay of Bengal (Huq, M.A., *Bangladesh Observer*, June 10, 1979) for which efforts are underway with mangrove afforestation (Choudhury, Z.M., in press; Karim, A., pers. comm.).

Table 1. Comparison of the Caribbean successional sequence of mangrove species (*Rhizophora mangle*, *Avicennia germinans*, *Laguncularia racemosa*, and *Conocarpus erecta*) cited by Davis (1940) with the general characters of secondary succession for tropical lowland forest synthesized by Budowski (1963).

Budowski's characters	Caribbean mangroves
'(1) The floristic composition of poineer communities is limited to a few species of wide natural distribution. There is little variation in the species represented in spite of different soil or climatic conditions.'	(1) Essentially true. Irrespective of which of the four species might occupy the seaward zone, the floristic composition is, by definition, limited. However, soil 'conditions' do appear to exert some influence on which species becomes dominant.
'(2) The number of strata in a community is highly indicative of its successional status. Few and well-defined strata reveal an early seral stage whereas several strata, difficult to separate, reveal an advanced stage of succession.'	(2) Inconclusive. In stable mono-specific zones, unevenaged regeneration in each zone makes it difficult to identify defined strata, suggesting an advanced successional stage which may not be a valid conclusion.
'(3) The absence of large stem diameters is a characteristic of early stages of succession.'	(3) True, but each species, when it appears as the pioneer, fits this character.
'(4) A dense undergrowth is characteristic of very early stages of development but not advanced stages or the climax.'	(4) True, for the same reasons given in (3). However, this is only characteristic of the mangrove fringe forest, but not necessarily the other forest types described by Lugo and Snedaker (1974).
'(5) The shape of the upper crowns is highly indicative. Early stages display uniform, thin light-green crowns. Older stages display many variations in crown forms and darker green color.'	(5) Not true. *R. mangle* consistently has irregular-shaped crown (except when it occurs in the riverine forest type) and has the darkest green leaves of any of the species. Both *A. germinans* and *L. racemosa* tend to have more regular-shaped crowns and lighter green leaf colors irrespective of the zone in which they might be found. *C. erecta* exists as a shrub with the lightest green coloration to the leaves of any of the four mangrove species.
'(6) Intolerance of the dominant species is characteristic of early stages and decreases towards the climax where most of the dominants are tolerant.'	(6) Inconclusive. Each of the species except *C. erecta*, appears to be variable in its shade tolerance under differing environmental conditions. Shade intolerance does, however, appear to be a general character of the *R. mangle* even though exceptions can be found. *C. erecta* may be the only consistently shade intolerant species of the four.
'(7) The evenaged condition is characteristic in early successional stages. There is a gradual change to an unevenaged condition with advance towards the climax.'	(7) True, but might be limited to stable or prograding shoreline environments. Unevenaged conditions are frequently induced by small light gaps, destructive storm events and localized areas of erosion/accretion irrespective of zones or species dominance.
'(8) Early poineer species characteristically have small seeds that are dispersed by wind, birds and bats. Old secondary or climax species mostly have large fruits and seeds, many of which are dispersed by gravity.'	(8) Not true (Rabinowitz, 1978c); the order appears to be reversed in mangroves. The primary dispersal agent for each species is water and may account for the discrepancy.
'(9) Deciduousness of many of the dominants in communities of intermediate status between the very early and the very advanced seral stages.'	(9) Not true. All of the mangrove species are evergreen.
'(10) Seeds of early pioneer species may remain dormant in the forest soil until favorable conditions such as clearing and fire trigger their development.'	(10) Probably not true for any of the mangroves although seedlings can persist for one or two years under the shade of a closed canopy. Subsequent growth appears to be 'triggered' or controlled by light availability.
'(11) Regeneration of the dominants is common in advanced stages but infrequent or absent in early poineer stages.'	(11) Inconclusive. Stable zones imply a regeneration of the dominants even in the fringing *R. mangle* zone.
'(12) Diameter and height growth is very rapid in early poineer stages.'	(12) True for each of the species when it appears as the pioneer or as the regeneration dominant in a light gap situation.

Table 1 *(continued)*.

Budowski's characters	Caribbean mangroves
'(13) Rapid reestablishment of an advanced stage of the original forest is favored by proximity of such a forest to the disturbed area; redevelopment of the original forest is more rapid in small clearings than in large ones.'	(13) True.
'(14) The presence of dominants having a very short life span is highly indicative of an early stage of succession.'	(14) Not true in the sense that the colonizing species in the mangrove environment are not annuals, perennials, etc. Life spans of each of the mangrove species appear to be comparable.
'(15) The presence of a large proportion of species with leaves of the macrophyll size class, is indicative of an early poineer stage. Climax species mostly have mesophyll leaves.'	(15) Not true. *R. mangle* has the largest leaves and *C. erecta* the smallest. Both *A. germinans* and *L. racemosa* are intermediate.
'(16) The hardness and weight of wood is highly indicative of successional position. The wood of trees representing early stages is soft and light whereas in species characteristic of advanced stages the wood is hard and heavy.'	(16) Not true. Record and Hess (1949) describe the wood of *R. mangle*, *A. germinans*, as 'hard and heavy', and *C. erecta* as being 'moderately heavy and hard'.
'(17) Climbers are highly indicative. In early stages of successional development, there are few species but many individuals and they are mostly herbaceous, often forming a tangle. In advanced stages of succession, they are large and woody with many species, but are not abundant.'	(17) Not true. Climbers only appear in mangrove forest directly exposed to freshwater runoff.
'(18) An increasing number of species and variety in life forms of epiphytes is characteristic of progressive development towards the climax.'	(18) Not true for the same reason given in (17). In addition, epiphytes are rare in mangroves of arid environment and most common in environments where rainfall exceeds potential evapotranspiration.
'(19) Certain species are highly indicative of the successional status of the community. Some can be correlated with past practices, notably exhaustive agricultural or fires. Others, notably palms, are indicative of long undisturbed conditions.'	(19) True as this is one of the basic arguments supporting successional zonation (cf. Davis, 1940) and anomalous conditions resulting from fire and hurricanes (Egler, 1950). When palms do occur in mangrove communities (e.g. *Nypa fruticans* Wurmb.) it is associated with the shoreline zone in the transition zone to freshwater swamp (Chapman, 1977).
'(20) On lateritic soils the presence of a community with dominants typical of habitats much drier than the rainfall would indicate for the region, points to past soil degradation, mainly compaction through extensive use of fires.'	(20) Marine laterites are unknown. However, an analogous condition is frequently observed on carbonate marls as in the Yucatan of Mexico, southeast Florida and the Bahama Islands, where each of the species exist in a dwarfed state.

vegetational patterns and species assemblages. On the west coast of Mexico, in the states of Sinaloa and Nyarit, others (Curray and Moore, 1964; Curray et al., 1969; Rollet, 1974; Connally, G.C., pers. comm.) present geomorphological evidence that shows how the coastal plain with its estuaries and beaches lead to a pattern of vegetation that fully captures the coastal history and resulting topographic relief. The aerial photographs provided by Rollet (1974) reveal a relatively precise correlation between vegetation and land-form pattern and, it is difficult to interpret the mangroves as having had any influence on the regional development of the land forms.

Although there is a substantial literature on coastal geomorphology, the works of Thom (1967) and Thom et al. (1975) are particularly informative as they show respectively, both in Mexico and Australia, the detailed interrelationship between geomorphic processes and the mangrove vegetation which results. Thom (1967) is able to relate species assemblages, distributions, and overall spatial organization to depositional and erosional histories, freshwater discharge, subsidence, compaction and sea level rise as related factors. On prograding shorelines experiencing the continuing accretion of sediments at a rate approximating mangrove colonization, it is reasonable to assume that a pattern of mangrove species zonation would result that fits the successional paradigm. In this regard, Thom (1967) does state that the *Rhizophora* can effectively accelerate land building processes during depositional sequences. Thus, mangroves are associated with land building only in a very restricted context, and, 'the self-maintenance of mangrove species in their preferred habitats continues until there is a critical change in habitat characteristics to induce vegetational change' (Thom, 1967).

The south Florida peninsula is a carbonate (karst) platform lacking delta-building rivers. The coastal area bordering Florida Bay is neither a significant source nor sink for non-carbonate clastic materials, and transport of biogenic-carbonate materials only results in localized areas of deposition and erosion. One dominant geologic influence is the progressive rise in sea level which Egler (1950) recognized and took into account in his consideration of mangrove zonation. Three theoretical possibilities of how rising sea level could affect mangrove zonation were described by Egler (1950): '(1) mangrove land formation predominates over submergence by the sea, and the zones shift seaward; (2) submergence predominates, and the zones shift landward; and (3) mangrove sedimentation approximately offsets submergence, in which case the mangrove belts widen, i.e. its seaward margin remains stationary, but its land margin advances inland.' Egler (1950) also argued that the deep beds of mangrove peat (Davis, 1946) did not represent land building, but rather, land retaining by mangroves during sea level rise. Although Egler's three theoretical possibilities each include mangrove zonation, it is only the first possibility that fits the paradigm of a colonizing zone initiating a successional sequence.

Although Egler (1950) did not comment on what would happen during a lowering of sea level (as this is not the case in south Florida), this author (SCS) has had the opportunity to observe a mangrove environment experiencing a lowering of sea level: the Marismas Nacionales in the State of Sinaloa, Mexico. There, a lowering of sea level is thought to be taking place due to tectonic uplifting of the coastal plain (Connally, G.C., pers. comm.). The dominant species in the shoreline zone in that environment is *Laguncularia racemosa*, (L.) Gaertn, inland from which, respectively occurs a zone dominated by *Avicennia* and a more inner zone of well-developed *L. racemosa*. The *Rhizophora* appear as scattered individuals throughout the intertidal area and as small mono-specific patches seemingly unrelated to local topography as they appear both on the shoreline and in the inner zones. The structure and species dominance of Marismas Nacionales mangrove forests have been described by Pool et al. (1977). In that environment, it is also interesting to note that *L. racemosa* is the dominant species colonizing recently-deposited sediments along shorelines and on shoals within the estuary.

From the geomorphological perspective, it is the size, configuration, topography and history of the coastal zone that determines the kinds and distributions of halophytes that can develop in the resulting habitats. A dominant factor that ties the physical environment to the character of the biological community which develops, is local tidal patterns with respect as to how frequently areas and zones are inundated. This is a highly variable function (although capable of stochastic description) due to tidal cycles which have diel, montly, seasonal, annual and 18.6 year frequencies, all superimposed on one another. In addition, storm and specific wind events can significantly alter the normal tidal cycle at what can be considered to be random intervals. Watson (1928) considered the pattern of tidal inundation (along with drainage) to be the primary factor involved in the distribution (e.g. zonation) of the mangrove species. He observed that some areas, or zones, become inundated more frequently than others and used this consideration to develop an empirical scheme of 'inundation classes' (Table 2) to explain the distribution of the Malaysian mangrove species (Watson, 1928). Others have used locally-determined inundation classes for the same purpose in other areas (Chapman, 1976). Snedaker and Lugo (1973) and

Table 2. The inundation classes of Watson (1928) developed for the area of Port Swettenham, Malaysia.

In-undation class	Flooded by	Height (ft.) above admiralty datum		Number of times flooded p.m.	
		from	to	from	to
1	All high tides	–	8	56	62
2	Medium high tides	8	11	45	59
3	Normal high tides	11	13	20	45
4	Spring high tides	13	15	2	20
5	Abnormal or equinoctial tides	15	–	–	2

Lugo and Snedaker (1974) extended this concept further by including the relative salinity of the inundating water and the direction of ingress versus egress of surface water. In their justification of the scheme they argued (Lugo and Snedaker, 1974) that over extensive areas of mangrove vegetation, it is only the shoreline 'fringe' forests that exhibit the classical pattern of species zonation observed by others.

4. Zonation as a physiological response to tide-maintained gradients

In the two previous sections (plant succession and geomorphology), it is only the viewpoint that allows distinction between the two interpretations of the same phenomenon in terms of the causal agent and the effect. From the life sciences viewpoint, the biological community of halophytes is presumed able to 'create' its own environmental gradients resulting in a succession of communities each adapted to a newly created habitat. From the perspective of the physical sciences, halophytic community distributions, both in space and time, are considered to be consequences of normally changing physical environments. But just as Davis (1940) criticized the historically-inadequate bases for successional zonation, and Egler (1950) rebuked the successional bases, so has there been an admonishment of physical control of the zonation of mangroves. Rabinowitz (1978c) reviewed part of the literature throughout 1974 and concluded that,

'The mode of action of these factors (i.e. physical) is not made explicit by most authors.' She alludes to a physiological explanation which would be required to adequately explain the 'mode of action'. In this regard, the majority of the literature focuses on salinity as the specific factor which links the physical environment through the physiology of the mangroves to the ultimate spatial organization.

The interrelationships among surface water hydrology (specifically, the 'tidal factors' of Chapman, 1976), salinity and the zonation of mangroves have received considerable attention, for example in the works of Watson (1928), De Haan (1931), Walter and Steiner (1936), Davis (1940), Chapman (1944), Chapman and Ronaldson (1958), Macnae (1968), Clarke and Hannon (1969, 1970, 1971), Baltzer (1969), and in the reviews of Walsh (1974), Lugo and Snedaker (1974), and Chapman (1976). In addition, site-specific correlations have been established for various parts of the world, for example the Malay Peninsula (Watson, 1928), Java (De Haan, 1931), South Africa (Day et al., 1953; Moog, 1963), south Florida (Davis, 1940; Egler, 1950), South Australia (Clarke and Hannon, 1971), and have all emphasized the correlation between salinity and the pattern of inundation, as well as the role of tides in maintaining characteristic salinity regimes. Although good correlations exist between salinity and mangrove species zonation, correlations are not proofs of direct cause-effect relationships. It would have to be shown that each mangrove species responds through its physiology to salinity and salinity gradients, in a specific manner, and that this response(s) results in species zonation.

Mangroves are generally considered to be facultative halophytes insofar as it appears that salt is not required for their development (Chapman, 1975, 1976). The basis for this consideration is the large number of observations of the different mangrove species growing, and apparently thriving, in freshwater (e.g. Hooker, 1875; Winkler, 1931; Chapman, 1944; Egler, 1948; Van Steenis, 1958, 1963; Gill, 1969; McMillan, 1971). This body of observations, however, is clouded by experimental efforts of a fewer number which ostensibly show that certain concentrations of specific salts are required for growth and development of mangroves (e.g., Stern and Voigt, 1959; Connor, 1969; Sidhu, 1975). Morrow and Nickerson (1973) present evidence which they claim '...indicates that *R. mangle* only survives in areas where salinity approximates that of seawater'. Whereas this seems to be an extreme example, the majority of the literature reporting salinities for a variety of mangrove species under field conditions indicates that they occur naturally in salinity regimes in excess of seawater (Walsh, 1974; Lugo and Snedaker, 1974; Chapman, 1976). The upper salinity limit, or tolerance, for mangroves for any extended period of time seems to be around 90‰ (McMillan, 1971, 1974; Connor, 1969; Clarke and Hannon, 1969, 1970; Kylin and Gee, 1970; Cintron et al., 1978). Cintron et al. (1978) have argued that the 90‰ upper limit is reasonable in light of the work of Scholander et al. (1965) which showed that the sap pressures in mangroves are high enough to enable them to extract freshwater from seawater with salt concentrations of about 2.5 times normal seawater. Although the literature appears contradictory, the weight of the evidence does support provisional acceptance that mangroves: (1) are facultative halophytes, (2) can exist over a range of salinities from 0+ to 90‰, and (3) that each species may have definable tolerances and optima under specific conditions.

Egler (1948) appears to have been the first to relate salinity tolerance to interspecific competition and thus lay another basis for explaining why mangrove species are zoned. He claimed that the *R. mangle* did well in a freshwater environment, but only in the absence of competitors. In a similar vein, West (1956) took the position that mangroves dominate in the coastal zone, not because they require salt, but rather because their potential competitors are less tolerant to salt (Chapman, 1975). This implies that along a salinity gradient, the halophytes are able to maximize their productive output at a higher metabolic efficiency in the saline portions to the exclusion of glycophytes adapted to lower salinities approximating freshwater. Clarke and Hannon (1971) provide experimental evidence suggesting that interspecific competition is probably 'intense' when associated with small or gradual changes in environmental gradients, but of lesser importance under either stable conditions or when there is an abrupt or large change in the gradients.

The elusive mode-of-action (Rabinowitz, 1978c) was partially elucidated in the work of Carter et al. (1973) who measured metabolic parameters and interstitial soil chlorinities under a variety of conditions in the Ten Thousand Islands in southwest Florida. They showed that respiration was positively correlated with interstitial chlorinity and inversely correlated with the ratio between surface water chlorinity and interstitial soil chlorinity. That work is also reviewed in Lugo and Snedaker (1974) and Hicks and Burns (1975). The importance of the interstitial-water surface-water chlorinity ratio is that it infers the degree of exchange between the sediment and overlying water; the smaller the ratio, the better the exchange and the less likelihood that high concentrations of chloride would develop. Based on independent work in the same area of Florida, Lugo et al. (1975) used simultaneous measurements of net daytime photosynthesis to reveal that the four Florida mangrove species were ranked from high to low in the order of *R. mangle* (highest), *A. germinans*, *L. racemosa*, and *Conocarpus erecta* L. (lowest). Lugo et al. (1975) thus considered that there was a metabolic basis for mangrove species zonation because the ranking recapitulates the sequence of species in zonation (Davis, 1940). They concluded that 'Within each zone the characteristic species has maximized, apparently, photosynthesis and thus exhibits a higher metabolic rate than any invading species, which would be at a competitive disadvantage due to its lower metabolic efficiency in that habitat' (Lugo et al., 1975). Support for this conclusion is also provided by Lind and Morrison (1974); they state that '...each species of the mangrove forest grows best under slightly different conditions which depend on factors such as the amount of water in the mud, the salinity, and the ability of the plant to tolerate shade. This means that the various species are not mingled together in a haphazard way, but occur in fairly distinct zonation.' In this specific regard, Clarke and Hannon (1971) tested a series of environmental-gradient experiments which showed that '...patterning resembling that found in the field can be obtained by subjecting combinations of the most prominent species to an artificial environmental gradient of elevation, salinity, waterlogging and soil type, and that the pattern coincides with predictions based on the laboratory performance of the species.' Their work suggests the existence of 'preferred' conditions for each of the species. From these reported works, it can be stated that each mangrove species probably has an optimum salinity, which at higher concentrations induces a proportionately higher respiration making it relatively less competitive, and a lower concentration is forced to compete with species better adapted to maximize photosynthesis. Salinity-induced high-respiratory demand has been cited as a metabolic cost of existence in a saline environment (Waisel, 1972; Queen, 1974). It is pertinent to note in this regard that Gale (1975) states that 'The increase of Rd (dark respiration) in response to salt is probably a truely adaptive and not a pathological reaction (which would merely reduce overall net CO_2 fixation).' He recognizes the potential teleological argument and cites published evidence of the 'salt respiration' phenomenon.

Although Carter et al. (1973) and Lugo et al. (1975) have shown, respectively, the relationship between interstitial chlorinity and mangrove metabolism and, mangrove metabolism and zonation (Cintron et al., 1978), the spatial and temporal variation in interstitial water chemistry make the 'concept' difficult to apply to all field conditions.[5] For example, under similar tidal regimes and surface water salinities, the interstitial soil water salinities can vary at least two orders of magnitude due to very localized conditions. Reasons which have been put forth to explain anomalously high or low salinities include: (1) density- driven accumulation of salt with sediment depth (Scholl, 1965; Lindberg and Harriss, 1973), (2) possible exposure to paleowater with high salt concentration (Scholl, 1962), (3) an unusual soil chemistry which can selectively concentrate chloride (Siever et al., 1961), (4) a high freshwater head or ground water dis-

charge (Lindberg and Harriss, 1973; Cintron et al., 1978), and (5) localized areas in which evapotranspiration exceeds the influx of diluting replacement water (Fosberg, 1961; Cintron et al., 1978). Thus, the salinity regime to which mangroves respond can vary spatially and temporally over a whole range of scales. Spot, or short-term measures would therefore not necessarily reveal the long-term mean to which the mangroves must adapt; it is an error in technique, not in concept. One wonders how much of the variation reported in the literature, as well as the superficially contradicting observations, are merely the result of the lack of long-term studies to define the long-term means and average conditions (?).

5. Zonation as a consequence of differential propagule dispersal

Recently, scientific attention has been given to the population dynamics of propagules and year-class-one seedlings as causal agents in the establishment and perpetuation of mangrove species zones. The principal propagule characteristics (e.g., size, weight, shape, buoyancy, longevity, numbers, location of source areas, etc.) are suggested to result in differential tidal sorting and depositing, and that this is a dominant factor in explaining the 'why' of zonation, or the lack thereof. The proposition is, in part, based on a dismissal of alternative explanations (Rabinowitz, 1978c)[6], and a series of field measurements and experiments in Panama. The pertinent results of that work can be summarized as follows: (1) mangrove genera which dominate at lower ground elevations close to openwater have large, heavy propagules, i.e. *Rhizophora* and *Pelliciera*, and those that dominate at higher elevations further inland have small propagules which require some five days on non-flooded substrate to become established, i.e. *Laguncularia* and *Avi-*

[5] In recently-formed hypersaline areas contiguous to arid salt flats, relatively stable salinity gradients develop which correlate well with mortality, dwarfing and healthy trees (Cintron et al., 1978). Such stable salinity gradients appear to be common in arid coastal zones (Por et al., 1977).

[6] Unfortunately, the pertinent literature on mangrove environmental physiology was not reviewed by the author, nor were related measurements incorporated in the field research efforts (Rabinowitz, 1975, 1978a, b, c). Therefore, the author's conclusions concerning processes which were neither reviewed nor researched are highly suspect.

cennia (Rabinowitz, 1978b), (2) seedling mortality rates were inversely correlated to propagule weight with seedling-number half-lives ranging from 85 days (*Avicennia*) to 604 days (*Pelliciera*) (Rabinowitz, 1978a), and (3) seedlings did not exhibit better growth under the canopies of the various adult species (Rabinowitz, 1978c). The latter was determined from the first year results of seedling plantings in reciprocal gardens prepared beneath the canopies of the various adult species in different parts of Panama where the author would find the most pure stands of the mature adult species. In addition to the prepared gardens (prop roots cut around perimeters and existing seedlings removed), wild seedling stocks were also tagged and monitored. From the information and data obtained, Rabinowitz (1978c) concluded that zonation was probably the result of differential tidal sorting and dispersion according to propagule size and the frequency of tidal inundation (Watson, 1928) of potential sites. Following establishment, it was suggested that competitive interaction between seedlings and adults dominated subsequent survival. Clearly, Rabinowitz falls within the varied group of zonation proponents who argue for environment domination by biological processes within the mangrove coastal zone. A concluding comment by Rabinowitz (1978c) is pertinent to the question of zonation: 'Physiologic specialization and optimal growth in a restricted band within the swamp may be precluded by fluctuating conditions that vary annually in concert over the entire swamp.'

Unlike the other three general explanations as to why mangroves so frequently appear in zones, the whole area of propagule dynamics, seedling establishment and community regeneration, is too poorly developed to permit the construction of a valid general hypothesis. In addition, it raises some difficult questions about community dynamics and plant succession. For example, Rabinowitz (1978b, c) has convincingly shown that propagule-size decreases, in the plant succession context, from large propagules in the 'colonizing, poineer' zone to small propagules in the more inland 'mature community' zones. Yet, other authors (e.g. Budowski, 1963) writing on the general aspects of secondary plant succession note that poineer, colonizing species

tend to have large numbers of small, easily dispersed seeds, and that the seeds/fruits of 'climax' species tend to be few in number, large and heavy. If this general relationship between successional maturity and propagule size is truly a universal characteristic, then the successional order of the mangrove zones would have to be interpreted as a diametric reversal on this basis alone (Table 1). That is, a species like *L. racemosa* better fits the model of a poineer species than does *R. mangle*. Some evidence in fact suggests that under certain conditions associated with new or recently exposed substrates, *L. racemosa* rapidly colonizes forming a single dominant (Snedaker and Pool, 1973; Savage, 1972) and in certain areas such as the west coast of Mexico it forms the mature coastal fringe (Pool et al., 1977). Much more needs to be known before this ecological anomaly can be resolved.

Pertinent to this discussion are some preliminary observations on a project initiated in south Florida in 1971 (Snedaker and Lugo, 1973). Snedaker and Pool (1973) clearcut a series of transects 5 m wide for the determination of compartmental biomass (Lugo and Snedaker, 1974) and the clearcut areas have been observed continuously since that time. The pattern of regeneration (Table 3) shows a variety of results, some unpredictable, which do not permit a general conclusion to be stated about either regeneration or successional sequences. For example, one area has been exposed to abundant propagules of each mangrove species, but has remained bare for eight years. The species assemblages of the regenerating community exhibit differing ratios than the surrounding undisturbed community, and differ remarkably from adjacent light gap areas which were quickly taken over by existing seedlings in the understory at the time the canopies opened. In general, we would tentatively conclude at this time, that: (1) all of the areas were and have been exposed to abundant propagules from each mangrove species in the area, (2) disturbed substrates versus undisturbed substrate in previously closed-canopy areas respond differently, (3) light appears to be the controlling factor in determining which of the regenerating species assume growth dominance (Clarke and Hannon, 1971), and (4) next to light, substrate condition

Table 3. Summary of observations made on regenerating clearcut areas (harvested in 1971) in the Ten Thousand Islands of south Florida. Original description of harvests in Snedaker and Pool (1973, summarized in Lugo and Snedaker, 1974); areas to be re-inventoried in 1981.

Mangrove forest type	Community description	Regeneration between 1971 and 1979
Successional fringe forest on dredge spoil island	*L. racemosa*, dominant; one small *R. mangle*	Area has remained bare since 1971 with exception of a few scattered *Batis maritima*
Overwash forest on small island	*R. mangle*, dominant; one large *L. racemosa*	The *L. racemosa* coppiced early and is the largest single plant; regeneration of *R. mangle* and *A. germinans* presently occurring and consisting of at least four age classes
Mature fringe forest	*R. mangle*, dominant; *L. racemosa*, codominant; *A. germinans* scattered understory seedlings	*L. racemosa* coppiced early and is the present dominant; area naturally reseeded with abundant *R. mangle* and *A. germinans*; *R. mangle* seedlings appearing as the early success of *A. germinans*
Riverine forest	*R. mangle* and *A. germinans* codominants; *L. racemosa* present as rare seedlings	*A. germinans* coppiced slowly but quickly reseeded along with propagules of *R. mangle* and *L. racemosa*; *Bostrichium* sp. rapidly covered remaining pneumatophores in response to light, now disappearing; no species dominance is evident
Riverine forest natural light gaps ca 5 m diameter	(Same as above)	Areas quickly colonized by *R. mangle* and *A. germinans* and existing now as codominants in regeneration; no evidence of dominant regeneration from newly arrived propagules

seems to be the second most important factor in determining the dominance of the regenerating species. Equally interesting in view of the metabolic basis for zonation developed by Lugo et al. (1975) are areas, which were once bare were seeded by *L. racemosa* resulting in 100% coverage after two years. These areas are now 8 to 12 years old and are being taken over by *R. mangle* which became established in the understory following occupation by *L. racemosa*.

Davis (1940) estimated the propagule crop per *R. mangle* tree as approximately 300 per year; it is suspected that similar, and probably higher seed crop numbers would characterize the other species as well. Based on our observations in south Florida, it would appear that no area or zone is ever propagule-limited due primarily to the fact that seed fall occurs during that portion of the year when there are severe storm events that scatter them over broad areas. Nor does this seasonal period of abnormally high water seem to preclude the establishment of each species in the seaward areas of deeper water. What percentage survival of each

year's propagule crop is necessary to offset natural mortality? Whatever it is, it appears that there are always far more propagules available in the south Florida area than are required to maintain the character of the species zones.

Despite an abundance of studies on propagules and their requirements for subsequent establishment and growth (e.g. Egler, 1948; Davis, 1940; Teas and Montgomery, 1968; Connor, 1969; McMillan, 1971; Banus and Kolehmainen, 1975; Sidhu, 1975; Steinke, 1975; Rabinowitz, 1975, 1978a, b, c) there seems to be inadequate reason to accept mangrove species zonation as being a consequence of propagule dispersal.

6. Discussion

In spite of the wealth of information and description on the zonation of mangroves, it is not possible to identify a consensus of scientific opinion explaining why these woody halophytes so frequently appear in discrete zones. At least two reasons account for this lack of a consensus. First, various authors have recognized the inadequacy and lack of rigorous scientific bases for the earlier interpretations and conclusions concerning zonation (e.g. Davis, 1940; Egler, 1950; Rabinowitz, 1978c). For example, the successional zonation hypothesis can be traced back to the late 1800's when Curtiss (1888) and Sargent (1893) wrote eloquent descriptions of the manner in which mangroves captured land from the sea. Without any supporting evidence, that presumption was repeated many times in the literature until the time of Davis (1938, 1940). Although Davis contributed much to our knowledge of mangroves, his evidence for the successional land-building role of the *R. mangle* was inferential and speculative as noted by Egler (1950). Not-with-standing, the presumed land-building role of mangroves is still being reported in the literature (e.g. Stephens, 1962; Savage, 1972; Carlton, 1974) with Davis (1938, 1940) frequently being cited as the authority. Second, many of the environmental gradients associated with the inter-tidal zone (e.g. topographic slope, hypsographic frequency of inundation, waterlogging, interstitial salinity, age of substrate, etc.) are simply auto-correlates of one another but which, independently and collectively, can be correlated with the zonation of the mangroves species. The simple correlation of species position along such compounded linear gradients is a useful predictive tool (Watson, 1928) but is not proof of cause-effect even when an associated function may be correctly identified (Rabinowitz, 1975, 1978a, b, c).

This review of the literature on the possible reasons for the zonation of mangroves has resulted in the identification of four separate bodies of scientific opinion. Of these, the geomorphology and environmental physiology studies appear to be the most relevant to the furtherance of our understanding of zonation and plant succession in the intertidal environment. Although the coastal geomorphologists have cast considerable doubt that mangroves can create land *de novo*, their work has provided a basis for understanding the physical mechanisms which create and maintain the environmental gradients to which mangroves appear to be responsive. In doing so, they have also opened up a new insight into mangrove succession. The newly developing field of seed and propagule population ecology exposes how little we know of the overall ecology of mangroves particularly in terms of the selection pressures which have guided their evolution from glycophytes to halophytes. The following synopsis summarizes what is believed to be the major points uncovered in this review.

It appears that many different kinds of mangroves can function as a poineer species through the ability of their propagules to rapidly colonize prograding shorelines and accreting shoals under a wide variety of local conditions. Certainly in the Caribbean area, the colonizing role is not restricted to *R. mangle* as was earlier believed. Although it is doubtful that mangroves can initiate the process leading to land building, it is apparent that their colonizing presence accelerates the local land-building processes. The ability for a variety of species to colonize new habitats suggests that propagule size and weight (Budowski, 1963) have little bearing on their designation as pioneers in the classical sense. Irrespective of which species first inhabits a specific site, there is a temporal tendency for each species to assume competitive dominance in its 'preferred' zone. (The term, 'preferred' used by Clarke and Hannon (1971) is used as an operational expression describing what, in fact, actually happens over time). This tendency is supported by both experiments (Clarke and Hannon, 1971) and field measurements (Lugo et al., 1975) and is consistent with general knowledge of the physiology of halophytes. Whether occupation of the preferred habitat along a gradient is guided by physical forces or results from interspecific competition (Clarke and Hannon, 1971), the general 'mode of action' (Rabinowitz, 1978c) appears to be the same; the species which can maximize its photosynthetic output with greatest metabolic efficiency dominates in competition with other species. The concept of a zone or environmental preference implies that each mangrove species does have a preferred optimum and a limit of tolerance related to the metabolic cost of existence along an environmental gradient (Clarke and Hannon, 1971; Waisel, 1972; Carter et al., 1973; Lugo and Snedaker, 1974; Queen, 1974; Gale, 1975; Lugo et al., 1975; Cintron et al., 1978). The '... fluctuating conditions that vary annually in concert over the entire swamp' (Rabinowitz, 1978c) probably only results in transitory shifts in interspecific competitive dominance which have two possible outcomes: (1) the changed condition may last long enough to result in competitive exclusion and domination by a previously subordinate competitor, or, (2) the 'fluctuating conditions' may oscillate around a more meaningful long-term average perpetuating the survival of the existing dominant and thus, the zone. The existence of fluctuating conditions does not intrinsically preclude 'Physiologic specialization and optimum growth in a restricted band within the swamp...' (Rabinowitz, 1978c), except possibly for first-year class seedlings which seem to be able to grow anywhere.

Literature cited

Baltzer, F. 1969. Les formations vegetales associees au delta de la Dumbea (Nouvelle Caledonie). Cah. ORSTOM, Ser. Geol. 1: 59–84.

Banus, M.D. and Kolehmainen, S.E. 1975. Floating, rooting and growth of red mangrove (Rhizophora mangle L.) seedlings: effect on expansion of mangroves in southwestern Puerto Rico. In Proc. Int. Symp. Biol. and Management of Mangroves, Vols. I and II, eds. G.E. Walsh, S.C. Snedaker and H.J. Teas, pp. 370–384. Int. Food Agric. Sci., Univ. Florida, Gainesville.

Budowski, G. 1963. Forest succession in tropical lowlands. Turrialba 13: 42–44.

Carlton, J.M. 1974. Land-building and stabilization by mangroves. Environmental Conservation 1: 285–294.

Carter, M.R., Burns, L.A., Cavinder, T.R., Dugger, K.R., Fore, P.L., Hicks, D.B., Revells, H.L. and Schmidt, T.W. 1973. Ecosystems Analysis of Big Cypress Swamp and Estuaries, EPA 904/9-74-002, U.S. Environmental Protection Agency, Region IV, Atlanta.

Chapman, V.J. 1944. 1939 Cambridge University expedition to Jamaica. I- A study of the botanical processes concerned in the development of the Jamaican shore-line. J. Linn. Soc. London Bot. 52: 407–447.

Chapman, V.J. 1968. Vegetation under saline conditions. In Saline Irrigation for Agriculture and Forestry, ed. H. Boyko, pp. 201–216. Dr W. Junk Publ., The Hague.

Chapman, V.J. 1970. Mangrove phytosociology. Trop. Ecol. 11: 1–19.

Chapman, V.J. 1975. Mangrove biogeography. In Proc. Int. Symp. Biol. and Management of Mangroves. Vols. I and II, eds. G.E. Walsh, S.C. Snedaker and H.J. Teas, pp. 3–22. Int. Food Agric. Sci., Univ. Florida, Gainesville.

Chapman, V.J. 1976. Mangrove Vegetation. Strauss & Cramer, Germany.

Chapman, V.J. ed. 1977. Ecosystems of the World. Vol. 1. Wet Germany.

Chapman, V.J. (ed.). 1977. Ecosystems of the World. Vol. 1. Wet Coastal Ecosystems. Elsevier Scientific Publ. Co., Amsterdam.

Chapman, V.J. and Ronaldson, J.W. 1958. The mangrove and saltgrass flats of Auckland Isthmus. N.Z. Dept. Sci. Ind. Res. Bull. No. 125.

Choudhury, A.M. 1968. Working Plan of Sundarban Forest Division for the Period from 1960–61 to 1979–80, Vol. 1. East Pakistan Government Press, Tejgaon, Dacca.

Choudhury, Z.M. 1978. Coastal development – a strategy. In Proc. UNESCO Sem. on Human Uses of the Mangrove Environment and Management Implications, eds. P. Kunstadter and S.C. Snedaker. Held in Dacca, Bangladesh, December 1978. (In preparation).

Cintron, G., Lugo, A.E., Pool, D.J. and Morris, G. 1978. Mangroves of arid environments in Puerto Rico and adjacent islands. Biotropica 10: 110–121.

Clarke, L.D. and Hannon, N.J. 1969. The mangrove swamp and salt marsh communities of the Sydney District. II– The holocoenotic complex with particular reference to physiography. J. Ecol. 57: 213–234.

Clarke, L.D. and Hannon, N.J. 1970. The mangrove swamp and salt marsh communities of the Sydney district. III– Plant growth in relation to salinity and waterlogging. J. Ecol. 58: 351–369.

Clarke, L.D. and Hannon, N.J. 1971. The mangrove swamp and salt marsh communities of the Sydney District. IV– The significance of species interaction. J. Ecol. 59: 535–553.

Coleman, J.M. 1969. Brahmaputra River: channel processes and sedimentation. Sediment. Geol. 3: 129–239.

Connally, G.G., Department of Geology, State University of New York, Buffalo, New York, U.S.A.

Connor, D.J. 1969. Growth of grey mangrove (Avicennia marina) in nutrient culture. Biotropica 1: 36–40.

Curray, J.R. and Moore, D.G. 1964. Holocene regressive littoral sand, Costa de Nayarit, Mexico. In Deltaic and Shallow Marine Deposits-Developments in Sedimentology, Vol. 1, ed. L.J.G.V. Van Straaten, pp. 76–82. Proc. 6th Sedimentological Congress 1963.

Curray, J.R. and Moore, D.G. 1971. Growth of the Bengal deep-sea fan and denudation in the Himalayas. Geol. Soc. Amer. Bull. 82: 563–572.

Curray, J.R., Emmel, F.J. and Crampton, P.J.S. 1969. Holocene history of a strand plain, lagoonal coast, Nayarit, Mexico. In Lagunas Costeras, un Simposio, eds. A. Ayala Castanares and F.B. Phleger, pp. 63–100. UNAM UNESCO 28–30 Nov. 1967.

Curtis, S.J. 1933. Working Plan for the Sunderbans Division (1931–51). Forest Dept., Bengal.

Curtiss, A.H. 1888. How the mangrove forms islands. Garden & Forest 1: 100.

Davis, J.H., Jr. 1938. Mangroves – makers of land. *Nature Mag.* 31: 551–553.

Davis, J.H., Jr. 1940. The ecology and geologic role of mangroves in Florida. In *Carnegie Inst. Washington Pub. No. 517*, pp. 303-412. Papers from the Tortugas Lab, Vol. 32.

Davis, J.H., Jr. 1946. The peat deposits of Florida, their occurrence, development and uses. *Fla. Geol. Surv. Bull.* No. 30. Tallahassee.

Day, J.H., Millard, N.A.H., and Broekhuysen, G.J. 1953. The ecology of South African estuaries. Part IV. The St. Lucia system. *Trans. R. Soc. South Africa* 34: 129.

De Haan, J.H. 1931. Het een en ander over de Tijlatjapsche vloedbosschen. *Tectona* 24: k39–75.

Egler, F.E. 1948. The dispersal and establishment of red mangrove in Florida. *Carib. Forest.* 9: 299–310.

Egler, F.E. 1950. Southeast saline Everglades vegetation, Florida, and its management. *Veg. Acta Geobot.* 3: 213–265.

Fosberg, F.R. 1961. Vegetation-free zone on dry mangrove coasts. *U.S. Geol. Surv. Prof. Paper* No. 424(D): 216–218.

Gale, J. 1975. Water balance and gas exchange of plants under saline conditions. In *Plants in Saline Environments*, eds. A. Poljakoff-Mayber and J. Gale, pp. 168–185. Springer-Verlag, New York.

Gifford, J.C. 1934. *The Keys and Glades of South Florida*. N.Y. Books Inc., New York.

Gill, A.M. 1969. Tidal trees: orient and occident. *Bull. Fairchild Trop. Gard.* 24: 7–10.

Harper, R.M. 1917. Geography of central Florida. *Florida State Geol. Surv. Annual Report* 13: 71–307.

Harshberger, J.W. 1914. The vegetation of South Florida. *Trans. Wagner Free Inst. Sci.* 7: 49–189.

Hicks, H.B. and Burns, L.A. 1975. Mangrove metabolic response to alterations of natural freshwater drainage to southwestern Florida estuaries. In *Proc. Int. Symp. Biol. and Management of Mangroves*, Vols. I and II, eds. G.E. Walsh, S.C. Snedaker and H.J. Teas, pp. 238–255. Int. Food Agric. Sci., Univ. Florida, Gainesville.

Hooker, D.J. 1875. *Flora of British India*, Vol. II. Reeve, London.

Ismail, M., Department of Botany, University of Dacca, Dacca-2, Bangladesh.

Karim, A., Department of Botany, University of Chittagong, Chittagong, Bangladesh.

Khan, S.A. 1966. Working Plan of the Coastal Zone Afforestation Division from 1963–64 to 1982–83. Government of West Pakistan, Agriculture Department, Lahore.

Kylin, A. and Gee, R. 1970. Adenosine triphosphatase activities in leaves of the mangrove *Avicennia nitida* Jacq. *Plant Physiol.* 45: 169–172.

Lind, E.M. and Morrison, M.E.S. 1974. *East African Vegetation*. Longman, London. (Quoted in Rabinowitz, 1978c).

Lindberg, S.E. and Harriss, R.C. 1973. Mechanisms controlling pore water salinities in a salt marsh. *Limnol. Oceanogr.* 18: 788–791.

Lugo, A.E. and Snedaker, S.C. 1974. The ecology of mangroves. *Ann. Rev. Ecology & Systematics* 5: 39–64.

Lugo, A.E., Evink, G., Brinson, M.M., Broce, A., and Snedaker, S.C. 1975. Diurnal rates of photosynthesis, respiration, and transpiration in mangrove forests of south Florida. In *Tropical Ecological Systems*, eds. F.B. Golley and E. Medina, pp. 335–350. Springer-Verlag, New York.

McMillan, C. 1971. Environmental factors affecting seedling establishment of the black mangrove on the central Texas coast. *Ecology* 52: 927–930.

McMillan, C. 1974. Salt tolerance of mangroves and submerged aquatic plants. In *Ecology of Halophytes*, eds. R.J. Reimold and W.H. Queen, pp. 379–390. Academic Press, New York.

Macnae, W. 1966. Mangroves in eastern and southern Australia. *Aust. J. Bot.* 14: 67–104.

Macnae, W. 1968. A general account of the fauna and flora of mangrove swamps and forests in the Indo-West-Pacific region. *Advances Marine Biol.* 6: 73–270.

Macnae, W. and Kalk, M. 1962. The ecology of the mangrove swamps of Inhaca Island, Mozambique. *J. Ecol.* 50: 19–34.

Moog, A.O.D. 1963. A preliminary investigation of the significance of salinity in the zonation of species in salt-marsh and mangrove swamp associations. *S. Afr. J. Sci.* 59: 81–86.

Morrow, L. and Nickerson, N.H. 1973. Salt concentrations in ground waters beneath *Rhizophora mangle* and *Avicennia germinans*. *Rhodora* 75: 102–106.

Odum, E.P. 1969. The strategy of ecosystem development. *Science* 164: 262-270.

Phillips, O.P. 1903. How the mangrove tree adds new land to Florida. *J. Geog.* 2: 10–21.

Pollard, C.L. 1903. Plant agencies in the formation of the Florida Keys. *Plant World* 5: 8–10.

Pool, D.J., Snedaker, S.C. and Lugo, A.E. 1977. Structure of mangrove forests in Florida, Puerto Rico, Mexico and Costa Rica. *Biotropica* 9: 195–212.

Por, F.D., Dor, I. and Amir, A. 1977. The mangal of Sinai: limits of an ecosystem. *Helgolander Wiss. Meeresunters* 30: 295–314.

Queen, W.H. 1974. Physiology of coastal halophytes. In *Ecology of Halophytes*, eds. R.J. Reimold and W.H. Queen, pp. 345–353. Academic Press, New York.

Rabinowitz, D. 1975. Planting experiments in mangrove swamps of Panama. In *Proc. Int. Symp. Biol. and Management of Mangroves*, Vols. I and II, eds. G.E. Walsh, S.C. Snedaker and H.J. Teas, pp. 385–393. Int. Food Agric. Sci., Univ. Florida, Gainesville.

Rabinowitz, D. 1978a. Mortality and initial propagule size in mangrove seedlings in Panama. *J. Ecol.* 66: 45–51.

Rabinowitz, D. 1978b. Dispersal properties of mangrove propagules. *Biotropica* 10: 47–57.

Rabinowitz, D. 1978c. Early growth of mangrove seedlings in Panama, and an hypothesis concerning the relationship of dispersal and zonation. *J. Biogeogr.* 5: 113–133.

Record, S.J. and Hess, R.W. 1949. *Timbers of the New World*. Yale Univ. Press, New Haven.

Richards, P.W. 1964. *The Tropical Rain Forest*. Cambridge Univ. Press.

Rollet, B. 1974. Introduction a l'étude des mangroves du Mexique. *Bios Forests Trop.* 156: 3–74.

Rollet, B., Division of Marine Sciences, UNESCO, 7 place de Fontenoy, 75700 Paris, France.

Sargent, C.S. 1893. The mangrove tree. *Garden & Forest* 6: 97–98, 101–103.

Savage, T. 1972. Florida mangroves as shoreline stabilizers. *Fla. Dept. Natur. Resources. Prof. Paper* No. 19.

Scholander, P.F., Hammel, H.T., Bradstreet, E.D. and Hemmingsen, E.A. 1965. Sap pressure in vascular plants. *Science* 148: 339–346.

Scholl, D.W. 1962. Sedimentary record of the holocene transgression across the southwestern margin of the Everglades, southern Florida. In *Proc. First Nat. Coastal and Shallow Water Res. Conf.* pp. 670–673.

Scholl, D.W. 1965. High interstitial water chlorinity in estuarine mangrove swamps, Florida. *Nature* (London) 207: 284–285.

Sidhu, S.S. 1975. Culture and growth of some mangrove species. In *Proc. Int. Symp. Biol. and Management of Mangroves*. Vols. I and II, eds. G.E. Walsh, S.C. Snedaker and H.J. Teas, pp. 394–401. Int. Food Agric. Sci., Univ. Florida, Gainesville.

Siever, R., Garrels, R.M., Kanwisher, J. and Berner, R.A. 1961. Interstitial waters of recent marine muds off Cape Cod. *Science* 134: 1071–1072.

Snedaker, S.C. and Lugo, A.E. 1973. The role of mangrove ecosystems in the maintenance of environmental quality and a high productivity of desirable fisheries. Bur. Sports Fisheris & Wildlife – A final report, contract no. 14-16-008-606. NTIS, Dept. of Commerce, Springfield, Va.

Snedaker, S.C. and Pool, D.J. 1973. Mangrove forest types and biomass. In *Bur. Sports Fisheries & Wildlife*, eds. S.C. Snedaker and A.E. Lugo. A final report, contract no. 14-16-008-606. NTIS, Dept. of Commerce, Springfield, Va.

Snedaker, S.C. and Stanford, R.L. 1976. Ecological studies of a subtropical terrestrial biome. Florida Power & Light Co., final report, Miami, Florida.

Steinke, T.D. 1975. Some factors affecting dispersal and establishment of propagules of *Avicennia marina* (Forsk.) Vierh. In *Proc. Int. Symp. Biol. and Management of Mangroves*, Vols. I and II, eds. G.E. Walsh, S.C. Snedaker and H.J. Teas, pp. 402–414. Int. Food Agric. Sci., Univ. Florida, Gainesville.

Stephens, W.M. 1962. Mangroves: trees that make land. *Sea Frontiers* 8: 491–496.

Stern, W.L. and Voigt, G.K. 1959. Effect of salt concentration on growth of red mangrove in culture. *Bot. Gaz.* 121: 36–39.

Teas, H.J. and Montgomery, F. 1968. Ecology of red mangrove seedling establishment. *Ass. S.E. Biol. Bull.* 15: 56–57.

Thom, B.G. 1967. Mangrove ecology and deltaic geomorphology: Tabasco, Mexico. *J. Ecol.* 55: 301–343.

Thom, B.G., Wright, L.D. and Coleman, J.M. 1975. Mangrove ecology and deltaic-estuarine geomorphology: Cambridge Gulf-Ord River, Western Australia. *J. Ecol.* 63: 203–232.

Vann, J.H. 1959. Landform-vegetation relationships in the Atrato delta. *Ann. Ass. Amer. Geogr.* 49: 345–360.

Van Steenis, C.G.G.J. 1958. Rhizophoraceae. *Flora Malesiana* 5: 431–493.

Van Steenis, C.G.G.J. 1963. Miscellaneous notes on New Guinea plants. VII. *Nova Guinea (Botany)* 12: 189.

Vaughan, T.W. 1909. The geologic work of mangroves in Southern Florida. *Smithson. Misc. Coll.* 52: 461–464.

Waisel, Y. 1972. *Biology of Halophytes.* Academic Press, New York.

Walsh, G.E. 1974. Mangroves: a review. In *Ecology of Halophytes*, eds. R.J. Reimhold and W.H. Queen, pp. 51–174. Academic Press, New York.

Walsh, G.E., Snedaker, S.C. and Teas, H.J. eds. 1975. *Proc. Int. Symp. Biol. and Management of Mangroves*, Vols. I and II. Inst. Food Agric. Sci., Univ. Florida, Gainesville.

Walter, H. 1971. *Ecology of Tropical and Subtropical Vegetation.* Van Nostrand-Reinhold, New York.

Walter, H. and Steiner, M. 1936. Die Okologie des Ost-Afrikanischen Mangroven. *Z. Bot.* 30: 65–193.

Watson, J.G. 1928. *Mangrove Forests of the Malay Peninsula.* (Malayan Forest. Rec. No. 6) Fraser and Neave, Ltd., Singapore.

Weaver, J.E. and Clements, F.E. 1938. *Plant Ecology*, 2nd edition. McGraw-Hill Book Co. Inc., New York.

Welch, B. 1962. Aspects of succession in shallow coastal waters of the Caribbean. Ph.D. dissertation. Department of Botany, Duke University.

West. R.C. 1956. Mangrove swamps of the Pacific coast of Columbia. *Ann. Ass. Amer. Geogr.* 46: 98–121.

Winkler, H. 1931. Einige Bemerkungen über Mangrove-Pflanzen und den *Amorphophallus titanum* im Hamburger Botanischer Garten. *Ber. Dtsch. Bot. Ges.* 49: 87–102.

Distribution and environmental control of productivity and growth form of *Spartina alterniflora* (Loisel.)

R. MICHAEL SMART

Environmental Laboratory, U.S.A.E. Waterways Experiment Station,
Vicksburg, MS 39180, U.S.A.

1. Introduction

Spartina alterniflora Loisel. is the most widespread of the North American tidal marsh plants. The species forms dense, monospecific stands and dominates the intertidal zone along the Atlantic Coast from southern Canada to northern Florida and along the Gulf of Mexico from Florida to southern Texas (Mobberley, 1956; Reimold, 1977). These *Spartina* marshes are among the most productive ecosystems in the world (Odum, 1971) and are considered major contributors to the high productivity of adjacent estuaries (de la Cruz, 1973). The high productivity of *S. alterniflora* is of particular interest in that it occurs over a wide geographical/climatological range and on sediments of high salinity and negative redox potential.

Another interesting aspect of the ecology of *S. alterniflora* is the occurrence of distinct growth forms in different zones of the salt marsh. In North American *Spartina* marshes a tall form occurs along creek banks and drainage channels. Immediately landward of the tall form an intermediate height form occurs and this form grades into the short form or stunted *S. alterniflora* in the interior of the marsh. These growth forms occur throughout the entire range of the species and result in large differences in productivity between different zones of the marsh. The cause of growth form differentiation has been variously attributed to environmental or genetic factors and is the subject of a considerable amount of ecological research.

The objectives of this chapter are to review the environmental factors controlling the distribution of *S. alterniflora*, to examine the physiological adaptations to the environment of the intertidal zone, and to critically assess contemporary hypotheses on the variation in productivity and growth form observed in tidal salt marshes. A conceptualized model will be used to assess possible interactions among growth-limiting variables and to identify avenues for further research.

2. Distribution

1. Temperature

The dominance of the lower intertidal zone between 30° and 50° N latitude (Reimold, 1977) illustrates a high degree of plasticity with respect to temperature required for photosynthetic and other metabolic processes (Mallot et al., 1975; Thomas and Long, 1978; Giurgevich and Dunn, 1979). The species is apparently restricted by low temperature at the northern end of its range (Hatcher and Mann, 1975) and by competitive exclusion by mangrove at the southern end (Davis, 1940). Recent evidence has indicated an inverse relationship between productivity and latitude (Keefe, 1972; Hatcher and Mann, 1975; Reimold, 1977), however considerable variation in productivity occurs among marshes in the same coastal region (Nixon and Oviatt, 1973a; Broome et al., 1975a), and between stands in

Tasks for vegetation science, Vol. 2 ed. by D.N. Sen and K.S. Rajpurohit

individual marshes (Good, 1965; Odum and Fanning, 1973; DeLaune et al., 1979). Anomalously high values of productivity with respect to latitude have also been shown (Nixon and Oviatt, 1973b). In spite of possible reductions in potential productivity with increasing latitude, temperature appears to be limiting only at the extreme northern end of the geographical range.

II. Substrate

The development of *Sartina* marshes is dependent on the occurrence of suitable substrate within the intertidal environment. The texture of the substrate does not appear to be particularly important as *Spartina* marshes develop on sand (Redfield, 1972; Kurz and Wagner, 1957), silt (Miller and Egler, 1950), or on fine-textured organic clays (Ranwell, 1964). The species is however restricted to relatively protected coastlines (Ranwell, 1973). According to Redfield (1972) instability of the substrate is more important in inhibiting plant establishment than mechanical damage by waves. Whatever the cuase, attempts to introduce *S. alterniflora* by transplanting in high energy environments have generally been unsuccessful below mean high water (Garbisch et al., 1975).

III. Tidal regime

Tidal regime (range, modality, and salinity) is the major factor limiting the distribution of *S. alterniflora* (Miller and Egler 1950; Adams, 1963). The species is generally restricted to the lower portion of the intertidal region (Redfield, 1972), although the actual extent will vary substantially with variations in tidal regime. Observations of the lower limit of occurrence vary. Johnson and York (1915), Chapman (1940), Hubbard (1969) and Redfield (1972) indicate that *Spartina* extends below mean sea level, while Kurz and Wagner (1957) determined lower limits ranging from below to slightly above mean sea level. In areas of extreme tidal range, such as the Bay of Fundy, the lower limit of occurrence will be substantially above mean sea level (Ganong, 1903). Hubbard (1969) indicates that submergence per se is not limiting the lower level of occurrence of *S.*

anglica as these plants withstood continuous submergence in clear seawater for over four months. The upper limit of the *S. alterniflora* zone is dependent on the slope and drainage characteristics as well as the tidal range (Kurz and Wagner, 1957; Chapman, 1974).

The factors controlling the distribution of *S. alterniflora* in the intertidal zone are highly complex due to interactions between the various aspects of the tidal regime (Ranwell, 1973; Chapman, 1974). The controlling factors include tidal range, modality, slope and aspect of the marsh in relation to prevailing winds, drainage characteristics, and salinity. Although the vertical range over which *S. alterniflora* will be dominant is impossible to predict, the species is apparently restricted to areas of moderate to high salinity, high moisture content, and frequent inundation. The absence of any one of these three factors may result in displacement.

3. Physiological adaptations

The observed high productivity of *S. alterniflora* is quite remarkable from a physiological viewpoint. Specific adaptations include the ability to rapidly colonize and stabilize freshly deposited sediments (Ranwell, 1964; Hubbard, 1965; Taylor and Burrows, 1968), a high degree of salinity tolerance (Ganong, 1903; Penfound and Hathoway, 1938; Adams, 1963; Smart and Barko, 1978), and an efficient network of aerenchyma tissues for the transport of oxygen to the roots (Teal and Kanwisher, 1966; Anderson, 1974).

I. Colonizing ability

Colonization of bare areas may occur by fragmentation or by seed (Goodman et al., 1959; Taylor and Burrows, 1968). Although the species is wind pollinated and wind dispersed, tidal currents may be of considerable importance in dispersal of seeds as well as fragments. Of interest in this regard, is that significant seed production occurs only along creek banks (Seneca, 1974). An afterripening period is required, and maximum germination occurs after cold storage for several months in estuarine water (Seneca, 1974).

The ability of *S. alterniflora* to rapidly colonize and stabilize bare areas is related to a high intrinsic growth rate and the rapid proliferation of stems through rhizomatous growth (Caldwell, 1957). Complete cover of newly seeded sites can be achieved in one year, and recently colonized sites are similar in productivity to mature marshes (Seneca, 1974). Sexual maturity can be reached in three to four months under favorable conditions, thus assuring an additional supply of seeds during early colonization. The root system is dense and highly branched (Anderson, 1974), and frequent branching is also evident in the rhizomes (Caldwell, 1957). The proliferation of dense assemblages of stems through tillering favors the deposition and subsequent entrapment of suspended sediments (Redfield, 1972). These sediments not only provide additional support for the plant but also provide an abundant source of nutrients (Ranwell, 1964).

II. Adaptations to Anaerobiosis

Anaerobic soils and sediments are characterized by a variety of biological and chemical reactions which do not occur in the presence of molecular oxygen. The physicochemical characteristics of these anaerobic soils and sediments have recently been reviewed (Ponnamperuma, 1972; Armstrong, 1975; Gambrell and Patrick, 1978). Of particular importance is the increased solubility of reduced forms of metals which can result in potentially toxic concentrations of iron and manganese. Additionally bacterial activity can generate potentially toxic quantities of ammonium or sulfide. The precipitation of trace metal sulfides is of considerable importance in moderating concentrations of these substances (Engler and Patrick, 1975). The actual concentrations of ferrous iron and sulfide attained in the interstitial water depend on the relative quantities of each. In freshwater sediments, iron is usually present in sufficient quantify to precipitate virtually all of the sulfide formed (Gambrell and Patrick, 1978). However salt marsh sediments receive quantities of sulfate from seawater and, in the presence of adequate organic substrates, sulfide production can result in precipitation of virtually all of the iron (DeLaune et al., 1976). Sulfide can

thus accumulate to potentially toxic levels in salt marsh sediments.

The ability of plants to exploit anaerobic sediments has been attributed to the provision of adequate root ventilation (Armstrong, 1975, 1978). Metabolic adaptations involving coupling of metabolic pathways or anaerobic metabolism (Crawford, 1978) may be of importance to occasionally flooded species or during dormant periods in true wetland plants, however these processes are insufficient to support active plant growth (Armstrong, 1978). The diffusion of oxygen through aerenchyma tissues fulfills the respiratory needs of the root system and also oxidizes potential toxicants in the rhizosphere (Armstrong, 1975, 1978). The aerenchyma network in *S. alterniflora* has been shown to be developed beyond the respiratory requirements of the plant and excess oxygen is lost to the sediment (Teal and Kanwisher, 1966). This radial loss of oxygen should provide a buffer against potentially toxic quantities of sulfide and/or iron. Evidence for this detoxification process is presented by Carlson (1979) who demonstrated that *S. alterniflora* roots moderate the accumulation of sulfides in salt marsh sediments.

III. Salinity tolerance

The maintenance of low osmotic potentials necessary for water uptake from saline soils generally requires the accumulation of osmotically-active substances in plant tissues. However these accumulated substances can result in impaired metabolism due to ion toxicity or electrolyte imbalance. High tissue concentrations of sodium chloride, for example are particularly inhibiting to enzymes of many halophytes (Flowers, 1972; Greenway and Osmond, 1972). Halophytes therefore must be capable of selective accumulation of high concentrations of ions or metabolites coupled with metabolic adaptation to these concentrations (Epstein, 1969; Jefferies, 1972). Maintenance of favorable ionic balance within halophyte shoots is accomplished through a variety of mechanisms depending on plant species. These mechanisms include salt secretion, compartmentalization, succulence, abscission of salt-saturated organs, and salt exclusion (Waisel, 1972).

The presence of secreted salts on the leaves of *S. alterniflora* is visible evidence of the importance of salt secretion in the species. The secretion process occurs in specialized organs called salt glands (Levering and Thompson, 1971; Anderson, 1974). In coastal marsh plants the predominant salt secreted is sodium chloride, with potassium and other non toxic electrolytes being retained (Waisel, 1972). There is some evidence that the secretion process in some species may be impaired at salinities approaching that of seawater (Waisel, 1961). However Smart and Barko (1980) demonstrated that *S. alterniflora* roots are capable of a high degree of selective ion uptake. They found that sodium was excluded and potassium was absorbed. This process greatly reduces the quantities of sodium which would normally be transported to leaves and subsequently secreted. Unfortunately salt exclusion by the roots can result in increasing salinization of the sediments.

IV. Water conservation

In view of the passive accumulation of salts in the rhizosphere of *S. alterniflora*, high salinity levels in salt marsh sediments may necessitate low rates of transpiration. Salinity has been demonstrated to depress transpiration in a number of halophyte species (Strogonov, 1962; Slatyer, 1970; Kaplan and Gale, 1972). Nestler (1977a) also observed increased resistance to transpiration in short form *S. alterniflora* plants which were exposed to higher salinities than the tall form. Decreased transpiration rates have also been shown to increase the time available for oxidation of reduced substances in the rhizosphere of wetland plants (Jones, 1971; Armstrong, 1975).

As early as 1903, Ganong recognized the xerophytic anatomy and ability for regulating transpiration which enabled *Spartina* to dominate the lower intertidal regions of salt marshes. He describes the thickly cutinized epidermis, the location of stomates in the furrows of the epidermis, and the ability of the leaves to curl, thus protecting the stomates. These observations are supported by Anderson (1974) who also describes the occurrence of accessory cells possessing branched papillae

which extend over the stomatal aperture.

An additional water conserving feature of *S. alterniflora* is the C_4 photosynthetic pathway. The adaptive significance of the C_4 pathway to growth under suboptimal moisture regimes has been stressed by Ludlow (1976). Huber and Sankhla (1976) state that the C_4 pathway may be an ecological necessity for plants growing in highly saline environments. Salinity has also been implicated in promoting a shift from C_3 to CAM (Shomer-Ilan and Waisel, 1973; Winter and Lüttge, 1976; Bloom, 1979) and from C_3 to C_4 (Sankhla and Huber, 1975). It is also of ecological interest to note that the majority of the more successful salt marsh plants of the coastal United States possess the C_4 pathway, while Hough and Wetzel (1977) were able to identify only one freshwater macrophyte species possessing the C_4 pathway. It appears likely that C_4 photosynthesis is of adaptive advantage to salt marsh plants and that this advantage may be associated with increased water economy.

Water conservation, to the extent that it involves stomatal closure, may also reduce photosynthesis. However it has been shown that partial stomatal closure reduces water losses to a greater extent than CO_2 fixation (Nobel, 1974). Additionally Kaplan and Gale (1972) have demonstrated that salinity induced large increases in mesophyll resistance to water vapor loss without a concomitant reduction in CO_2 fixation by the halophyte *Atriplex halimus*. Gale (1975) attributes this response to a decrease in hydraulic conductivity of the mesophyll cell walls. The relevance of these findings, in addition to xeromorphology and C_4 photosynthesis, in *S. alterniflora* has not been examined.

4. Variations in productivity and growth form

The occurrence of distinct growth forms of *S. alterniflora* has generated a great deal of research as to the possible environmental or biological factors responsible. These growth forms occur throughout the entire range of the species and result in large differences in productivity between different zones of the marsh. In most North American *Spartina* marshes a tall form occurs along creek banks and drainage channels. Immediately landward of the tall form an intermediate height form usually occurs and this form grades into the short form or stunted *S. alterniflora* in the interior of the marsh. A number of hypotheses have been proposed to explain these variations in productivity and growth form in individual salt marshes, and some of these are supported by considerable experimental data. However the repeated occurrence of this phenomenon over a wide range of climate, salinity, nutrients, and tidal regimes suggests that a single factor may be responsible. The objective of this section is to evaluate the current hypotheses regarding growth limitation in *S. alterniflora*.

I. Ecotypic differentiation

Morphological differences among height forms of *S. alterniflora* characterizing different zones of the marsh are so distinct and extensive that these forms were once considered different varieties or subvarieties (cf. Mobberley, 1956). However, Mobberley, in a monograph on the genus *Spartina*, rejected these classifications and regarded the varietal names (*glabra*, *pilosa*, etc.) as synonomous with *alterniflora*.

Stalter and Batson (1969) used a reciprocal transplant technique in an attempt to illucidate factors controlling plant zonation in a South Carolina salt marsh. Short form *S. alterniflora* transplanted into a stand of the tall form remained short after six months. Tall form *S. alterniflora* transplanted into the short zone likewise remained taller than the surrounding plants. They concluded that the two forms were ecotypes. However, in a laboratory investigation, Mooring et al. (1971) demonstrated that seedlings originating from the two forms responded similarly when grown under the same conditions. They concluded that the two forms were ecophenes and were responding to differences in salinity in the different zones of the marsh. In a later investigation Shea et al. (1975) observed reciprocal transplants over a period of eighteen months and demonstrated that the transplants eventually grew to resemble the surrounding plants. They also examined aspects of the biochemistry of the two forms and concluded that they were geneti-

cally similar. Although Seneca (1974) presented evidence of possible genetic differences among *S. alterniflora* populations along the Atlantic and Gulf Coasts, there is little evidence for genetic differentiation within individual salt marshes.

II. Iron limitation

Adams (1963) observed signs of iron chlorosis in hydroponically grown *S. alterniflora* and suggested that the species had a high iron requirement. Chlorosis was more evident in the more rapidly growing plants held in freshwater, perhaps indicating a depletion of stored reserves of iron. Although chlorosis was mitigated by either foliar applications of iron or by the combination of high salinity (which reduced growth) and high iron concentrations in the culture solution, there was no significant increase in plant growth which was attributable to iron additions. However, on the basis of higher soluble iron levels in the sediments of the tall *S. alterniflora* zone, Adams concluded that differences in iron availability were responsible for the observed differences in growth form. Taylor (1939) and Mooring et al. (1971) have also observed iron chlorosis in solution cultures of *S. alterniflora*. Mooring et al. (1971) alleviated iron chlorosis in solution cultures only through frequent foliar applications of ferrous sulfate, indicating an inability of *S. alterniflora* to assimilate iron from the culture solution. Kneller et al. (1975) suggested that optimal growth occurred when solution cultures were maintained under reducing conditions which favor the more soluble divalent form of iron. Similarly, Smart (unpublished data) observed poor growth and chlorosis in cultures of *S. alterniflora*. Normal pigmentation and growth occurred only after the onset of anaerobiosis and negative redox potentials. Broome et al. (1975b) in a field investigation, demonstrated that applications of iron to either the foliage or sediments of *S. alterniflora* did not result in increased biomass or plant height. The characteristic chlorotic appearance of these plants was likewise unaffected by additions of iron but was ameliorated by nitrogen fertilization. Broome et al. (1975b) therefore concluded that availability of iron was not responsible for the reduced growth of short

131

S. alterniflora. Evidence for iron limitation has only been obtained from aerated solution culture experiments and has not been observed under more natural anaerobic culture conditions (Kneller et al., 1975; Smart and Barko, 1980) or under field conditions (Broome et al., 1975b), thus the occurrence of different growth forms of *S. alterniflora* is considered unrelated to the availability of iron.

III. Anaerobiosis

In spite of the well developed aerenchyma system in *S. alterniflora* (Teal and Kanwisher, 1966) , recent research suggests impaired growth under anaerobic conditions. Linthurst (1979) grew *S. alterniflora* plants under differing degrees of aeration. These conditions included permanent flooding, diurnal inundation and drainage, and permanent flooding with aeration accomplished by bubbling air through the sediment. He observed increased growth and nutrient uptake with increasing degree of aeration and attributed these responses to associated changes in redox potential, pH, and ion availability. The causal mechanism promoting increased growth was unclear, and in some cases large differences in growth were associated with relatively minor differences in sulfide concentrations and redox potentials. In another investigation, Linthurst and Seneca (1980) demonstrated increased growth of *S. alterniflora* as a result of either aeration or nitrogen addition. The major difficulty in interpreting the results of these studies is that the method used does not allow for the separation of mixing effects from those of aeration. Thus the increased growth and nutrient uptake observed in these studies may be attributable to overcoming diffusional limitations normally associated with nutrient supply.

In an investigation of dieback sites in a *S. anglica* marsh, Goodman and Williams (1961) were unable to find direct evidence of damage attributable to anaerobiosis. Mahall and Park (1976a), in a detailed study of oxygen diffusion rates in a *S. foliosa* marsh, concluded that oxygen availability was unrelated to plant zonation or productivity. Smart and Barko (1980) grew *S. alterniflora* on a variety of anaerobic sediments under permanently flooded conditions. Plant growth was affected by salinity and nitrogen availability but was unaffected by negative sediment redox potentials (unpublished data).

If the growth inhibiting effects of anaerobiosis (Linthurst, 1979; Linthurst and Seneca, 1980) were due to sulfide toxicity, additions of iron to salt marsh sediments might stimulate growth due to sulfide precipitation (Ponnamperuma, 1972). Conversely if the effects were due to iron toxicity, additions of iron would be expected to depress growth due to increased concentrations of the toxicant. Additions of iron to salt marsh sediments did not significantly affect plant growth or the uptake of sulfur or iron (Broome et al., 1975b), indicating that these substances may be of little importance in affecting the growth form of *S. alterniflora.*

IV. Salinity

Salinity has been considered to play a major role in the zonation of inland salt marshes (Bolen, 1964; Ungar, 1965; Ungar et al., 1969), and although the tidal factor has received considerably more attention with regard to plant zonation in tidal marshes (Chapman, 1974), a number of investigators have considered salinity to play a major role (Penfound and Hathoway, 1938; Reed, 1947; Kurz and Wagner, 1957; Ranwell et al., 1964). Kurz and Wagner (1957) also repeatedly observed an inverse relationship between height of *S. alterniflora* and chlorinity of the soil solution. Reed (1947) observed salinities in the short *S. alterniflora* zone that were twice as high as those in the tall form zone. More recently, Good (1965) indicated that the short height form of *S. alterniflora* was associated with higher and more variable salinities than was the tall height form. Nestler (1977a) observed stable salinity gradients increasing with distance from tidal creeks in a Georgia *S. alterniflora* marsh. He attributed the low productivity of the short form to high salinity. Smart and Barko (1978) indicated that increased sediment salinity significantly reduced growth of *S. alterniflora.* Results of solution culture studies have also demonstrated reductions in growth of *Spartina* with increasing salinity (Gosselink, 1970; Mooring et al., 1971; Phleger, 1971; Mahall and Park, 1976b;

Parrondo et al., 1978). These results indicate that salinity may exert a controlling influence on the productivity of S. alterniflora.

V. Nitrogen limitation

A number of studies have shown growth responses of short S. alterniflora to nitrogen fertilization (Sullivan and Daiber, 1974; Valiela and Teal, 1974; Gallagher, 1975). Nitrogen additions to the tall form of S. alterniflora occurring along creek banks have demonstrated a growth response only in a Louisiana salt marsh (Patrick and DeLaune, 1976). The increased productivity of short S. alterniflora after nitrogen fertilization has been considered indicative of nitrogen limitation in these plants. However, as pointed out by Mendelssohn (1979a) and Smart and Barko (1980), additions of nitrogen can result in changes in microbial decomposition and nutrient mineralization rates as well as changes in ionic composition, and pH of the interstitial water. These secondary changes, either singly or in combination, may be responsible for the increased growth of S. alterniflora after nitrogen fertilization. Smart and Barko (1980) avoided these problems by using a plant assay to assess the limiting nutrient status of S. alterniflora. Their technique is based on the minimum tissue concentration (critical concentration) required for normal growth and metabolism (Gerloff, 1969, 1975; Bates, 1971). Critical concentrations were determined from sand culture experiments and these values were compared with tissue concentrations of plants grown on a variety of sediments. They demonstrated nitrogen limitation of plant growth on all fifteen sediments examined. Squiers and Good (1974) used changes in standing crop and tissue nitrogen to demonstrate nitrogen limitation in a New Jersey S. alterniflora marsh. Mendelssohn (1979a) used assays of glutamate dehydrogenase activity to examine nitrogen metabolism of S. alterniflora in a North Carolina salt marsh. Root glutamate dehydrogenase activities were lower in short S. alterniflora than in tall or intermediate forms (Mendelssohn, 1979a) and nitrogen fertilization produced a growth response in the short height zone (Mendelssohn, 1979b). However, the short form S. alterniflora plants were exposed to interstitial water ammonium concentrations that were an order of magnitude greater then those occurring in the tall form sediments. Thus it is paradoxical that these plants were nitrogen limited in the presence of high concentrations of available nitrogen. Chalmers (1979) obtained similar results for a short S. alterniflora marsh in Georgia. She also measured high inorganic nitrogen concentrations in the interstitial water, but fertilization with nitrogen-containing sewage sludge resulted in increased growth compared to non-fertilized controls. Maye (1972), in a survey of salt marsh sediments in Georgia, also noted high concentrations of ammonium in the interstitial waters of sediments in the zones of thick (short form) S. alterniflora.

The high concentrations of interstitial water ammonium present in short S. alterniflora zones do not imply a greater nitrogen supply to this zone in comparison with the tall zone, but they do suggest an inability of the short S. alterniflora plants to utilize the quantities of ammonium present. Perhaps, as Mendelssohn (1979a) suggested, some factor is preventing the uptake of ammonium in the short S. alterniflora zone.

Additional evidence for an inhibition of nitrogen uptake is the characteristic export of nitrogen (particularly ammonium) from S. alterniflora marshes to adjacent estuaries (Axelrad et al., 1976; Heinle and Flemer, 1976; Stevenson et al., 1977; Valiela et al., 1978). Undisturbed ecosystems are considered to develop 'tight' nutrient cycles and usually export small quantities of essential nutrients (Likens and Bormann, 1972, 1974; Hobbie and Likens, 1973). Limiting nutrients in particular are most closely conserved (Vitousek and Reiners, 1975). If nitogen is limiting to the salt marsh, its export in significant quantities is a characteristic of a disturbed ecosystem. The nature of this disturbance will be discussed in a later section.

VI. Nitrogen fixation

Anaerobic sediments are likely to be deficient in nitrogen due to nitrification-denitrification processes (DeLaune et al., 1976; Reddy et al., 1976; Patrick and Reddy, 1976). Nitrogen fixed at the sediment surface is subject to these same losses prior to reaching the rhizosphere, thus minimizing its contribution to plant uptake. High rates of nitrogen fixation have recently been measured in the rhizosphere of wetland plants (Patriquin and Keddy, 1978). This nitrogen would presumably be available for immediate assimilation by the plant, and has been shown to account for significant portions of the nitrogen requirement of some wetland plants (Tjepkema and Evans, 1976; Zuberer and Silver, 1978). However high concentrations of ammonium depress rates of nitrogen fixation in the rhizosphere (Patriquin and Keddy, 1978) as well as in the bulk sediment (Van Raalte et al., 1974; Hanson, 1977). For these reasons nitrogen fixation and subsequent assimilation may be slight in short *S. alterniflora* zones. However, differences in rates between tall and short form zones (Valiela and Teal, 1979) are insufficient to account for the differences in productivity between these zones.

VII. Tidal subsidy

Odum and Fanning (1973) and Odum (1974) equated tidal flushing in salt marshes with irrigation of agricultural crops. Odum (1974) suggested that productivity of *S. alterniflora* in a Georgia salt marsh was correlated with the degree of tidal subsidy (frequency of inundation). Steever et al. (1976) further illustrated the importance of this phenomenon with tidal and productivity data obtained from salt marshes along the Atlantic Coast. Similarly, intertidal stands of the freshwater emergent, *Zizania aquatica*, were shown to be more productive than those not subject to tidal flooding (Whigham and Simpson, 1977). Conner and Day (1976) also related productivity of Cypress swamps to water flow.

The reasons for the above relationships are unclear, however a number of studies have demonstrated decreased sediment water fluxes in short *S. alterniflora* zones relative to the more productive stands along creekbanks (Gardner, 1973; Odum and Riedeburg, 1976; Nestler, 1977b). Smart and Barko (1978, 1980) suggested that increased fluxes of tidal water through salt marsh sediments might increase productivity by flushing excluded salts from

the sediments and by providing additional nitrogen supplies. The potential influence of increased tidal flushing on sediment salinity and nitrogen availability will be discussed in a later section.

5. Nitrogen-salinity interactions

There is some evidence that productivity and growth form of *S. alterniflora* are affected by both nitrogen availability and salinity. In a study of Rhode Island salt marshes (Nixon and Oviatt, 1973a), both nitrogen concentration and salinity of the tidal water were correlated with productivity of *S. alterniflora*. Broome et al. (1975a) observed a highly significant negative correlation between soil salinity and yield of nitrogen limited *S. alterniflora* in North Carolina salt marshes. Valiela et al. (1976) state that nitrogen limited *S. alterniflora* in a Massachusetts salt marsh repeatedly shows greening and increased growth following substantial rains. They did not report sediment salinity changes, but lowered salinities following heavy rains have been documented for *Spartina* marshes in England (Ranwell et al., 1964). If lowered salinities also occurred in the Massachusetts salt marsh studied by Valiela et al. (1976), the increased growth may have been a result of lessened salinity stress. In a later investigation, Valiela et al. (1978) state that four years of nitrogen fertilization in the short *S. alterniflora* zone had failed to produce plants equivalent to those occurring naturally along creekbanks. They suggest the possibility of a secondary limiting factor such as salinity. Similarly the highest rate of nitrogen fertilization (8 mol. m^{-2}) employed by Mendelssohn (1979b) did not increase productivity of short *S. alterniflora* stands to the level of the unfertilized creekbank stands. Mendelssohn (1979a) also suggests that some factor (such as salinity) may be inhibiting the uptake and/or assimilation of nitrogen by the short *S. alterniflora* plants.

The possible interaction between nitrogen and salinity in limiting the productivity of *S. alterniflora* is further substantiated by the findings of Smart and Barko (1980). These authors demonstrated the importance of ion exclusion-selection processes in regulating the salt content of plant tissues. Although the molar ratio of sodium to potassium in seawater is approximately 45:1, *S. alterniflora* tissues contain only one to two moles of sodium per mole of potassium. Preferential absorption of potassium, coupled with sodium exclusion, results in increased salinity and Na:K in sediment interstitial waters. The magnitude of the increase is apparently dependent on the rate of transpiration, as increased growth (and leaf area) results in a proportional increase in salinization (Fig. 1). Thus growth stimulation resulting from nitrogen fertilization may be self-limiting due to increasing salinity stress. Evidence for this is presented by Chalmers (1979) who observed increased salinity following nitrogen fertilization.

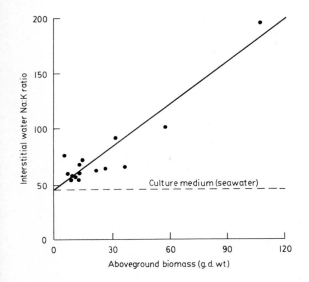

Fig. 1. The effect of aboveground growth of *Spartina alterniflora* on the ion balance of sediment interstitial water.

I. Competitive ion effects

The ability of *S. alterniflora* to maintain potassium absorption under increasingly high Na:K ratios is unknown; however, the depressed potassium levels (Table 1) in the belowgrond tissues of plants exposed to high Na:K (195:1) may indicate a limited ability in this regard. Rains and Epstein (1967) demonstrated depressed potassium uptake in the halophyte *Avicennia marina* at a similar ratio (200:1). In sediments repeatedly flushed by tidal

Table 1. The effect of interstitial water salinity ($g.kg^{-1}$) and Na:K ratio on the ion content ($mmol\ g.d.\ wt^{-1}$) of aboveground and belowground portions of *Spartina alterniflora*. Adapted from Smart and Barko (1980).

| Interstitial water | | Aboveground | | Belowground | |
Salinity	Na:K	Na	K	Na	K
23	88	0.33	0.15	0.30	0.30
32	195	0.30	0.16	0.30	0.15

water, such as those along creek banks, excluded salts are continually leached and these high Na:K ratios are unlikely to be attained. However in the hydraulically stagnant short *S. alterniflora* zone (Gardner, 1973; Odum and Riedeburg, 1976; Nestler, 1977b), excluded sodium may accumulate to the detriment of the plant community.

This salinization process may also affect the nitrogen economy of the plant by greatly increasing the quantity of sodium in the interstitial water. The resulting high concentration of sodium competing with ammonium may exceed the capacity of the roots for selective uptake of ammonium. Increased nitrogen assimilation (and subsequently increased plant growth) repeatedly observed following nitrogen fertilization may be simply a mass action response overriding the inhibition of nitrogen uptake. In this regard it should be noted that the amounts of nitrogen used in many fertilization studies are extremely high. For example, the highest rate employed by Mendelssohn (1979b) is equivalent to the nitrogen requirement of an aboveground standing crop of 14,000 g.d. wt m^{-2} (based on the critical nitrogen concentration determined by Smart and Barko, 1980).

In opposition to the above, however, examination of sodium and ammonium concentrations of interstitial waters associated with *S. alterniflora* roots reveals a high affinity for ammonium uptake in spite of high concentrations of sodium (Smart and Barko, 1980). For example, *S. alterniflora* reduced rhizosphere ammonium to levels of 0.01 mM in the presence of 350 mM sodium. It should be noted that these ammonium levels are considerably below those reported by Mendelssohn (1979a) and are in opposition to his conclusion that ammonium uptake and/or assimilation is inhibited

135

in short *S. alterniflora*. The method of interstitial water extraction used by Mendelssohn differed from that used by Smart and Barko (1980) and may more closely approximate the bulk interstitial water than that which is in intimate contact with *S. alterniflora* roots. If this is correct then the characteristically high levels of ammonium present in short *S. alterniflora* sediments (Maye, 1972; Chalmers, 1979; Mendelssohn, 1979a) may be indicative of decreased ammonium transport to the roots rather than to decreased uptake or assimilative capacity.

II. Nitrogen efficiency

The nitrogen efficiency of a plant can be examined on the basis of dry weight production per unit nitrogen uptake or on the ability of a plant to extract nitrogen from a given soil (Gerloff, 1977). In an examination of the former aspect of nitrogen efficiency in *S. alterniflora*, Smart and Barko (1980) determined a high relative requirement for nitrogen. The significance of this high nitrogen requirement in relation to the proposed osmoregularoty role of nitrogen-containing compounds such as proline (Stewart and Lee, 1974) and glycinebetaine (Storey and Wyn Jones, 1977) is unknown, however recent evidence indicates that *S. alterniflora* can accumulate proline under highly saline conditions (Cavalieri and Huang, 1979). Whether this accumulation performs an osmoregulatory function, or is simply a manifestation of impaired metabolism, remains to be tested.

The second aspect of nitrogen efficiency, the ability to extract nitrogen from a given soil, is of obvious ecological importance for plants growing in a nitrogen limiting environment. The uptake of nitrogen by the salt marsh plants *Distichlis spicata* (L.) and *S. alterniflora* is compared with that of the freshwater emergent, *Cyperus esculentus* (L.) in Table 2. All species were grown concurrently on the same sediment type and were harvested at peak mass. The salt marsh species were subjected to a 'tidal' salinity of 15 g. kg^{-1} while the freshwater species was maintained in deionized water. Both growth and nitrogen uptake by the salt marsh plants were reduced relative to the freshwater spe-

Table 2. Comparison of the efficiency of nitrogen uptake between a freshwater emergent, *Cyperus esculentus* and two salt marsh plants, *Distichlis spicata* and *Spartina alterniflora*. All plants were concurrently grown on the same sediment type and growth was nitrogen limited.

Species	Plant part	Biomass (g.d. wt)	Tissue N (mmol g.d. wt^{-1})	N Uptake (mmol)
Cyperus [*] *esculentus*	Shoots	177	0.30	53.1
	Roots	76	0.42	31.9
	Tubers	154	0.27	41.8
	Total	407		126.8
Spartina [+] *alterniflora*	Shoots	108	0.45	48.6
	Roots/ rhizomes	135	0.32	43.4
	Total	243		92.0
Distichlis [+] *spicata*	Shoots	107	0.56	59.6
	Roots/ rhizomes	87	0.36	31.7
	Total	194		91.3

[*] Data from Barko and Smart (1979).
[+] Previously unpublished data from the study of Smart and Barko (1980).

cies. *D. spicata* and *S. alterniflora* removed 91 and 92 mmol. of nitrogen respectively while *C. esculentus* removed 127 mmol. This difference was not associated with impaired uptake by the salt marsh species as all three species reduced rhizosphere ammonium concentrations to low levels. Reduced transport of ammonium to the roots, due to low rates of transpiration by the salt marsh plants, may have been involved (unpublished data) and this aspect will be discussed in a later section.

III. Nutrient transport mechanisms

Plant roots contact nutrients through the processes of root extension, diffusion, and mass-flow. Root extension, or growth into unexploited volumes of soil, generally accounts for less than ten percent of the plant's nutrient requirement (Barber et al., 1963). In short *S. alterniflora* this value may be even less due to the restricted depth of root penetration (Gallagher and Plumpey, 1979). Diffusional transport is involved in the uptake of 'fixed' elements such as potassium or phosphorus (Barber et al., 1963). and mass-flow is responsible for significant transport of mobile elements. Nutrient transport by mass-flow can be substantial; and, where it is ineffective, due to continually high humidity, plant growth is severely limited (Odum, 1968). In salt marsh soils much of the ammonium formed may be mobile due to displacement from the cation exchange complex by the high concentrations of sodium and other cations in seawater. Ammonium thus released to the interstitial water would be subject to transport via mass-flow. The magnitude of this transport would be dependent on the rate of transpiration. However, as discussed above, increased transport of salts into the rhizosphere would negate the nutritional benefit derived from rapid transpiration. Mass-flow of nutrients in salt marshes may thus be of limited importance.

Another mechanism for nutrient transport in marshes is associated with sediment water table fluctuations. This aspect of nutrient transport has not been investigated but could provide considerable transport of nutrients in sediments subjected to tidally-induced water table fluctuations. As suggested by Smart and Barko (1978, 1980), these sediment water movements could redistribute excluded salts in addition to replenishing rhizosphere ammonium concentrations.

6. General hypothesis

The preceding discussion has been an attempt to reduce the predominant hypotheses on growth limitation of *S. alterniflora* to their lowest common denominators in order to synthesize a general hypothesis which incorporates most of the available data. This preliminary attempt indicates an interrelationship among salinity, nitrogen availability, and tidal energy subsidy in controlling productivity and growth form. The possible nature of this relationship is discussed below.

A feedback diagram relating transpiration and ammonium transport is shown in Fig. 2. Under low salinity conditions, transpiration drives a flux of ammonium to the rhizosphere (Loop 1) where it is assimilated. Resultant growth stimulation and leaf proliferation exerts a positive effect on transpiration, which eventually results in elevated salinity as well as the ratio of sodium to potassium (Loop 2). Increased salinity stress reduces transpiration and subsequently the rate of ammonium and salt

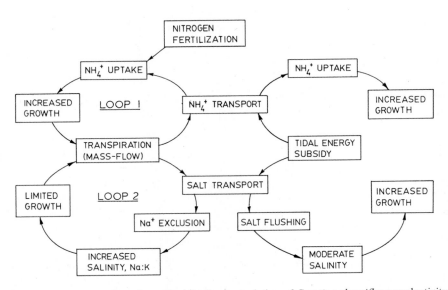

Fig. 2. Conceptualization of proposed feedback regulation of *Spartina alterniflora* productivity.

transport to the rhizosphere. These feedback loops are evident in the data of Nestler (1977a), Smart and Barko (1978, 1980), and Chalmers (1979).

Nitrogen fertilization eliminates the dependence on transpiration (mass-flow) and diffusion for supplying nitrogen. However, the increased growth due to fertilization increases transpiration, thus promoting salinity increases and subsequent stress. This is evident in Fig. 1, and in the data of Chalmers (1979) and may explain the repeated failures of nitrogen fertilization to increase productivity to the level of tall form *S. alterniflora* (Sullivan and Daiber, 1974; Valiela and Teal, 1974; Gallagher, 1975; Mendelssohn, 1979b). Indeed Valiela et al. (1978) indicate that nitrogen fertilization over four successive years did not accomplish this.

Tidal energy subsidy results in a flushing action in the sediments of the creekbank and ditchbank *S. alterniflora* zones. This flushing has been equated with crop irrigation (Odum, 1974) and Smart and Barko (1978, 1980) suggested that it removes accumulated salts and replenishes ammonium concentrations, thus performing a dual function of moderating salinity and driving nutrient transport. This mechanism may be responsible for the enhanced productivity attributed to tidal energy subsidy (Odum, 1974; Steever et al., 1976).

Growth of short form *S. alterniflora*, although limited by nitrogen in the classical sense, may actually be controlled by the continual salinization of the sediments. Although additions of nitrogen stimulate growth, the nature of the response is moderated by salinity. Moreover decreases in the level of sediment salinity elicit a response similar to that of nitrogen addition (Nestler, 1977a).

The survival of short *S. alterniflora* depends on its ability to extract nitrogen with minimum expenditure of water, thus avoiding intolerably high salt burdens. This feat is accomplished through a variety of mechanisms, including the development of a large belowground biomass for acquisition/storage of nitrogen (Valiela et al., 1976; Smith et al., 1979). In this regard over 50% of the seasonal maximum nitrogen content may be present in the overwintering shoots (Squiers and Good, 1974; Gallagher et al., 1980). Belowground storage accounts for an additional unknown percentage and root uptake of nitrogen may be slight. Nitrogen economy may be enhanced by accelerated leaf senescence (unpublished data), and by the absence of seed production (Seneca, 1974). Additional adaptive responses include decreased plant stature and leaf area index as well as increased resistance to transpiration (Nestler, 1977a; Giurgevich and Dunn, 1979). Additions of nitrogen, while stimulating plant growth, do not significantly modify these morphological adaptations to salinity. Future investigations should consider the role of the tidal factor in salinity tolerance and nitrogen nutrition as well as the physiological, morphological, and phenological adaptations enabling *S. alterniflora* to dominate the intertidal zone.

Acknowledgements

The author acknowledges the able technical assistance of John W. Barko in the collection and analysis of the data. Dana L. Smart prepared the figures. This chapter was based on research supported by the United States Army Engineer Dredged Material Research Program.

Literature cited

Adams, D.A. 1963. Factors influencing vascular plant zonation in North Carolina salt marshes. *Ecology* 44: 445–456.

Anderson, E.E. 1974. A review of structure in several North Carolina salt marsh plants. In *Ecology of Halophytes*, eds. R.J. Reimold and W.H. Queen, pp. 307–340. Academic Press, New York.

Armstrong, W. 1975. Waterlogged soils. In *Environment and Plant Ecology*, ed. J.R. Etherington, pp. 181–218. John Wiley and Sons, New York.

Armstrong, W. 1978. Root aeration in the wetland condition. In *Plant Life in Anaerobic Environments*, eds. D.D. Hook and R.M. Crawford, pp. 269–297. Ann Arbor Science, Ann Arbor.

Axelrad, D.M., Moore, K.A. and Bender, M.E. 1976. *Nitrogen Phosphorus, and Carbon Flux in Chesapeake Bay Marshes*. Virginia Water Resources Research Center Bulletin 79, Blacksburg.

Barber, S.A., Walker, J. M. and Vasey, E. G. 1963. Mechanisms for the movement of plant nutrients from the soil and fertilizer to the plant root. *J. Agr. Food Chem.* 11: 204–207.

Barko, J.W. and Smart, R.M. 1979. The nutritional ecology of *Cyperus esculentus*, an emergent aquatic plant, grown on different sediments. *Aquat. Bot.* 6: 13–28.

Bates, T.E. 1971. Factors affecting critical nutrient concentrations in plants and their evaluation: A review. *Soil Sci.* 112: 116–130.

Bloom, A.J. 1979. Salt requirement for Crassulacean acid metabolism in the annual succulent, *Mesembryanthemum crystallinum. Plant Physiol.* 63: 749–753.

Bolen, E.G. 1964. Plant ecology of spring-fed salt marshes in western Utah. *Ecol. Monogr.* 34: 143–166.

Broome, S.W., Woodhouse, W.W., Jr. and Seneca, E.D. 1975a. The relationship of mineral nutrients to growth of *Spartina alternifloa* in North Carolina: I. Nutrient status of plants and soils in natural stands. *S.S.S.A. Proc.* 39: 295–301.

Broome, S.W., Woodhouse, W.W., Jr. and Seneca, E.D. 1975b. The relationship of mineral nutrients to growth of *Spartina alterniflora* in North Carolina: II. The effects of N, P and Fe fertilizers. *S.S.S.A. Proc.* 39: 301–307.

Caldwell, P. A. 1957. The spatial development of *Spartina* colonies growing wihout competition. *Ann. Bot.* 21: 203–214.

Carlson, P.R. 1979. The role of *Spartina alterniflora* in the sulfur cycle of the salt marsh. *Abstracts, 42nd annual meeting, Am. Soc. Limnol. Oceanogr.*

Cavalieri, A.J. and Huang, A.H.C. 1979. Evaluation of proline accumulation in the adaptation of diverse species of marsh halophytes to the saline environment. *Amer. J. Bot.* 66: 307–312.

Chalmers, A.G. 1979. The effects of fertilization on nitrogen distribution in a *Spartina alterniflora* salt marsh. *Estuarine Coastal Mar. Sci.* 8: 327–337.

Chapman, V.J. 1940. Succession on New England salt marshes. *Ecology* 21: 279–282.

Chapman, V.J. 1974. *Salt Marshes and Salt Deserts of the World.* Cramer, Lehre.

Conner, W.H. and Day, J.W., Jr. 1976. Productivity and composition of a baldcypress-water tupelo site and a bottomland hardwood site in a Louisiana swamp. *Amer. J. Bot.* 63: 1354–1364.

Crawford, R.M.M. 1978. Metabolic adaptations to anoxia. In *Plant Life in Anaerobic Environments*, eds. D.D. Hook and R.M.M. Crawford, pp. 119–137. Ann Arbor Science.

Davis, J.H. 1940. The ecology and geologic role of mangrove in Florida. *Carn. Inst. Wash. Publ.* 517: 303–412.

de la Cruz, A.A. 1973. The role of tidal marshes in the productivity of coastal waters. *Assoc. Southeastern Biol. Bull.* 20: 147–156.

De Laune, R.D., Buresh, R.J. and Patrick, W.H., Jr. 1979. Relationship of soil properties to standing crop biomass of *Spartina alterniflora* in a Louisiana marsh. *Estuarine Coastal Mar. Sci.* 8: 477–487.

Delanne, R.D., Patrick, W.H., Jr. and Brannon, J.M. 1976. *Nutrient Transformations in Louisiana Salt Marsh Soils.* Sea Grant Publication LSU-T-76-009. Baton Rouge.

Engler, R.M. and Patrick, W.H., Jr. 1975. Stability of sulfides of manganese, iron, zinc, cooper and mercury in flooded and nonflooded soil. *Soil Sci.* 119: 217–221.

Epstein, E. 1969. Mineral metabolism of halophytes. In *Ecological Aspects of the Mineral Nutrition of Plants*, ed. R.H. Rorison, pp. 345–355. Blackwell Scientific, Oxford.

Flowers, T.J. 1972. The effect of sodium chloride on enzyme activities from four halophyte species of Chenopodiaceae. *Phytochemistry* 11: 1881–1886.

Gale, J. 1975. Water balance and gas exchange of plants under saline conditions. In *Plants in Saline Environments*, eds. A. Poljakoff-Mayber and J. Gale, pp. 168–185. Springer-Verlag, New York.

Gallagher, J.L. 1975. Effect of an ammonium nitrate pulse on the growth and elemental composition of natural stands of *Spartina alternifora* and *Juncus roemerianus. Amer. J. Bot.* 62: 644–648.

Gallagher, J. L., Reimold, R.J., Linthurst, R.A. and Pfeiffer, W.J. 1980. Aerial production, mortality, and mineral accumulation dyamics in *Spartina alterniflora* and *Juncus roemerianus* in a Georgia salt marsh. *Ecology* 61: 303–312.

Gallagher, J.L. and Plumley, F.G. 1979. Underground biomass profiles and productivity in Atlantic Coastal marshes. *Amer. J. Bot.* 66: 156–161.

Gambrell, R.P. and Patrick, W.H., Jr. 1978. Chemical and microbiological properties of anaerobic soils and sediments. In *Plant Life in Anaerobic Sediments*, eds. D.D. Hook and R.M.M. Crawford, pp. 375–423. Ann Arbor Science, Ann Arbor.

Ganong, W.F. 1903. The vegetation of the Bay of Fundy salt and diked marshes: An ecological study. *Bot Gaz.* 36: 161–186.

Garbisch, E.W., Jr., Woller, P.B., Bostian, W.J. and McCallum, R.J. 1975. Biotic techniques for shore stabilization. In *Estuarine Research*, ed. L.E. Cronin, pp. 405–426. Academic Press, New York.

Gardner, L.R. 1973. The effect of hydrologic factors on the pore water chemistry of intertidal marsh sediments. *Southeastern Geology* 15: 17–28.

Gerloff, G.C. 1969. Evaluating nutrient supplies for the growth of aquatic plants in natural waters. In *Eutrophication: Caues,Consequences, Correctives*, ed. G.E. Likens, pp. 537–555. National Academy of Sciences, Washington.

Gerloff, G.C. 1975. *Nutritional Ecology of Nuisance Aquatic Plants.* U.S. Environmental Protection Agency, Corvallis.

Gerloff, G.C. 1977. Plant efficiencies in the use of nitrogen, phosphorus and potassium. In *Proceedings of the Workshop on Plant Adaptation to Mineral Stress in Problem Soils*, ed. M.J. Wright, pp. 161–173. Cornell University Press, Ithaca.

Giurgevich, J.R. and Dunn, E.L. 1979. Seasonal patterns of CO_2 and water vapor exchange of the tall and short height forms of *Spartina alterniflora* Loisel. in a Georgia salt marsh. *Oecologia* 43: 139–156.

Good, R.E. 1965. Salt marsh vegetation, Cape May, New Jersey. *Bull. New Jersey Acad. Sci.* 10: 1–11.

Goodman, P.J., Braybrooks, E.M. and Lambert, J.M. 1959. Investigations into 'dieback' in *Spartina townsendii* agg. I. The present status of *Spartina townsendii* in Britain. *J. Ecol.* 47: 651–677.

139

Goodman, P.J. and Williams, W.T. 1961. Investigations into 'dieback' in *Spartina townsendii* agg. III. Physiological correlates of 'dieback'. *J. Ecol.* 49: 391–398.

Gosselink, J.G. 1970. Growth of *Spartina patens* and *S. alterniflora* as influenced by salinity and source of nitrogen. Louisiana State University, *Coatal Studies Bulletin* 5: 97–110.

Greenway, H. and Osmond, C.B. 1972. Salt responses of enzymes from species differing in salt tolerance. *Plant Physiol.* 49: 256–259.

Hanson, R.B. 1977. Comparison of nitrogen fixation activity in tall and short *Spartina alterniflora* salt marsh soils. *Appl. Env. Microbiol.* 33: 596–602.

Hatcher, B.G. and Mann, K.H. 1975. Above-ground production of marsh cordgrass *(Spartina alterniflora)* near the northern end of its range. *J. Fish. Res. Bd. Canada* 32: 83–87.

Heinle, D.R. and Flemer, D.A. 1976. Flows of materials between poorly flooded tidal marshes and an estuary. *Mar. Biol.* 35: 359–373.

Hobbie, J.E. and Likens, G.E. 1973. The output of phosphorus, dissolved organic carbon and fine particulate carbon from Hubbard Brook watersheds. *Limnol. Oceanogr.* 18: 734–742.

Hough, R.A. and Wetzel, R.G. 1977. Photosynthetic pathways of some aquatic plants. *Aquat. Bot.* 3: 297–313.

Hubbard, J.C.E.1965. *Spartina* marshes in southern England. VI. Pattern of invasion in Poole Harbour. *J. Ecol.* 53: 799–813.

Hubbard, J.C.E. 1969. Light in relation to tidal immersion and the growth of *Spartina townsendii* (s.l.). *J. Ecol.* 57: 795–804.

Huber, W. and Sankhla, N. 1976. C$_4$ pathway and regulation of the balance between C$_4$ and C$_3$ metabolism. In *Water and Plant Life*, eds. O.L. Lange, L. Kappen, and E.D. Schulze, pp. 335–363. Springer-Verlag, New York.

Jefferies, R.L. 1972. Aspects of salt marsh ecology with particular reference to inorganic plant nutrition. In *The Estuarine Environment*, ed. R.S.K. Barnes, pp. 61–85. Applied Science Publishers Ltd., London.

Johnson, D.S. and York, H.H. 1915. The relation of plants to tide levels. *Carn. Inst. Wash. Publ.* 206.

Jones, H.E. 1971. Comparative studies in plant growth and distribution in relation to waterlogging. II. An experimental study of the relationship between transpiration and the uptake of iron in *Erica cinerea* L. and *E. tetralix* L. *J. Ecol.* 59: 167–178.

Kaplan, A. and Gale, J. 1972. Effect of sodium chloride salinity on the water balance of *Atriplex halimus*. *Aust. J. Biol. Sci.* 25: 895–903.

Keefe, C.W. 1972. Marsh production: A summary of the literature. *Contr. Mar. Sci.* 16: 165–181.

Kneller, K., Parrondo, R.T. and Gosseling, J.G. 1975. Influence of iron source and concentration on growth of *Spartina alterniflora* Loisel. Unpublished manuscript. Louisiana State University, Baton Rouge.

Kurz, H. and Wagner, K. 1957. Tidal marshes of the Gulf and Atlantic Coasts of northern Florida and Charleston, South Carolina. *Florida State University Studies* 24, Tallahassee.

Levering, C.A. and Thompson, W.W. 1971. Salt glands of *Spartina foliosa*. *Planta* 97: 183–196.

Likens, G.E. and Bormann, F.H. 1972. Nutrient cycling in ecosystems. In *Ecosystems: Structure and Function*, ed. J. Weins, pp. 25–67. Oregon State University Press, Corvallis.

Likens, G.E. and Bormann, F.H. 1974. Linkages between terrestrial and aquatic ecosystems. *Bioscience* 24: 447–456.

Linthurst, R.A. 1979. The effect of aeration on the growth of *Spartina alterniflora* Loisel. *Amer. J. Bot.* 66: 685–691.

Linthurst, R.A. and Seneca, E.D. 1980. Aeration, nitrogen and salinity as determinants of *Spartina alterniflora* Loisel. growth response. *Estuaries* (in press).

Ludlow, M.M. 1976. Ecophysiology of C$_4$ grasses. In *Water and Plant Life*, eds. O.L. Lange, L. Kappen, and E.D. Schulze, pp. 364–386. Springer-Verlag, New York.

Mahall, B.E. and Park, R.B. 1976a. The ecotone between *Spartina foliosa* Trin and *Salicornia virginica* L. in salt marshes of San Francisco Bay. III. Soil aeration and tidal immersion. *J. Ecol.* 64: 811–819.

Mahall, B.E. and Park, R.B. 1976b. The ecotone between *Spartina foliosa* Trin and *Salicornia virginica* L. in salt marshes of San Francisco Bay. II. Soil water and salinity. *J. Ecol.* 64: 793–809.

Mallot, P.G., Davey, A.J., Jefferies, R.L. and Hutton, M.J. 1975. Carbon dioxide exchange in leaves of *Spartina anglica* Hubbard. *Oecologia* 21: 351–358.

Maye, P.R., III. 1972. Some important inorganic nitrogen and phosphorus species in a Georgia salt marsh. Environmental Resources Center Report No. 0272, Georgia Institute of Technology, Atlanta.

Mendelssohn, I.A. 1979a. Nitrogen metabolism in the height forms of *Spartina alterniflora* in North Carolina. *Ecology* 60: 574–584.

Mendelssohn, I.A. 1979b. The influence of nitrogen level, form, and application method on the growth response of *Spartina alterniflora* in North Carolina. *Estuaries* 2: 106–112.

Miller, W.R. and Egler, F.E. 1950. Vegetation of the Westquetequock-Pawcatuck tidal-marshes, Connecticut. *Ecol. Monogr.* 20: 143–172.

Mobberley, D.G. 1956. Taxonomy and distribution of the genus *Spartina*. *Iowa State College J. Sci.* 30: 471–574.

Mooring, M.T., Cooper, A.W. and Seneca, E.D. 1971. Seed germination response and evidence for height ecophenes in *Spartina alterniflora* from North Carolina. *Amer. J. Bot.* 58: 48–55.

Nestler, J. 1977a. Interstitial salinity as a cause of ecophenic variation in *Spartina alterniflora*. *Estuarine Coastal Mar. Sci.* 5: 707–714.

Nestler, J. 1977b. A preliminary study of the sediment hydrology of a Georgia salt marsh using Rhodamine WT as a tracer. *Southeastern Geol.* 18: 265–271.

Nixon, S.W. and Oviatt, C.A. 1973a. Analysis of local variation in the standing crop of *Spartina alterniflora*. *Bot. Mar.* 26: 103–109.

Nixon, S.W. and Oviatt, C.A. 1973b. Ecology of a New England salt marsh. *Ecol. Monogr.* 43: 463–498.

Nobel, P.S. 1974. *Introduction to Biophysical Plant Physiology*.

W.H. Freeman and Co., San Francisco.

Odum, E.P. 1971. *Fundamentals of Ecology*. W.B. Saunders Company, Philadelphia.

Odum, E.P. 1974. Halophytes, energetics, and ecosystems. In *Ecology of Halophytes*, eds. R.J. Reimold and W.H. Queen, pp. 599–602. Academic Press, New York.

Odum, E.P. and Fanning, M.E. 1973. Comparison of the productivity of *Spartina alterniflora* and *Spartina cynosuroides* in Georgia coastal marshes. *Bull. Georgia Acad. Sci.* 31: 1–12.

Odum, E.P. and Riedeburg, C.H. 1976. A dye study of interstitial water flow in tidal marsh sediments. Abstracts, 39th annual meeting, *Am. Soc. Limnol. Oceanogr.*

Odum, H.T. 1968. Work circuits and systems stress. In *Symposium on Primary Production and Mineral Cycling in Natural Ecosystems*, ed. H.E. Young, pp. 81–138. University of Maine Press, Orono.

Parrondo, R.T., Gosselink, J.G. and Hopkinson, C.S. 1978. Effects of salinity and drainage on the growth of three salt marsh grasses. *Bot. Gaz.* 139: 102–107.

Patrick, W.H., Jr. and De Laune, R.D. 1976. Nitrogen and phosphorus utilization by *Spartina alterniflora* in a salt marsh in Barataria Bay, Louisiana. *Estuarine Coastal Mar. Sci.* 4: 59–64.

Patrick, W.H. Jr. and Reddy, K.R. 1976. Nitrification-denitrification reactions in flooded soils and water bottoms: Dependence on oxygen supply and ammonium diffusion. *J. Environ. Qual.* 5: 469–472.

Patriquin, D.G. and Keddy, C. 1978. Nitrogenase activity (acetylene reduction) in a Nova Scotian salt marsh: Its association with angiosperms and the influence of some edaphic factors. *Aquat. Bot.* 4: 227–224.

Penfound, W.T. and Hathoway, E.S. 1938. Plant communities in the marshlands of southeastern Louisiana. *Ecol. Monogr.* 8: 1–56.

Phleger, C.F. 1971. Effect of salinity on growth of a salt marsh grass. *Ecology* 52: 908–911.

Ponnamperuma, F.N. 1972. The chemistry of submerged soils. *Adv. Agron.* 24: 29–96.

Rains, D.W. and Epstein, E. 1967. Preferential absorption of potassium by leaf tissue of the mangrove *Avicennia marina*: An aspect of halophytic competence in coping with salt. *Aust. J. Biol. Sci.* 14: 519–540.

Ranwell, D.S. 1964. *Spartina* salt marshes in southern England. III. Rates of establishment, succession and nutrient supply at Bridgwater Bay, Somerset. *J. Ecol.* 52: 95–105.

Ranwell, D.S. 1973. Ecology of salt marshes and sand dunes. Chapman and Hall, London.

Ranwell, D.S., Bird, E.C.F., Hubbard, J.C.E. and Stebbings, R.E. 1964. *Spartina* salt marshes in southern England. V. Tidal submergence and chlorinity in Poole Harbour. *J. Ecol.* 52: 627–641.

Reddy, K.R., Patrick, W.H., Jr. and Phillips, R.E. 1976. Ammonium diffusion as a factor in nitrogen loss from flooded soils. *S.S.S.A.J.* 40: 528–533.

Redfield, A.C. 1972. Development of a New England salt marsh. *Ecol. Monogr.* 42: 201–237.

Reed, J.F. 1947. The relation of the *Spartinetum glabrae* near Beaufort, North Carolina, to certain edaphic factors. *Amer. Midl. Nat.* 38: 605–614.

Reimold, R.J. 1977. Mangals and salt marshes of eastern United States. In *Ecosystems of the World Vol. 1. Wet Coastal Ecosystems*, ed. V.J. Chapman, pp. 157–166. Elsevier Scientific, New York.

Sankhla, N. and Huber, W. 1975. Regulation of balance between C_3 and C_4 pathway: Role of abscisic acid. *Z. Pfl. Physiol* 74: 267–271.

Seneca, E.D. 1974. Germination and seedling response of Atlantic and Gulf Coasts populations of *Spartina alterniflora*. *Amer. J. Bot.* 61: 947–956.

Shea, M.L., Warren, R.S. and Niering, W.A. 1975. Biochemical and transplantation studies of the growth form of *Spartina alterniflora* on Connecticut salt marshes. *Ecology* 56: 461–466.

Shomer-Ilan, A. and Waisel, Y. 1973. the effect of sodium chloride on the balance between the C_3- and C_4- carbon fixation pathways. *Physiol. Plant.* 29: 190–193.

Slatyer, R.O. 1970. Comparative photosynthesis, growth and transpiration of two species of *Atriplex*. *Planta* 93: 175–189.

Smart, R.M. and Barko, J.W. 1978. Influence of sediment salinity and nutrients on the physiological ecology of selected salt marsh plants. *Estuarine Coastal Mar. Sci.* 7: 487–495.

Smart, R.M. and Barko, J.W. 1980. Nitrogen nutrition and salinity tolerance of *Distichlis spicata* and *Spartina alterniflora*. *Ecology* 61: 630–638.

Smith, K.K., Good, R.E. and Good, N.F. 1979. Production dynamics for above and belowground components of a New Jersey *Spartina alterniflora* tidal marsh. *Estuarine Coastal Mar. Sci.* 9: 189–201.

Squiers, E.R. and Good, R.E. 1974. Seasonal changes in the productivity, caloric content, and chemical composition of a population of salt-marsh cord-grass (*Spartina alterniflora*). *Chesapeake Sci.* 15: 63–71.

Stalter, R. and Batson, W.T. 1969. Transplantation of salt marsh vegetation, Georgetown, South Carolina. *Ecology* 50: 1087–1089.

Steever, E.Z., Warren, R.S. and Niering, W.A. 1976. Tidal energy subsidy and standing crop production of *Spartina alterniflora*. *Estuarine Coastal Mar. Sci.* 4: 473–478.

Stevenson, J.C., Heinle, D.R. Flemer, D.A., Small, R.J., Rowland, R.A. and Ustach, J.F. 1977. Nutrient exchanges between brackish water marshes and the estuary. In *Estuarine Processes Vol. II. Circulation, Sediments and Transfer of Materials in the Estuary*, ed. M. Wiley, pp. 219–240. Academic Press, New York.

Stewart, G.R. and Lee, J.A. 1974. The role of proline accumulation in halophytes. *Planta* 120: 279–289.

Storey, R. and Wyn Jones, R.G. 1977. Quaternary ammonium compounds in plants in relation to salt resistance. *Phytochemistry* 16: 447–453.

Strogonov, B.P. 1962. *Physiological Basis of Salt Tolerance of Plants*. English Edition, I.P.S.T., Jerusalem.

Sullivan, M.J. and Daiber, F.C. 1974. Response in production of cord-grass, *Spartina alterniflora*, to inorganic nitrogen and phosphorus fertilizer. *Chesapeake Sci.* 15: 121–123.

Taylor, N. 1939. Salt tolerance of Long Island salt marsh plants. *New York State Museum Circ.* 23.

Taylor, M.C. and Burrows, E.M. 1968. Studies on the biology of *Spartina* in the Dee estuary, Cheshire. *J. Ecol.* 56: 795–809.

Teal, J.M. and Kanwisher, J.W. 1966. Gas transport in the marsh grass, *Spartina alterniflora. J. Exp. Bot.* 17: 355–361.

Thomas, S.M. and Long, S.P. 1978. C_4 Photosynthesis in *Spartina townsendii* at low and high temperatures. *Planta* 142: 171–174.

Tjepkema, J.D. and Evans, H.J. 1976. Nitrogen fixation associated with *Juncus balticus* and other plants of Oregon wetlands. *Soil Biol. Biochem.* 8: 505–509.

Ungar, I.A. 1965. An ecological study of the vegetation of the Big Salt Marsh, Staffort County, Kansas. *University of Kansas Science Bull.* 46: 1–99.

Ungar, I.A., Hogan, W. and McClelland, M. 1969. Plant communities of saline soils at Lincoln, Nebraska. *Amer. Midl. Natur.* 82: 564–577.

Valiela, I. and Teal, J.M. 1974. Nutrient limitation in salt marsh vegetation. In *Ecology of Halophytes*, eds. R.J. Reimold and W.H. Queen, pp. 547–563. Academic Press, New York.

Valiela, I. and Teal, J.M. 1979. The nitrogen budget of a salt marsh ecosystem. *Nature* 280: 652–656.

Valiela, I., Teal, J.M. and Deuser, W.G. 1978. The nature of growth forms in the salt marsh grass *Spartina alterniflora. Amer. Natur.* 112: 461–470.

Valiela, I., Teal, J.M. and Persson, N.Y. 1976. Production and dynamics of experimentally enriched salt marsh vegetation: Belowground biomass. *Limnol. Oceanogr.* 21: 245–252.

Valiela, I., Teal, J.M., Volkmann, S., Shafer, D. and Carpenter, E.J. 1978. Nutrient and particulate fluxes in a salt marsh ecosystem: Tidal exchanges and inputs by precipitation and ground water. *Limnol. Oceanogr.* 23: 798–812.

Van Raalte, C.D., Valiela, I., Carpenter, E.J. and Teal, J.M. 1974. Inhibition of nitrogen fixation in salt marshes measured by acetylene reduction. *Estuarine Coastal Mar. Sci.* 2: 301–305.

Vitousek, P.M. and Reiners, W.A. 1975. Ecosystem succession and nutrient retention: A hypothesis. *Bioscience* 25: 376–381.

Waisel, Y. 1961. Ecological studies on *Tamarix aphylla* (L.) Karst. III. The salt economy. *Plant and Soil* 13: 356–364.

Waisel Y. 1972. *Biology of Halophytes*. Academic Press, New York.

Whigham, D.F. and Simpson, R.L. 1977. Growth, mortality, and biomass partitioning in freshwater tidal wetland populations of wild rice (*Zizania aquatica* var. *aquatica*). *Bull Torrey Bot. Club* 104: 347–351.

Winter, K. and Lüttge, U. 1976. Balance between C_3 and CAM pathway of photosynthesis. In *Water and Plant Life*, eds. O.L. Lange, L. Kappen, and E.D. Schulze, pp. 323–334. Springer-Verlag, New York.

Zuberer, D.A. and Silver, W.S. 1978. Biological dinitrogen fixation (acetylene reduction) associated with Florida mangroves. *Appl. Env. Microbiol.* 35: 567–575.

CHAPTER 3

Germination ecology of halophytes

IRWIN A. UNGAR

Department of Botany, Ohio University,
Athens, Ohio 45701, USA

1. Salt tolerance

Germination is an important stage in the life cycle of species growing in saline environments because it determines the soil conditions later stages in the life cycle will be exposed to. Laboratory investigations of seed germination indicate that seeds of most halophytic species reach their maximum germination in distilled water (Seneca, 1969; Onnis and Bellettato, 1972; Breen et al. 1977; Okusanya, 1977; Ungar, 1977a, 1978a; Zid and Boukhris, 1977; Dietert and Shontz, 1978). Seed germination in saline environments ordinarily occurs during the spring or in a season with high precipitation, when soil salinity levels are usually reduced (Ward, 1967; Chapman, 1974; McMahon and Ungar, 1978). Several studies have shown that seeds of many halophytes remain dormant when exposed to low water potentials (Ungar, 1962, 1974a, b, 1975, 1978a; Williams and Ungar, 1972). With a reduction in salinity stress, it was found that germination would occur at levels equivalent to that of the original distilled water controls (Table 1).

In general both halophytes (Waisel and Ovadia, 1972; Chapman, 1974; Okusanya, 1977; Zid and Boukhris, 1977; Breen et al., 1977; Joshi and Iyengar, 1977; Ungar, 1978a, b) and glycophytes (Saini, 1972; Albregts and Howard, 1973; Ryan et al., 1975; Varshney and Baijal, 1977) respond in a similar manner to increased salinity stress; with both a reduction in the total number of seeds germinating and a delay in the initiation of the germination process. The precise salinity concentration causing a delay and reduction in the number of seeds germinating depends upon the salt tolerance of each individual species.

Darwin (1857) found that seeds from 87 species of plants varied in their ability to tolerate soaking in seawater. Twenty-five percent of the species studied did not remain viable after a 23 day soaking period. However, seeds of *Beta vulgaris* were found to retain their viability after 100 days immersion in seawater. More recent investigations with halophytes have demonstrated that seeds of several species, including *Atriplex halimus*, *Salicornia europaea*, *Crithmum maritimum*, and *Sporobolus virginicus*, remain dormant at low water potentials. These seeds do not lose their viability and will germinate when returned to a distilled water treatment (Zid and Boukhris, 1977; Breen et al., 1977; Okusanya, 1977; Ungar, 1977a). McMahon and

Table 1. Germination percentage of *Hordeum jubatum* seeds in distilled water after soaking for ten days in salt solutions (Ungar, 1974).

Days	Original NaCl concentration (%)			
	2	3	4	5
5	0	0	0	0
10	73	71	61	59
15	93	90	90	92
20	97	97	97	98
25	97	97	97	98

Tasks for vegetation science, Vol. 2 ed. by D.N. Sen and K.S. Rajpurohit
© *1982, Dr W. Junk Publishers, The Hague. All rights reserved. ISBN 90 6193 942 9*

Ungar (1978) showed that the germination of *Atriplex triangularis* seeds under field conditions was related to reduced soil salinity levels in early spring. Germination also occurred sporadically until early summer when periods of precipitation reduced the relative salt content of surface soils. Laboratory results of Zid and Boukhris (1977) indicate that seeds of *Atriplex halimus*, which did not germinate when exposed to 4% or 5% NaCl and were subsequently returned to distilled water, germinated at levels equal to those of the original distilled water controls, 85 to 99%. After being placed in distilled water, *Crithmum maritimum* seeds also recovered from treatments with half-strength seawater and germinated at levels equal to that of the orginal controls. Germination of *Sporobolus virginicus* was completely inhibited by 2% NaCl (Breen et al., 1977). Viability of seeds of this species was not reduced even after a 7 week storage period in 12% NaCl. A number of other examples of similar recovery responses from osmotically induced dormancy have been reported in the literature (Ungar, 1978a).

The fact that this recovery response is common among halophytes implies that it might be of some ecological significance within highly saline environments, reflecting a physiological response that is strongly selected for during the evolution of these species. The importance of this dormancy response to high salt stress in seeds of coastal species is that it inhibits seeds from germinating while floating in seawater. Establishment would be impossible under these conditions (Lesko and Walker, 1969). Another important effect of the release of dormancy with the alleviation of salt stress is that it determines the salinity level at the period of seedling development, which is probably one of the most sensitive periods in the life cycle of halophytes. The shallow rooted seedlings develop close to the soil surface and are exposed to the most severe salinity stress. Delay in the time of germination until a period of reduced soil salinity stress assures that seedlings will not die immediately and provides them with some chance for survival to maturity. It has been reported in the literature that adult plants are more salt tolerant than are seeds (Millington et al., 1951; Mayer and Poljakoff-Mayber, 1975). Extremely

high salt tolerance in terms of germination of seeds could prove to be deleterious at later stages of development. Seeds may have been selected for germination at reduced salinities to assure survival of the developing seedlings. Apparent reductions in salt tolerance at the germination stage could be due to the fact that seeds in the surface soils are exposed to higher salinity levels than are roots of an actively growing plant. The ability of seeds to remain dormant at extremely low water potentials indicates that they may in reality be more salt tolerant than actively growing plants. The criteria used to interpret salt tolerance is a critical factor here.

Waite and Hutchings (1978) found that increasing the density at which seeds were planted caused a general increase in the percentage germination of *Plantago coronopus* seeds at salinities ranging up to 10% of seawater. This density related acceleration of germination may be the result of an ecological strategy, assuring high exploitation of safe sites or it may result from exogenous secretions of seeds which facilitate germination.

Seeds of *Salicornia europaea*, a highly salt tolerant halophyte, remained dormant at salinities above 5% NaCl (Ungar, 1977a). Variations in soil salinity levels in the field probably extend the germination period of this species. When microedaphic conditions were such that germination was not possible, *S. europaea* seeds apparently remained stored in the soil until the following growing season (Ungar, 1974a). Preliminary investigaions by Ungar (unpublished data) indicated that an average of 30 seeds were present in a 10 g sample of surface soil collected from *Salicornia* communities in September. Viability of seeds was tested on 1.5 × 9 cm samples of soil and germination data indicated relatively high viability for these seeds. Seedling numbers from the *Salicornia* community soil samples ranged from 10 to 175 seedlings. *Atriplex triangularis* did not occur growing in the *Salicornia* community, but was overrepresented in the seed flora. The reduced number of seeds in the submerged zone correlates well with lower plant densities in this area (Table 2). These data indicate that long-term seed storage could be an important factor influencing changes in the floristic composition of zonational salt marsh communities. Milton (1939),

Table 2. Seed banks in a highly saline, *Salicornia europaea* community.[a]

Species	Community plants/100 cm^2			Seeds No./10 g soil			Seedlings No./ 1.5 × 9 cm volume		
	SS	ES	TS	SS	ES	TS	SS	ES	TS
Salicornia europaea	47	359	551	35	119	95	6	20	51
Atriplex triangularis	0	0	0	17	30	39	0	4	13
Hordeum jubatum	0	0	0	0	0	0	0	1	1

[a] SS = submerged *Salicornia* on pan, ES = *Salicornia* bordering pan, TS = Tall *Salicornia* bordering *Atriplex* zone.

in his investigations with buried viable seed content in the Dovey estuarine marshes of Wales, found that some species were not proportionately represented in the seed bank in terms of community composition. Species such as *Glyceria maritima* and *Agrostis* sp. were more strongly represented in the seed bank, while other more abundant species such as *Armeria maritima* and *Festuca rubra* were under-represented. Communities occurring closest to cultivated areas contained many weed species in their seed bank, including *Stellaria media*, *Urtica dioica*, and *Rumex obtusifolius*.

2. Specific ion toxicity

Darwin (1857) demonstrated that seeds vary in their ability to tolerate soaking in seawater. More recent investigations with pasture grasses by Varshney and Baijal (1977) indicate that specific ion toxicities occur when seeds were exposed to various salts, in the following order of descending toxicity: $NaHCO_3 > Na_2CO_3 > NaCl > CaCl_2$. Other studies with glycophytes indicate that species vary in their response to specific salts (Yadava et al., 1975; Younis and Hatata, 1971; Ryan et al., 1975). Strogonov (1964, 1974) has also descibed the importance of specific ions on the behavior and distribution of halophytes. Ungar and Capilupo (1969) and Macke and Ungar (1971), using inorganic and organic osmotic agents to examine specific ion effects with halophytes, *Suaeda depressa* and *Puccinellia nuttalliana*, indicate that these salts act osmotically rather than through a specific ion effect (Table 3). Ungar and Hogan (1970) have also shown the same effect with the less salt tolerant *Iva annua*.

Uhvits (1946) and Redmann (1974) studies with *Medicago sativa* demonstrated that NaCl was more toxic than organic solvents at the germination stage. Redmann (1974) found that sodium sulfate and magnesium chloride were more toxic to *M. sativa* than isotonic solutions of NaCl. Studies with steppe species from Washington by Choudhuri (1968)

Table 3. Percentage germination (± standard error) of *Puccinellia nuttalliana* seeds after 25 days at varied osmotic concentrations (Macke and Ungar, 1971).

	Osmotic potential (-bars)				
	4	8	12	16	25 days
Na_2So_4	83.6 ± 2.0	79.2 ± 2.0	40.0 ± 2.8	1.6 ± 0.8	0.0
$NaHCO_3$	88.0 ± 1.6	87.2 ± 1.6	31.0 ± 5.1	2.8 ± 1.2	0.0
NaCl	90.0 ± 3.2	78.8 ± 2.8	34.0 ± 4.2	5.2 ± 0.8	0.0
Ethylene glycol	72.4 ± 2.8	77.2 ± 4.0	10.0 ± 1.6	2.0 ± 1.2	0.0
Distilled water	86.0 ± 2.0				

indicate that sodium carbonate was more toxic at the germination stage than other inorganic salts or oganic solvents. Sodium chloride was found to be least inhibitory to seed germination. Prisco et al. (1978) found that NaCl and Na_2SO_4 were equally inhibitory to germination of *Sorghum bicolor*. The difference in response between halophytes and some glcophytes indicates a possibly fundamental difference in response to specific ions. The more general osmotic response of halophytes is probably of survival value to widely occurring inland halophytic species of North America since they are exposed to different combinations of ions under field conditions (Ungar, 1972, 1974a).

3. Hormones

Three taxa in the genus *Suaeda*, *S. maritima* var. *macrocarpa*, *S. maritima* var. *flexilis*, and *S. depressa* were found to contain reduced endogenous cytokinin concentrations when seeds were treated with 0.85 M NaCl (Ungar and Boucaud, 1975; Boucaud and Ungar, 1976) (Fig. 1). Germination was stimulated by treatments with 10^{-3} M GA_3, but not with kinetin. It may be true as suggested by Kahn (1971) that cytokinins act as intermediary growth regulators, mediating the process of germination. Exogenous kinetin treatments did not stimulate germination in the salinity treatments containing the above-mentioned species of *Suaeda*, but GA_3 served to alleviate part of the dormancy induced by salt stress. However, it may be that kinetin was not the proper form of cytokinin to alleviate this dormancy. Several researchers have reported that kinetin treatments stimulate seeds of *Lactuca sativa* and *Triticum aestivum* to germinate at lower water potentials than untreated controls (Kaufmann and Ross, 1970; Odegbaro and Smith, 1969). These results conflict with studies of halophytes including *Spergularia media*, *Salicornia europaea*, *Suaeda* sp. by Ungar and Binet (1975), Boucaud and Ungar (1976), Ungar (1977a), and Ungar (1978a) in which kinetin did not stimulate germination, but GA_3 was found to be stimulating in each case. These data indicate that more than one mechanism may be involved in breaking dormancy that is induced by low water potentials. The response of halophytes to GA_3 indicates a somewhat different mechanism may be involved here than in the case of the less salt tolerant glycophytes.

4. Protein and amino acids

Hadas (1976) found that reducing the water potential of the media in which seeds of *Cicer arietinum* and *Vicia faba* were germinated caused a decrease in germination rates and suggested that a reduction in enzymatic activity could accompany a reduction in meristematic activity and radicle emergence. Cotyledons of *Vigna sinensis*, which were exposed to −4.3 bars osmotic stress, accumulated over ten times more soluble amino acids after a nine day germination period than did the distilled water controls. Prisco and Vieira (1976) concluded that the chief inhibitory effect of salinity on seed protein reserve mobilization was due to the inhibition of the translocation of hydrolysis products rather than to an inhibition of protease activity. It was found that in vivo protease activity was not inhibited by 0.1 M NaCl treatments, even though this salt concentration strongly inhibited in vitro protease activity (Filho and Prisco, 1978). It was suggested that the cytoplasmic organization of proteins in vivo protected enzymes from NaCl toxicity, while in vitro

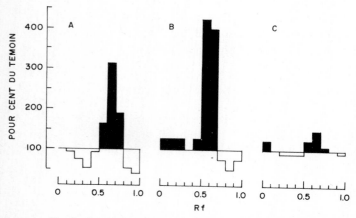

Fig. 1. Cytokinin activity from the acid hydrolyzed fraction of seeds of *Suaeda maritima* var. *macrocarpa*; (A) air-dry seeds, (B) seeds imbibed for two days in distilled water. (C) seeds imbibed for two days in 0.85 M NaCl (Ungar and Boucaud, 1975).

responses may have been an artifact produced in aqueous solutions. Studies with the salinity tolerance of enzymes from shoots of halophytes and glycophytes indicate almost no differences in response to salt stress that was related to the tolerance of the plants from which enzyme extracts were made (Austenfeld, 1974, 1976; Flowers, 1975; Fowers et al., 1976; Flowers et al., 1977; Cavalieri and Huang, 1977). This prompted Flowers (1972a, 1972b), Greenway and Osmond (1972), and Hall et al. (1974) to conclude that the cellular enzymes of halophytes may not be in direct contact with high salt concentrations. Salts may be sequestered in vacuoles, membrane bound vesicles may contain enzymes or salt tolerant enzymes may exist in vivo.

Increase in salt stress has been associated with a rise in endogenous proline concentrations in plants (Goas, 1971; Goas et al., 1970; Treichel, 1975; Stewart and Lee, 1974). Stewart and Lee (1974) and Flowers (1972a, 1975) suggest that proline accumulation in the cytoplasm serves as a counterbalance to excess salts which may be stored in the cell vacuole, bringing about osmotic adjustment. Bar-Nun and Poljakoff-Mayber (1977) found that concentrations of 100 μM proline alleviated the inhibitory effects of 120 mM NaCl on *Pisum sativum* seed germination. Phenylalanine and aspartic acid (100 μM) also stimulated germination of seeds in 120 mM NaCl.

5. Temperature-salinity interactions

Ungar (1978a) has indicated that interactions between temperature, salinity, and seed germination optima exist in halophytes. Recent work by Okusanya (1977) with *Crithmum maritimum* indicates that alternating temperatures of 5° C and 15° C, 5° C and 25° C, and 15° C and 25° C produced significantly higher germination percentages at all salinities tested than did constant temperatures. No seeds germinated at constant temperature treatments of 5° C or 25° C. Maximum germination was found in treatments containing 0 to 20% seawater. Higher seawater concentrations, 30% to 50%, were inhibitory. Okusanya (1977) concluded that high seed germination percentages in the spring were due

to both reductions in soil salinity and the marked diurnal fluctuations in temperature under field conditions.

Zohar et al. (1975) found an interaction between temperature and osmotic stress on the level of germination of *Eucalyptus occidentalis* seed. When seeds were exposed to -9.1 bars at 15° C only 13% of the seeds germinated, while at 30° C 73% of the seeds germinated. Distilled water treatments only varied 10% in their final germination percentages at these temperatures. Seeds of *E. occidentalis* germinated in the dark under osmotic stress had low germination percentages, 10%, removal of dark stress yielded 25% germination, removal of osmotic stress, 46%, and removal of both stresses, 77%, indicating an additive effect. These results support other earlier data reporting temperature-salinity effects, on the germination of halophyte seeds (Binet and Combes, 1961; Langlois 1961, 1966; Malcolm, 1964; Binet, 1964, 1968; Ungar, 1965; Springfield, 1966, 1970; Ungar and Hogan, 1970; Rivers and Weber, 1971; Onnis and Mazzanti, 1971). The exact nature of these synergistic effects are not clearly understood. It may be that osmotic adjustment is more rapid at one temperature regime than at another, causing an apparent change in salt tolerance. High temperatures may stimulate more rapid ion uptake and in this way cause an apparent increase in specific ion toxicity or osmotic inhibition of enzymatic processes. Further studies looking into endogenous changes in salt concentrations of seed and the influence of this factor on enzymatic activity would assist in explaining these interactions.

6. Ecotypes

Variations in the salt tolerance of seeds from closely related plant species or varieties and from different populations or ecotypes within a particular plant species have been reported in the literature (Dewey, 1960; Workman and West, 1967, 1969; Bazzaz, 1973; Kingsbury et al., 1976). These data indicate that the source from which seeds were obtained may be very critical in determining their germination response when exposed to saline conditions.

147

Evidence is available which indicates that the level of salt tolerance varies within different populations of seeds from a single species or within species complexes within a genus. It is most probable that genetic selection has taken place for increased salt tolerance in the evolution of at least some taxa found growing in both saline and non-saline environments (Clarke and West, 1969; Bazzaz, 1973; Kingsbury et al., 1976).

Marked differences in the level of seed germination response to varying salinity concentrations were reported for taxa within the genus *Agropyron* (Dewey, 1960). The most highly salt tolerant species, *Agropyron intermedium*, had 74.6% germination in 1.8% NaCl, while the least salt tolerant, *A. subsecundum*, had no germination under these conditions. Populations of *A. intermedium* varied in their response to 1.8% NaCl concentrations, ranging in their germination response from 31.4% in var. *Greenar* to 74.6% germination in the var. *Amur*. These results indicate that some evolutionary selection for salt tolerance has taken place in the response of seeds from different populations.

Clarke and West (1969) found that the date of seed collection for *Kochia americana* was significantly correlated with the response of seeds to salinity treatments. Seeds collected on 29 September 1967 had a maximum salt tolerance and germinated in up to 10.0% NaCl, while collections from 11 November 1967 tolerated up to 8.0% NaCl. Some of the differences in germination response could be related to the soil salinity concentrations at the collection localities. On the other hand, Workman and West (1967, 1969) also reported ecotypic differentiation in four populations of *Eurotia lanata*, but the levels of salt tolerance within populations at the germination stage did not correlate well with field soil salinity levels. Actually, seeds collected from the LaSal population were the most salt tolerant, germinating in up to 4.0% NaCl concentrations, and yet the source of these seeds were plants growing in the least saline location.

Bazzaz (1973) investigated the salt tolerance of seeds collected from three *Prosopis farcta* populations inhabiting southern saline and northern non-saline soils of Iraq. Seeds were treated with NaCl solutions ranging from 0 to −18 bars. Little difference was observed in the germination response of seeds from these three populations in the 0 to −9 bar salinity range. The northern Mosul population was more sensitive to salinity stress than the two sources of seed collected from saline soils. Germination dropped to 27% in the Mosul population when treated with −18 bars NaCl, in comparison to 64% germination in the seeds from the Basra population. These data indicate that interpopulation variation exists in the response of seeds to salinity stress. Radicle length and seedling respiration of plants from the three populations also differed in their response to salinity treatments.

Cavers and Harper (1967) studied the adaptations to salinity of *Rumex crispus* populations from maritime shingle beaches and compared their behavior with populations of inland *R. crispus* from north Wales. Seeds from both populations were soaked in seawater for 28 days and then returned to distilled water to germinate, resulting germination percentages were over 86% for both populations. Exposure of seeds to an eighty day soaking period in seawater completely inhibited germination in both populations. Seed collections from maritime plants proved to be less sensitive to salinity treatments of 25% seawater than were inland plants, but maritime seeds were more sensitive to the higher, 50% seawater concentrations. No single sharp difference in response to seawater was found between the maritime and inland populations of *R. crispus*, but seeds of the maritime form remained floating longer in seawater and after soaking in seawater they germinated more rapidly than did the inland form when seeds were returned to distilled water.

Dotzenko and Dean (1959) reported that seeds obtained from five varieties of *Medicago sativa* varied considerably in their response to −12 bars osmotic pressure. Germination of seeds for these different cultivars ranged from a low of 1.2% in variety Buffalo to 78.4% in the variety Caliverde when they were exposed to −12 bars osmotic stress.

Halophytes such as *Uniola paniculata* have broad distributional ranges on the Atlantic and Gulf coasts. Seneca (1972) found that seed sources from ten populations throughout this range had different stratification requirements. Over 75% of the seeds from two populations on the Atlantic coast of Florida germinated without a stratification treatment, while germination of seeds collected from North Carolina and Virginia populations had lower germination percentages, ranging from 6% to 26%. Germination of the more northern populations were strongly stimulated by cold treatments. There was no difference in the salt tolerance response of seedlings from the ten populations studied.

No significant differences in seed germination responses to salinities ranging from 0 to 1.75% NaCl were found for seed from clones of *Typha latifolia*, *T. angustifolia*, and intermediates (McMillan, 1959). Apparently, there was no specific selection for salt tolerance at the germination stage in this hybrid complex of *Typha*. *Typha angustifolia* was more salt tolerant during the growth stage of development than was *T. latifolia* and hybrid clones appeared to be intermediate in their response to salinity.

Wells (1959) studied the germination response of seeds from two closely related desert species of tobacco, *Nicotiana attenuata* and *N. trigonophylla* and found that they were strongly inhibited by osmotic stress of -11.8 bars. At -7.1 bars *N. trigonophylla* had four times greater germination, 68%, than did *N. attenuata*, 17%. Soils in the habitat of *N. trigonophylla* were higher in salinity, -7.1 bars at saturation than were soils from those habitats containing *N. attenuata*, -1.8 bars, indicating that the germination and seedling growth responses of *N. attenuata* may have evolved as adaptations to the saline soil conditions in which this species was found growing. Kingsbury et al. (1976) reported that populations of the winter annual, *Lasthenia glabrata*, varied in their salt tolerance at the germination stage. The northern subspecies had the greatest salt tolerance at the germination stage of development.

Binet and Boucaud (1968) demonstrated that two varieties, possibly ecotypes of *Suaeda maritima*, var. *macrocarpa*, and var. *flexilis*, as well as *S. splendens*, and *S. fruticosa* were highly salt tolerant. *Suaeda maritima* var. *macrocarpa* had no dormancy and germinated rapidly, but the other taxa within the genus *Suaeda* apparently had some level of dormancy. A mechanical dormancy induced by the testa was the primary factor influencing germination and it could be alleviated by scarification treatments. Boucaud (1962) found that the most dormant form of *S. maritima* was var. *vulgaris*, with 86% of the seeds not germinating after a 60 day stratification treatment. A comparison of seed proteins for the vars. *macrocarpa*, *flexilis*, and *vulgaris* by disc gel electrophoresis indicated that they were closely related, with similarity coefficients ranging from 88.4% to 97.6% (Ungar and Boucaud, 1975). A close relationship, based on protein banding patterns, between other taxa of *Suaeda*, *S. maritima* var. *macrocarpa* and *S. salsa*, was reported by Hekmat-Shoar et al. (1978). Billard and Binet (1975) have also found high similarities using disc gel electrophoresis for seed proteins between three taxa of *Atriplex* from the Normandy coast.

Salinity levels of 0.85 M NaCl caused a sharp reduction in cytokinin concentrations in both *S. maritima* var. *macrocarpa* and var. *flexilis* (Boucaud and Ungar, 1976). Both mechanically induced dormany and dormancy caused by osmotic stress were relieved by treatments with 10^{-3} M gibberellic acid (Boucaud and Ungar, 1973, 1976).

Some evidence is available indicating that natural selection is working in highly saline habitats to select for the most highly salt tolerant biotypes. The data supporting this concept are not overwhelmingly strong, and some reports indicated a rather random response in salt tolerance at the germination stage for populations found in a wide range of saline habitats. It might be more beneficial for seed to germinate at reduced salinities to insure growth and reproduction of offspring. In inland saline habitats low salinities at the germination stage could insure completion of the life cycles of annual species, whereas, if seeds were highly salt tolerant germination would occur under conditions that might be too severe for growing plants. Many marine plants produce seed that will not germinate while floating in seawater, thus insuring that floating seed will not produce seedlings that cannot anchor in soil.

7. Seed dimorphism and polymorphism

One possible selective advantage for the development of somatic dimorphism or polymorphism in seeds is in the regulation of resource allocation by plants to the different seed types, which permits a direct response to changing environments (Harper et al., 1970, Harper, 1977). Various morphological forms that have evolved in species demonstrating some type of seed polymorphism may have different ecological roles, represented by differences in dispersal behavior and dissimilarities in germination requirements, assuring that seed germination occurs over extended periods of time and under varied microenvironmental conditions (Cavers and Harper, 1967; Harper et al., 1970; Ungar, 1971, 1978b; Maun and Cavers, 1971a, b; Baskin and Baskin, 1976; Harper, 1977; Flint and Palmblad, 1978). Germination polymorphism is of significant advantage to species growing in variable environments because it provides alternate temporal and spatial germination situations. Thus, preventing a single local change in microenvironmental conditions from eliminating an entire plant population.

A number of plant species found growing in saline environments have developed some form of seed dimorphism or polymorphism (Beadle, 1952; Salisbury, 1958; König, 1960; Dalby, 1962; Sterk, 1969a, 1969b; Ungar, 1971, 1978b, Drysdale, 1973; Grouzis et al., 1976). One advantage that plants with polymorphic seeds have in unstable environments, with changing soil salinity stress, could be that this morphological characteristic is associated with physiological differences which provide for multiple germination periods. An entire population of plants might be eliminated by increased salinity stress and with only a single highly synchronized germination period no seed reserve would be available. Both *Atriplex triangularis* and *Salicornia europaea* have prolonged germination periods under field conditions (McMahon and Ungar, 1978; Ungar et al., 1979). These extended germination periods, running from February through June and sometimes later into the summer, are probably related to both the response of dimorphic seeds in these species and fluctuating soil salinity stress under field conditions, which at times may be beyond the salt tolerance of these organisms (Ungar, 1971; McMahon and Ungar 1978; Ungar, 1978a,b). The larger seed produced by both *S. europaea* and *A. triangularis* are apparently less dormant and germinate more rapidly than smaller seeds. Grouzis et al. (1976) found that a stratification treatment was stimulatory to the germination of seeds of *Salicornia patula* under saline conditions, and that small seeds were more sensitive to salinity stress, needing both a light and stratification treatment to break seed dormancy.

Each flower of *S. europaea* produces a single seed, and in an inland North American population, the median flower at a node produced a large seed (0.78 mg), while the two lateral flowers produced small seeds (0.24 mg) (Ungar, 1978b). Figure 2 indicates the pattern of seed dimorphism for this inland North American population of *S. europaea*. Ungar (1978b) found that small seeds of *S. europaea* had both a light and scarification requirement similar to that found in *S. patula* described by Grouzis et al. (1976). However, the major difference in germination response was in the increased rate of germination rather than in final germination percentages (Table 4). Small seeds of *S. europaea* had lower germination percentages than large seeds and were apparently less salt tolerant (Fig. 3).

Early germination is of selective advantage in an environment which is usually characterized by ris-

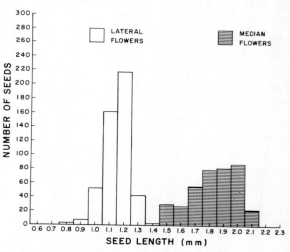

Fig. 2. Variation in length between seeds of *Salicornia europaea* that originate on median and lateral flowers (Ungar, 1978b).

Table 4. Germination percentages for small seeds of *Salicornia europaea* under varying conditions of light, nutrients, and cold stratification (Ungar, 1978b).

Treatment	No stratification				21-day stratification			
	Days				Days			
	7	14	21	28	7	14	21	28
Distilled water								
Light	9	17	18	22	80	84	84	84
Dark	5	11	16	19	14	22	47	48
Nutrient solution								
Light	10	20	20	26	75	80	80	80
Dark	6	18	19	21	—[a]	—[a]	—[a]	—[a]

[a] no data collected

ing salinity stress from spring through summer. The roots of the earliest germinating seeds have a better opportunity to penetrate the most highly saline upper few centimeters of soil prior to extreme changes in salinity stress and this increases the chance for plants to survive. Slower growing seedlings, developing from small seeds, are at a competitive disadvantage because there roots remain in the suface soil for a longer period and seedlings of these plants would succumb more rapidly than seedlings arising from large seeds, with sharp increases in salt stress. Baker (1974) demonstrated that *Atriplex patula* var. *hastata* plants arising from larger seeds had faster root and shoot growth than seedlings from small seeds which germinated at the same time. After two months of growth there was no difference in the size of seedlings developing from the two seed types. Similar observations have been made with *Atriplex triangularis* (McMahon, Riehl and Ungar, unpublished data).

8. Conclusions

The behavior of halophyte seeds under field conditions is still not well understood. More research is needed to determine the length of time seeds are stored in the soil and how this seed bank affects community diversity. Seed dimorphism is prevalent in several halophytic taxa, but whether this is a direct response to environmental stimuli is not known. Although we know that salt stress induces dormancy in seeds, we do not know the exact nature of this inhibition. It is not clearly understood whether the direct effect of salt stress is due to an endogenous low water potential or specific ion interference or a combined interference with hormonal balance, inhibition of enzyme production or activity, disturbance of membrane integrity, direct inhibition of DNA transcription or RNA translation, or a combination of these factors. Future research should be aimed at solving some of these very interesting problems in halophyte seed germination.

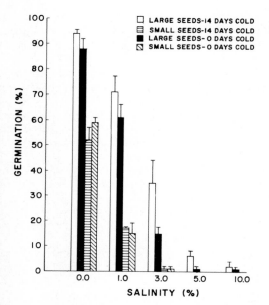

Fig. 3. The influence of 0-day and 14-day stratification periods on the percentage germination of large and small *S. europaea* seeds at different salinities (Ungar, 1978b).

Acknowledgement.

This research was supported in part by National Science Foundation research grants DEB-76-00444, DEB-79-27236 and by Ohio University Research Committee Grant No. 578.

Literature cited

Albregts, E.E. and Howard, C.M. 1973. Influence of temperature and moisture stress from sodium chloride salinization on okra emergence. *Agron. J.* 65: 836–837.

Austenfeld, F-A. 1974. Der Einfluss des NaCl und anderer Alkalisalze auf die Niträtreduktaseaktivitat von *Salicornia europaea* L. Z. *Pflanzenphysiol.* 71: 288–296.

Austenfeld, F-A. 1976. The effect of various alkaline salts on the glycolate oxidase of *Salicornia europaea* and *Pisum sativum* in vitro. *Physiol. Plant.* 36: 82–87.

Baker, H.G. 1974. The evolution of weeds. *Ann. Rev. Ecol. Syst.* 5: 1–24.

Bar-Nun, N. and Poljakoff-Mayber, A. 1977. Salinity stress and the content of proline in roots of *Pisum sativum and Tamarix tetragyna*. *Ann. Bot.* 41: 173–179.

Baskin, J.M. and Baskin, C.C. 1976. Germination dimoprhism in *Heterotheca subaxillaris* var. *subaxillaris*. *Bull. Torr. Bot. Club* 103: 201–206.

Bazzaz, F.A. 1973. Seed germination in relation to salt concentration in three populations of *Prosopis farcta*. *Oecologia* 13: 73–80.

Beadle, N.C.W. 1952. Studies in halophytes. I. The germination of the seed and establishment of the seedlings of five species of *Atriplex* in Australia. *Ecology* 33: 49–62.

Billard, J-P. and Binet, P. 1975. Physio-écologie des *Atriplex* des milieux sableux littoraux. *Bull. Soc. Bot Fr.* 122: 51–64.

Binet, P. 1964. La germination des semences des halophytes. *Bull. Soc. Fr. Physiol, Végét.* 10: 253–263.

Binet, P. 1968. Dormances et aptitude a germer en milieu sale chez les halophytes. *Bull. Soc. Fr. Physiol. Végét.* 14: 115–124.

Binet, P. and Boucaud, J. 1968. Dormance, levée de dormance et aptitude a germer en milieu sale dans le genre *Suaeda* Forsk. *Bull. Soc. Fr. Physiol. Végét.* 14: 125–132.

Binet, P. and Combes, M.R. 1961. Action de la temperature et de l'eau de mer sur la germination des graines de *Cochlearia anglica*. L. *C.R. Acad. Sci. Paris* 253: 895–897.

Boucaud, J. 1962. Étude morphologique et écophysiologique de la germination de trois variétés de *Suaeda maritima* Dum. *Bull. Soc. Linn. Norm.* 3: 63–74.

Boucaud, J. and Ungar, I.A. 1973. The role of hormones in controlling the mechanically induced dormancy of *Suaeda* spp. *Physiol. Plant.* 29: 97–102.

Boucaud, J. and Ungar, I.A. 1976. Hormonal control of germination under saline conditions of three halophytic taxa in the genus *Suaeda*. *Physiol. Plant.* 37: 143–148.

Breen, C.M., Everson, C., and Rogers, K. 1977. Ecological studies on *Sprorobolus virginicus* (L.) Kunth with particular reference to salinity and inundation. *Hydrobiologia* 54: 135–140.

Cavalieri, A.J. and Huang, A.H.C. 1977. Effect of NaCl on the in vitro activity of malate dehydrogenase in salt marsh halophytes of the U.S. *Physiol. Plant.* 41: 79–84.

Cavers, P.B. and Harper, J.L. 1967. The comparative biology of closely related species living in the same area. IX. *Rumex*: the nature of adaptation to the seashore habitat. *J. Ecol.* 55: 73–82.

Chapman, V.J. 1974. *Salt Marshes and Salt Deserts of the World*. J. Cramer, Bremerhaven, West Germany.

Choudhuri, G.N. 1968. Effect of soil salinity on germination and survival of some steppe plants in Washington. *Ecology* 49: 465–471.

Clarke, L.D. and West, N.E. 1969. Germination of *Kochia americana* in relation to salinity. *J. Range Manag.* 22: 286–287.

Dalby, D.H. 1962. Chromosome number, morphology and breeding behavior in the British Salicorniae. *Watsonia* 5: 150–162.

Darwin, C. 1857. On the action of sea-water on the germinton of seeds. *J. Linn. Soc.* 1: 130–140.

Dewey, D.R. 1960. Salt tolerance of twenty-five strains of *Agropyron*. *Agron. J.* 52: 631–635.

Dietert, M.F. and Shontz, J.P. 1978. Germination ecology of a Maryland population of saltmarsh bulrush (*Scirpus robustus*). *estuaries* 1: 164–170.

Dotzenko, A.D. and Dean, J.G. 1959. Germination of six alfalfa varieties at three levels of osmotic pressure. *Agron. J.* 51: 308–309.

Drysdale, F.R. 1973. Variation of seed size in *Atriplex patula* var. *hastata* (L.) Gray. *Rhodora* 75: 106–110.

Filho, E.G. and Prisco, J.T. 1978. Effects of NaCl salinity in vivo and in vitro on the proteolytic activity of *Vigna sinensis* (L.) Savi cotyledons during germination. *Revta. brasil. Bot.* 1: 83–88.

Flint, S.D. and Palmblad, I.G. 1978. Germination dimorphism and developmental flexibility in the ruderal weed *Heterotheca grandiflora*. *Oecologia* 36: 33–43.

Flowers, T.J. 1972a. Salt tolerance in *Suaeda maritima* (L.) Dum. The effect of sodium chloride on growth, respiration, and soluble enzymes in a comparative study with *Pisum sativum* L. *J. Exptl. Bot.* 23: 310–321.

Flowers, T.J. 1972b. The effect of sodium chloride on enzyme activities from four halophyte species of Chenopodiaceae. *Phytochemistry* 11: 1881–1886.

Flowers, T.J. 1975. Halophytes. In *Ion Transport in Cells and Tissues*, eds. D.A. Baker and J.L. Hall, pp. 309–334. North Holland Publ. Co., Amsterdam.

Flowers, T.J. and Hall, J.L. 1976. Properties of membranes from the halophyte *Suaeda maritima*. II. Distribution and properties of enzymes in isolated membrane fractions. *J. Exptl. Bot.* 27: 673–689.

Flowers, T.J., Troke, P.F. and Yeo, A.R. 1977. The mechanism of salt tolerance in halophytes. *Ann. Rev. Plant. Physiol.* 28: 89–121.

Goas, M. 1971. Métabolisme azoté des halophytes. *C.R. Acad. Sci. Paris* 272: 414–417.

Goas, M., Goas, G. and Larher, F. 1970. Métabolisme azoté des halophytes: utilisation de l'acide glutamique ^{14}C-3-4 par les jeunes plantes d'*Aster tripolium* L. *C.R. Acad. Sci. Paris* 271: 1763–1766.

Greenway, H. and Osmond, C.B., 1972. Salt response of enzymes from species differing in salt tolerance. *Plant Physiol.* 49: 256–259.

Grouzis, M., Berger, A. and Heim, G. 1976. Polymorphisme et germination des graines chez trois espèces annuelles du genre *Salicornia*. *Oecologia Plant.* 11: 41–52.

Hadas, A. 1976. Water uptake and germination of leguminous seeds under changing external water potential in osmotic solutions. *J. Exptl. Bot.* 27: 480–489.

Hall, J.L., Yeo, A.R., and Flowers, T.J. 1974. Uptake and localization of rubidium in the halophyte *Suaeda maritima*. *Z. Pflanzenphysiol.* 71: 200–206.

Harper, J.L. 1977. *Population Biology of Plants.* Academic Press, New York.

Harper, J.L. Lovell, P.H. and Moore, K.G. 1970. The shapes and sizes of seeds. *Ann. Rev. Ecol. Syst.* 1: 327–356.

Hekmat-Shoar, H., Billard, J-P. and Boucaud, J. 1978. Étude chimiotaxonomique des proteines séminales de 4 Suaeda d'Iran. Relations phylogéniques au sein du genre Suaeda Forsk. *C.R. Acad. Sci. Paris* 286: 53–56.

Joshi, A.J. and Iyengar, E.R.R. 1977. Germination of *Suaeda nudiflora* Moq. *Geobios* 4: 267–268.

Kahn, A.A. 1971. Cytokinins: permissive role in seed germination. *Science* 171: 853–859.

Kaufmann, M.R. and Ross, K.J. 1970. Water potential, temperature, and kinetin effects on seed germination in soil and solute systems. *Amer. J. Bot.* 47: 413–419.

Kingsbury, R.W., Radlow, A., Mudie, P.J., Rutherford, J. and Radlow, R. 1976. Salt stress in *Lasthenia glabrata*, a winter annual composite endemic to saline soils. *Can. J. Bot.* 54: 1377–1385.

König, D. 1960. Beitrage zur Kenntnis de deutschen Salicornien. *Mitt. Flor-soziol. Arbeits.* 8: 5–58.

Langlois, J. 1961. Aspects morphologiques et ecophysiologiques de la germination de trois variétés de *Salicornia herbacea* L. *Bull. Soc. Linn. Norm.* 2: 160–174.

Langlois , J. 1966. Étude comparée de l'aptitude a germer des graines de *Salicornia stricta* Dumort, *Salicornia disarticulata* Moss, and *Salicornia radicans* Moss. *Rev. Gén. Botanique* 73: 25–39.

Lesko, G.L. and Walker, R.B. 1969. Effect of seawater on seed germination in two Pacific atoll beach species. *Ecology* 50: 730–734.

Macke, A. and Ungar, I.A. 1971. The effect of salinity on seed germination and early growth of *Puccinellia nuttalliana*. *Can. J. Bot.* 49: 515–520.

Malcolm, C.V. 1964. Effects of salt, temperature, and seed scarification on germination of two varieties of *Arthrocnemum halocnemoides*. *J. Roy. Soc. West. Aust.* 47: 72–74.

Mann, M.A. and Cavers, P.B. 1971a. Seed production and dormancy in *Rumex crispus*. I. The effects of removal of cauline leaves at anthesis. *Can. J. Bot.* 49: 1123–1130.

Maun, M.A. and Cavers, P.B. 1971b. Seed production and dormancy in *Rumex crispus*. II.The effects of removal of various proportions of flowers at anthesis. *Can. J. Bot.* 49: 1841–1848.

Mayer, A.M. and Poljakoff-Mayber, A. 1975. *The Germination of Seeds*. Pergamon Press, New York.

McMahon, K.A. and Ungar, I.A. 1978. Phenology, distribution and survival of *Atriplex triangularis* Willd. in an Ohio salt pan. *Amer. Midl. Nat.* 100: 1–14.

McMillan, C. 1959. Salt tolerance within a *Typha* population *Amer. J. Bot.* 46: 521–526.

Millington, A.J., Burvill, G.H. and Marsh, B. 1951. Soil salinity investigations. Salt tolerance, germination and growth tests under controlled salinity conditions. *J. Agric. West Australia* 28: 198–210.

Milton, W.E.J. 1939. Occurrence of buried viable seed in soils at different elevations and on a salt marsh. *J. Ecol.* 27: 149–159.

Odegbaro, O.A. and Smith, O.E. 1969. Effect of kinetin, salt concentration and temperature on germination and early seedling growth of *Lactuca sativa* L. *Amer. Soc. Hort. Sci.* 94: 167–170.

Okusanya, O.T. 1977. The effect of sea water and temperature on the germination behavior of *Crithmum maritimum*. *Physiol. Plant.* 41: 265–267.

Onnis, A. and Bellettato, R. 1972. Dormienza e alotolleranza in due specie spontanee di *Hordeum* (*H. murinum* L. e *H. marinum* Huds). *G. Bot. Ital.* 106: 101–113.

Onnis, A. and Mazzanti, M. 1971. *Althenia filiformis* Petit: Azione della temperatura e dell'acqua di mare sulla germinazione. *G. Bot. Ital.* 105: 131–143.

Prisco, J.T. and Vieira, G.H.F. 1976. Effects of NaCl salinity on nitrogenous compounds and proteases during germination of *Vigna sinensis* seeds. *Physiol. Plant.* 36: 317–320.

Prisco, J.T., Souto, G.F. and Ferreira, L.G.R. 1978. Overcoming salinity inhibition of sorghum seed germination by hydration-dehydration treatment. *Plant Soil* 49: 199–206.

Redmann, R.E. 1974. Osmotic and specific ion effects on the germination of alfalfa. *Can. J. Bot.* 52: 803–808.

Rivers, W.G. and Weber, D.J. 1971. The influence of salinity and temperature on seed germination in *Salicornia bigelovii*. *Physiol. Plant.* 24: 73–75.

Ryan, J., Miyamoto, S. and Stroehlein, J.L. 1975. Salt and specific ion effects on germination of four grasses. *J. Range Management* 28: 61–64.

Saini, G.R. 1972. Seed germination and salt tolerance of crops in coastal alluvial soils of New Brunswick, Canada. *Ecology* 53: 524–525.

Salisbury, E.J. 1958. *Spergularia salina* and *Spergularia marginata* and their heteromorphic seeds. *Kew Bull.* 1: 41–51.

Seneca, E.D. 1969. Germination response to temperature and salinity of four dune grasses from the outer banks of North Carolina. *Ecology* 50: 45–53.

Seneca, E.D. 1972. Germination and seedling responses of Atlantic and Gulf coasts populations of *Uniola paniculata*.

Amer. J. Bot. 59: 290–296.

Springfield, H.W. 1966. Germination of fourwing saltbush seeds at different levels of moisture stress. *Agron. J.* 58: 149–150.

Springfield, H.W. 1970. Germination and establishment of fourwing saltbush in the southwest. *U.S.D.A. Forest Serv. Res. Paper* RM55: 1–48.

Sterk, A.A. 1969a. Biosystematic studies of *Spergularia media* and *S. marina* in the Netherlands. I. The morphological variability of *S. media. Acta Bot. Neerl.* 18: 325–338.

Sterk, A.A. 1969b. Biosystematic studies on *Spergularia media* and *S. marina* in the Netherlands. III. The variability of *S. media* and *S. marina* in relation to the environment. *Acta Bot. Neerl.* 18: 561–577.

Stewart, G.R. and Lee, J.A. 1974. The role of proline accumulation in halophytes. *Planta* 120: 279–289.

Strogonov, B.P. 1964. *Physiological Basis of Salt Tolerance of Plants.* IPST, Jerusalem.

Strogonov, B.P. 1974. *Structure and Function of Plant Cells in Saline Habitats.* John Wiley and Sons, New York.

Treichel, S. 1975. Der Einfluss von NaCl auf die Prolinekonzentration verschiedener Halophyten. *Z. Pflanzenphysiol.* 76: 56–68.

Uhvits, R. 1946. Effect of osmotic pressure in water absorption and germination of alfalfa seeds. *Amer. J. Bot.* 33: 278–285.

Ungar, I.A. 1962. Influence of salinity on seed germination in succulent halophytes. *Ecology* 43: 763–764.

Ungar, I.A. 1965. An ecological study of the vegetation of the Big Salt Marsh, Stafford County, Kansas. *Univ. Kansas Sci. Bull.* 46: 1–98.

Ungar, I.A. 1971. *Atriplex patula* var. *hastata* (L.) Gray seed dimorphism. *Rhodora* 73: 548–551.

Ungar, I.A. 1972. The vegetation of inland saline marshes of North America, north of Mexico. In *Grundfragen und Methoden in der Pflanzensoziologie,* ed. R. Tüxen, pp. 397–411. Dr. W. Junk Publ., The Hague.

Ungar, I.A. 1974a. Inland halophytes of the United States. In *Ecology of Halophytes,* eds. R. Reimold and W. Queen, pp. 235–305. Academic Press, New York.

Ungar, I.A. 1974b. The effect of salinity and temperature on seed germination and growth of *Hordeum jubatum. Can. J. Bot.* 52: 1357–1362.

Ungar, I.A. 1977a. Salinity, temperature, and growth regulator effects on seed germination of *Salicornia europaea* L. *Aquatic Bot.* 3: 329–335.

Ungar, I.A. 1977b. The effects of salinity and hormonal treatments on ion uptake in *Salicornia europaea. Bull. Soc. Bot. Fr.* (in press).

Ungar, I.A. 1978a. Halophyte seed germination. *Bot. Rev.* 44: 233–264.

Ungar, I.A. 1978b. Seed dimorphism in *Salicornia europaea* L. *Bot. Gaz.* 140: 102–108.

Ungar, I.A., Benner, D.K. and McGraw, D.C. 1979. The distribution and growth of *Salicornia europaéa* on an inland salt pan. *Ecology* 60: 329–336.

Ungar, I.A. and Binet, P. 1975. Factors influencing seed dormancy in *Spergularia media* (L.) C. Presl. *Aquatic Bot.* 1: 45–55.

Ungar, I.A. and Boucaud, J. 1975. Action des fortes teneurs en NaCl sur l'evolution des cytokinines au cours de la germination d'un halophyte: le *Suaeda maritima* (L.) Dum. var. *macrocarpa* Moq. *C.R. Acad. Sci. Paris* 281: 1239–1242.

Ungar, I.A. and Capilupo, F. 1969. An ecological life history study of *Suaeda depressa* (Pursh) Wats. *Adv. Front. Plant Sci.* 23: 137–158.

Ungar, I.A. and Hogan, W. 1970. Seed germination in *Iva annua* L. *Ecology* 51: 150–154.

Varshney, K.A. and Baijal, B.D. 1977. Effect of salt stress on seed germination of some pasture grasses. *Comp. Physiol. Ecology* 2: 104–107.

Waisel, Y. and Ovadia, S. 1972. Biological flora of Israel 3. *Suaeda monoica* Forsk. ex J.F. Gmel. *Israel J. Botany* 21: 42–52.

Waite, S. and Hutchings, M.J. 1978. The effects of sowing density, salinity and substrate upon the germination of seeds of *Plantago coronopus* L. *New Phytol.* 81: 341–348.

Ward, J.M. 1967. Studies in ecology of a shell barrier beach. III. Chemical factors of the environment. *Vegetatio* 15: 77-112.

Wells, P.V. 1959. An ecological investigation of ´two desert tobaccos. *Ecology* 40: 626–644.

Williams, M.D. and Ungar, I.A. 1972. The effect of environmental parameters on the germination, growth, and development of *Suaeda depressa* (Pursh) Wats. *Amer. J. Bot.* 59: 912–918.

Workman, J.P. and West, N.E. 1967. Germination of *Eurotia lanata* in relation to temperature and salinity. *Ecology* 48: 659–661.

Workman, J.P. and West, N.E. 1969. Ecotypic variation of *Eurotia lanata* populations in Utah. *Bot. Gaz.* 130: 26–35.

Yadava, R.B.R., Mehra, K.L., Magoon, M.L., Sreenath, P.R. and Pandey, R.M. 1975. Varietal differences in salt tolerance during seed germination of guar. *Indian J. Plant Physiol.* 18: 16–19.

Younis, A.F. and Hatata, M.A. 1971. Studies on the effects of certain salts on germination, on growth of root, and on metabolism. I. Effects of chlorides sulphates of sodium, potassium, and magnesium on germination of wheat grains. *Plant Soil* 13: 183–200.

Zid, E. and Bourkhris, M. 1977. Some aspects of salt tolerance of *Atriplex halimus* L. Multiplication, growth, mineral composition. *Oecol. Plant.* 12: 351–362.

Zohar, Y., Waisel, Y. and Karschon, R. 1975. Effects of light, temperature, and osmotic stress on seed germination of *Eucalyptus occidentalis* Endl. *Aust. J. Bot.* 23: 391–397.

CHAPTER 4

Aspects of salinity and water relations of Australian chenopods

M.L. SHARMA
Division of Land Resources Management, CSIRO,
Wembley, W.A. 6014, Australia

1. Introduction

Many plants belonging to the family Chenopodiaceae are indigenous to arid and semi-arid regions of the world. Various species of this family are dominant components of the shrub-steppe which occupy about six percent of the Australian land surface area of 4.34×10^6 km². However, their contribution to pastoral production is much more than their share of the occupied area. Chenopod plant communities lie mainly south of the Tropic of Capricorn in a region receiving predominantly winter rain in the range of 150 to 500 mm per annum. The most extensive areas occur in South Australia, western New South Wales and south-eastern Western Australia (Fig. 1). In Eurasia and the Great Basin of North America, the main chenopod-dominated areas are some 10–20° further from the equator than in Australia, and are subject to severe winters. In some cases, chenopod dominated vegetation is confined to highly saline situations.

From the viewpoint of providing forage for livestock, and controlling land erosion, the two most important genera are *Atriplex* and *Maireana*** as they are considered to be the mainstay of the pastoral industry of some arid parts of Australia (Jackson, 1958). Chenopods are noted for low transpiration rate, high water use efficiency and their resistance to drought and salinity (Sharma, 1977). These features in addition to their unusually

* Genus previously called *Kochia* (Wilson, 1975).

high protein content are the major reasons for their pastoral significance. The potential for growing these species under routine agronomic conditions, and their high yield potentials are encouraging signs for them to be adapted in many arid regions and on saline wastelands (Malcolm, 1969). Because of their ability to accumulate salts, they could be used to reclaim salted lands.

Despite the general acceptance of drought and salt tolerance of chenopods, comparatively little is known about their relative endurance under natural conditions. The mechanisms underlying these are not wel onderstood. This chapter reviews evidence of such tolerances and aspects of salinity and water relations of some of the most studied members of the family. It is not intended to be an extensive review, but rather an attempt to synthesize information which the author considers important in understanding the behaviour and performance of this unique group of plants under adverse soil and environmental conditions.

2. Salinity aspects of chenopods

I. Salt tolerance

Chenopods are considered highly salt tolerant as evidenced by their natural occurrence on salted-lands in dry environments. In culture solutions, they are reported to grow at very high electrolyte concentrations (e.g. Beadle, 1952; Black, 1960;

Tasks for vegetation science, Vol. 2 ed. by D.N. Sen and K.S. Rajpurohit
© *1982, Dr W. Junk Publishers, The Hague. All rights reserved. ISBN 90 6193 942 9*

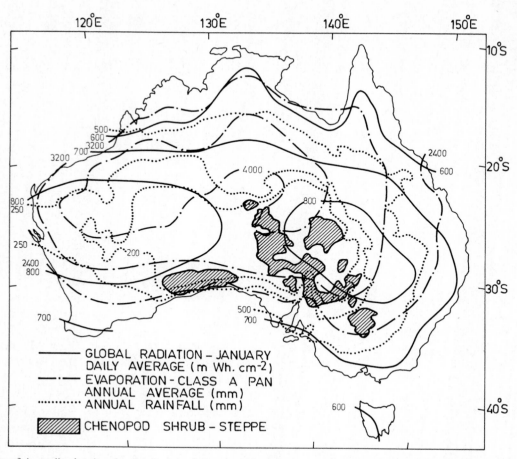

Fig. 1. Map of Australia showing the distribution of chenopod shrubsteppe (From Leigh, 1972). Average annual rainfall (mm) and evaporation from Class A pan (mm) and average global radiation (mW cm^{-2} d^{-1}) for a summer month (January) are also shown.

Chatterton et al., 1969). Black (1960) reported that seedlings of *Atriplex vesicaria* were successfully established in solutions of up to 1 M NaCl ($\approx$ −45 bars osmotic potential). Chatterton and McKell (1969) showed that seedlings of *Atriplex polycarpa* could grow in solutions of 39,000 ppm NaCl ($\approx$ −27 bars). However, germination of most *Atriplex* declines at higher osmotic potentials. For example, Beadle (1952) reported a severe reduction in germination of several *Atriplex* species at a osmotic potential of −9 bars, while Chatterton and McKell (1969) reported complete failure of germination of *A. polycarpa* at about −16 bars. Germination response of two common Australian *Atriplex* species to osmotic and matric potential effects is shown in Fig. 2. It appears, *Atriplex* differ considerably in their germination response to salinity, but their

germination is higher at equivalent levels of osmotic compared with matric potential. The increased salt tolerance of seedlings and mature plants is probably due to salt regulatory mechanisms within the leaf. This will be discussed later.

Atriplex are also reported to be tolerant to high concentrations of boron, which could be a problem in many saline conditions. Chatterton et al. (1969) reported little effect of boron on growth of *Atriplex polycarpa* in culture solutions containing up to 80 ppm boron, a concentration several hundred-times greater than the level of boron needed to satisfy trace element requirements.

II. Salt accumulation within plant

Halophytes survive at high electrolyte concen-

156

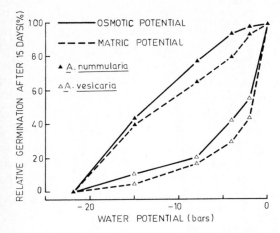

Fig. 2. Relative germination percentages of two *Atriplex* species over a range of 0 to −22 bars osmotic and matric potential. The osmotic and matric potentials were simulated by osmotic solutions of sodium chloride and polyethylene glycol (PEG – 20,000 mol wt.) (redrawn from Sharma, 1973).

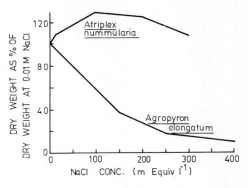

Fig. 3. Growth of a 'salt accumulator', *Atriplex nummularia* and a salt excluder, *Agropyron elongatum* at a range of sodium chloride concentrations (from Greenway and Osmond, 1970).

trations either by salt exclusion or by salt accumulation. The characteristics of salt accumulators (Hayward and Wadleigh, 1949) are: rapid salt uptake, regulation of internal salt concentration and tolerance of metabolic machinery to high electrolyte concentration. By considering these criteria, Greenway and Osmond (1970) suggested that *Atriplex* species fall into this category, although they did not find the isolated enzymes of *Atriplex* had any higher tolerance to electrolytes than those of salt sensitive *Phaseolus*.

Most Australian *Atriplex* and other members of the family Chenopodiaceae, are xerophytic halophytes that accumulate large quantities of electrolytes, mainly Na^+ and Cl^-, in their leaves and redistribute them into specific parts. Another example of a salt accumulator is *Tamarix* which concentrates salt into anatomically distinct parts called salt glands and secretes salt crystals to the surface of the leaf (Thomson and Liu, 1967). In *Atriplex*, the salt is concentrated in salt bladders or vesiculated hairs (Osmond et al., 1969; Pallaghy, 1970; Mozafar and Goodin, 1970).

Comparison of a salt accumulator, *Atriplex nummularia* with a salt excluder *Agropyron elongatum* (Fig. 3) indicates that *A. nummularia* grows more vigorously at high (0.1–0.2 M) than at low (0.01–0.1 M) sodium chloride concentrations.

Atriplex species contain high salt concentrations in their leaves even when grown at low electrolyte concentrations (1–10 m equiv l^{-1}), and this internal concentration increases steadily with increase in external concentration to 600 m equiv l^{-1} (Black, 1956; Ashby and Beadle, 1957). Black (1960) found that salt levels of mature leaves of *Atriplex vesicaria* maintained a gradient to the culture solution concentration of about −12 bars over the full concentration range of 0–1 M NaCl. The salt concentration of roots remained relatively unchanged at varying electrolyte concentrations (Fig. 4).

Suitability of *Atriplex* as dryland forage may be reduced because of their high salt content (Wood, 1925; Beadle, et al., 1957) which may lead to unpalatability (Leigh and Mulham, 1971) and to an

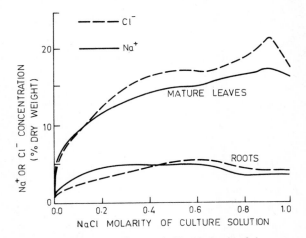

Fig. 4. Comparative sodium and chloride levels of the mature leaves and roots of *Atriplex vesicaria*, grown at varying concentrations of sodium chloride (drawn from Black, 1960).

increased water requirement by the grazing animal (Wilson, 1966). Earlier reports indicate that the leaf salinity of *Atriplex* is not related to soil salinity. For example, Beadle et al. (1957) observed that the chloride concentration in leaves of several *Atriplex* did not vary significantly even when the chloride content of soils varied from 19–9,200 ppm. Even at low soil salinity, *Atriplex* leaves contain large quantities of salt which fluctuate considerably with the season, being highest during summer and lowest in winter (Wood, 1925; Lachover and Tadmor, 1965). However, no satisfactory explanation has been advanced for such seasonal fluctuations. Wood (1925) suggested that the increase in salt content of leaves during summer may be due to an increase in the salt content of soil and also to a gradual accumulation of salt in the leaves, while Lachover and Tadmor (1965) implied that these increases were caused by decreases in soil water content.

In field conditions, Sharma et al. (1972) observed a twofold increase in Na^+ and Cl^- concentration of mature leaves of *A. vesicaria* and *A. nummularia* from wet winter conditions to dry summer conditions (Fig. 5). Salt concentration of leaves increased with increase in the electrolyte concentration of soil water. With a decrease in average osmotic potential of soil water from −2.4 to −7.4 bars, there was a corresponding increase in Cl^- concentration of *A. nummularia* leaves from 5.8 to 11.7%. However, this increase in leaf salinity did not occur when the osmotic potential of soil water was kept at high levels by irrigation. The Na^+ and Cl^- concentrations of *Atriplex* leaves were found to be correlated with the average osmotic potential of soil water and total water in the profile. *Sharma et al.* (1972) postulated that the build up of salinity in *Atriplex* leaves during summer is due not only to increased electrolyte concentration in the growing medium and to high transpiration rates (Greenway, 1965), but may also be associated with drought tolerance of these plants.

III. Ionic balance

There is usually a preferred uptake and transport of monovalent cations (Na^+ and K^+) over Ca^{++} in chenopods. For example, Osmond (1966) observed that total cation ($Na^+ + K^+ + Ca^{++}$) concentration in leaves of *Atriplex*, grown in solutions of 100 m equiv 1^{-1} of NaCl or KCl, was not significantly different from that when plants were grown in a solution of 100 m equiv 1^{-1} of $CaCl_2$. At high solution concentrations of $NaCl + KCl + CaCl_2$, uptake of Ca^{++} was depressed compared with

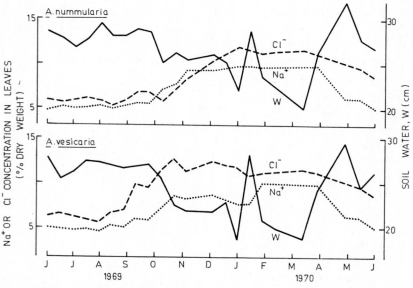

Fig. 5. Sodium and chloride concentrations in mature leaves of *Atriplex nummularia* and *A. vesicaria* and corresponding soil water content of Billabong clay during June 1969 to May 1970 (redrawn from Sharma et al., 1972).

uptake in $CaCl_2$ alone. This reflects a higher mobility of monovalent cations.

Sodium is an essential micro-nutrient for *Atriplex* plants (Brownell, 1965) and this requirement in culture solutions is satisfied at less than 0.01 mM Na^+. Growth responses to Na^+ are also commonly observed (Black, 1956; Ashby and Beadle, 1957). However, the preference of Na^+/K^+ during uptake is not fully understood. Most other species show preference to K^+ over Na^+ (Collander, 1941). A Na^+/K^+ ratio of 5.2 was found in the leaves of *A. nummularia* compared with that of 2 in *Hordeum vulgare*, when both were grown in standard culture solutions containing 0.1 M NaCl (Greenway and Osmond, 1970). Even higher Na^+/K^+ ratios are possible for xerophytic *Atriplex* (Ellern et al., 1974). When grown in culture solution, the K^+ concentration of *Atriplex* plants decreases to a lower level with increasing NaCl concentration. The uptake of sufficient K^+ to meet nutritional requirement is maintained even when the Na^+/K^+ ratio in culture solution is 100 (Black, 1960), and similarly, Na^+ uptake is maintained in solutions of very high KCl concentration (Osmond, 1966; Mozafar et al., 1970).

It is possible that Na^+ partially substitutes for deficient K^+ situations. Thus, high capacity for Na^+ uptake in the chenopods may be an essential part of the osmoregulatory mechanism rather than nutritional. However, even at high levels of K^+ preferential Na^+ uptake still occurs (Mozafar et al., 1970). It is unclear whether the selectivity of ion uptake is in young cells or in the ion transport from the export tissues (Greenway and Osmond, 1970).

In some 'salt accumulators', solutes might consist entirely of inorganic anions such as Cl^-, which would be balanced by mobile cations such as Na^+ and K^+. However, in *Atriplex* there is usually an excess uptake of $Na^+ + K^+$, over Cl^-, and this is balanced by oxalates (Osmond, 1967; Mozafar and Goodin, 1970; Ellern et al., 1974). The high concentration of oxalates in leaves may be partly responsible for the unpalatability of some of the chenopods, although this has been proven only in a few cases (Jones et al., 1970). Ellern et al. (1974) discount the likelihood that the oxalate content of *Atriplex* species is a serious poisoning hazard to stock.

IV. Partitioning of salt within plant

Several reports (e.g. Black, 1956, 1960; Charley, 1959; Sharma and Tongway, 1973; Wallace et al., 1971) show that in arid *Atriplex*, the accumulation of salt mainly occurs in the leaves, while the other components of plants contain rather small concentrations (Table 1). By contrast, Black (1956) noted that for *Atriplex hastata*, a species from salt marsh habitats, the salt concentration of roots was higher than in the shoots. This is an important difference, suggesting that uptake and partitioning of ions in *A. hastata* differs from that of halophytes from the arid regions (Black, 1960; Greenway and Osmond, 1970).

A common feature of *Atriplex* leaves is possession of bladder-like epidermal hairs, which are called salt bladders or vesicles or trichomes. These vesicles comprise large (100-200 μ diam.) vacuolated cells, resting on stalk cells. They are extremely rich in salt and play an important role in regulating electrolyte levels in the leaves (Osmond et al., 1969). The electrolyte content of the vesicles may represent about half of the total salt in the leaf (Pallaghy, 1970). This compartmentalization prevents excessive accumulation of salt in the mesophyll cells. Mozafar and Goodin (1970) reported

Table 1. Sodium chloride* concentration in various parts of field grown mature plants of two Atriplex species (Sharma and Tongway, 1973).

Plant parts		Percent sodium chloride	
		A. nummularia	*A. vesicaria*
Leaves			
	young	23.68	21.46
	mature	18.52	15.05
	senescing	15.58	12.01
Stems			
	fine	0.58	0.59
	main	0.28	0.22
Roots			
	fine	0.69	0.58
	main	0.41	0.33

* Chlorides are expressed as sodium chloride assuming that all the chlorides are associated with sodium.

that in *Atriplex halimus*, the solute concentration of vesicles increased considerably when the plants were grown at increasing concentrations, although the salt concentration of the expressed sap of leaves did not change appreciably. The estimated osmotic potential of the vesicles due to sodium and potassium chlorides ranged from −110 to −559 bars in the control and salt treated (0.1 M NaCl + 0.1 M KCl) plants respectively.

Accumulation of salt in the vesicles occurs against a concentration gradient and is stimulated by light (Osmond et al., 1969). This indicates that it is an active process. Possible sites of active electrolyte transport in the mesophyll-stalk-epidermal-bladder cell system are elegantly demonstrated in Fig. 6. The vesicles are denser in young than in mature leaves. How they grow and replace the older cells is not fully understood. It is possible that, as salt content of these cells reaches a critical level,

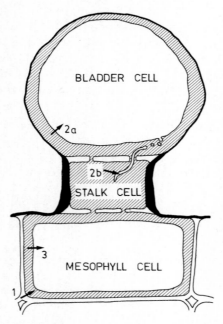

Fig. 6. Possible sites of active transfer of chloride from the external solution to the bladder vacuole. (1) Active transfer of chloride ions across the plasma membrane of the mesophyll cells followed by diffusive movement through the symplasm to the bladder vacuole. (2a) Active chloride uptake to the bladder vacuole from the cytoplasm. (2b) Active transfer of chloride to the membrane system of the stalk cytoplasm and movement to the bladder vacuole by means of the small vesicles found in the bladder cytoplasm. (3) Active transport of chloride across the tonoplast of the mesophyll cells (redrawn from Pallaghy, 1970.)

they collapse and may form a mat on the leaf surface. Alternatively, as new vesicles develop from the epidermis they push up and break the stems of the older ones thus depositing their contents upon the leaf surface. The highly reflective salt crust of leaves usually observed in the field evidently forms when vesicles collapse and dry. Charley (1959) observed that the salt leached by rain from leaves of *A. vesicaria* growing in the field, may amount to about half of the total salt content of these plants. However, Sharma et al. (1972) did not observe any significant leaching of salt from two *Atriplex* species when leaves were washed for 24 hours. Similarly, Mozafar and Goodin (1970) observed that prolonged washing did not decrease the salt concentration in the vesicles of *A. halimus*.

Regulation of solute concentration in the leaves of many chenopods, through a dilution effect, is achieved by increased succulence. It involves an elongation of hypodermal cells. With increasing salinity, higher succulence is reported to be induced in a salt marsh halophyte, *A. hastata* than in an arid halophyte, *A. vesicaria* (Black 1956, 1958), although subsequent studies report salinity-induced succulence in many arid species (e.g. Gale and Poljakoff-Mayber, 1970; Gates, 1972). Thus, both salt secretion and succulence are operative in chenopods in regulating the salt concentration in leaves, otherwise salts might accumulate in the cell walls and abolish turgor of individual cells, as has been suggested for non-halophytes by Oertli (1968).

3. Water relations of chenopods

I. Drought tolerance

The natural occurrence of various chenopods in low rainfall areas where high levels of radiation and temperature are encountered, testifies to their drought tolerance (Fig. 1). The drought tolerance of these plants has long been recognized (Wood, 1923), as they are reported to remain leafy and viable under highly stressed environmental conditions (Gates and Muirhead, 1967), although quantitative data on the magnitude and extent of drought tolerance became available much later.

The performance and survival of chenopods under exceedingly low soil water potentials has been demonstrated experimentally both in laboratory as well as in field. While most agricultural plants exhibit permanent wilting at a soil water potential of −15 bars, chenopods are able to survive when the water potential of the entire soil profile decreases to much lower than −15 bars (e.g. Jones, 1969; Sharma, 1976a).

Under laboratory conditions, Palmer et al. (1964) measured transpiration of *A. nummularia* at soil water potentials much lower than −15 bars, and the permanent wilting occurred at about −60 bars. Moore et al. (1972) reported that *Atriplex confertifolia* could transpire even when plant water potential had reached as low as −115 bars. Under controlled environment, Slatyer (1972) reported transpiration and photosynthesis of *Atriplex spongiosa* and *A. hastata* at a plant water potential as

low as −35 bars. *A. polycarpa* was able to recover when exposed to a plant water potential of −69 bars, while mesophytic plants died (Sankary and Barbour, 1972). In nature *A. polycarpa* exhibited plant water potentials as low as −58 bars.

Although no data on soil and plant water potentials are given, field studies comparing drought tolerance of *A vesicaria* and *Maireana sedifolia* by Carrodus and Specht (1965) suggest that both species were able to reduce soil water to water potentials much lower than −15 bars. Jones (1969) observed that over a two year period, the soil profile beneath *A. vesicaria* pasture was at soil matric potentials much lower than −15 bars for approximately 48 percent of the time during the first year, and 64 percent of the time during the second year.

Soil water potential profiles developed under two *Atriplex* communities, *A. vesicaria* and *A. nummularia* for a drying period are shown in Fig. 7. At

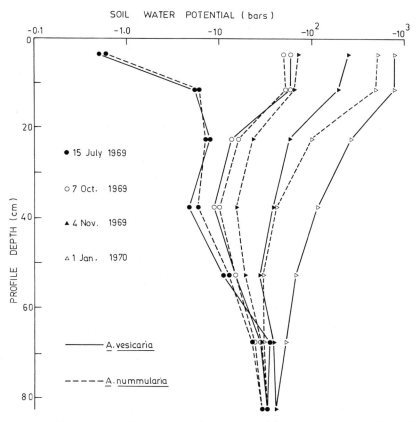

Fig. 7. Soil water potential profiles under *Atriplex vesicaria* and *A. nummularia* during a drying cycle from July 1969 to January 1970 (redrawn from Sharma, 1976a).

depths below 60 cm the soil water potential was always < −15 bars and did not vary appreciably. Both species were able to survive exceedingly low water potentials. For example on January 1, 1970, one of the driest periods, the water potentials observed in 7.5–15 cm, 15–30 cm and 30–45 cm depth intervals under *A. vesicaria* were −800, −250 and −120 bars, and under *A. nummularia* −500, −100 and −40 bars respectively. Under these harsh environmental conditions, when all the intervening vegetation was dead or became dormant, both *Atriplex* species retained at least 30% of their original leaf material. The lowest plant water potentials (early morning) were −110 bars for *A. vesicaria* and −46 bars for *A. nummularia*.

II. Water use and efficiency of water use

Early studies (Wood, 1923) indicated that relative transpiration rates of *Atriplex* and *Maireana* species were considerably low and that these rates were not appreciably affected by environmental factors. Wood suggested that this not only helps these plants to survive under drought but also leads to high water use efficiency. Because of relatively low transpiration rates and suggested high photosynthetic efficiency, various members of Chenopodiaceae have been reported (Richardson and Trumble, 1928; Trumble, 1932) to have lower transpiration ratios (amount of water transpired per unit of dry matter produced) than lucerne and other vegetation (Table 2).

Subsequent studies (e.g. Palmer et al., 1964; Slatyer, 1970a,b; Moore et al., 1972) have confirmed low transpiration rates in *Atriplex*. Unique anatomical features (West, 1970) and characteristic of some arid plants which have the 'C$_4$' photosynthetic pathway (Osmond et al., 1969) suggests that they have a very efficient photosynthetic apparatus compared with most mesophytic plants which have the Calvin cycle 'C$_3$' pathway. The C$_4$ plants are associated with high light and high temperatures and they generally occur in tropical and arid environments. Osmond (1970) lists various differentiating features of C$_3$ and C$_4$ plants, emphasizing that C$_4$ plants are equipped with a much more efficient apparatus for CO$_2$ assimilation. All

Table 2. Transpiration ratios of two cultivated chenopods, *Atriplex vesicaria* and *A. semibaccatum* in comparison with other vegetation.

Vegetation	Transpiration ratio		
Atriplex vesicaria	354 ⎤	Richardson and	
Danthonia penicillata	622 ⎦	Trumble (1928)	
Atriplex semibaccatum	359	(361)* ⎤	Trumble (1932)
Lucerne (alfalfa)	966	(756) ⎦	
Corn	317 ⎤		
Alfalfa	626		
Brome grass	880	Lee (1942)**	
Evergreen trees	140		
Deciduous trees	825 ⎦		

* Values in brackets are the average of three different seasons.
** From Physical Climatology by W.D. Sellers, 1965. Univ. of Chicago Press, Chicago, pp. 91.

the endemic Australian *Atriplex* species of arid and semi-arid regions, tested so far, have the C$_4$ pathway (Hatch et al., 1972). This may help to explain the higher water use efficiency of these plants (Table 2).

Under controlled environmental conditions, Slatyer (1970a) compared photosynthesis, growth and transpiration of *A. spongiosa*, a C$_4$ species with those of *A. hastata*, a C$_3$ species. *A. spongiosa* maintained higher photosynthetic rates than *A. hastata* while transpiration rates were consistently lower in *A. spongiosa*. The water use efficiency, both on a single-leaf and whole-plant basis, was much greater for *A. spongiosa*. Figure 8 not only shows substantially greater efficiency of *A. spongiosa* than *A. hastata*, but also the relative constancy of this ratio during the experimental period. The photosynthetic and transpiration rates of these species were also compared under water stress (Fig. 9) by Slatyer (1972). Again, the water use efficiency of *A. spongiosa* was substantially greater than that of *A. hastata* due primarily to high rates of photosynthesis but partially to lower rates of transpiration.

Although the C$_4$ photosynthetic pathway is generally thought to confer higher photosynthetic rates, particularly in arid environments (Larcher, 1975), recent work (e.g. Mooney et al., 1976; Ehleringer and Björkman, 1977) shows that some C$_3$ desert plants have high photosynthetic rates

162

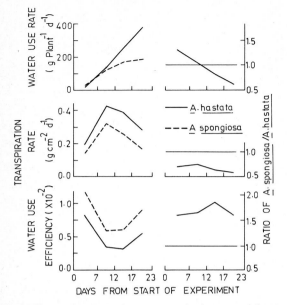

Fig. 8. Changes in water use rates, transpiration rates and water use efficiency of *Atriplex spongiosa* and *A. hastata* during the experimental period. The experiments were performed under controlled environments when water supply was non-limiting (redrawn from Slatyer, 1970a).

missonis, but the water use efficiency of *A. leucophylla* was the highest.

Caution must be exercised in extending short-term laboratory results to field conditions. In regard to productivity of a community, community structure and the distribution of leaf area may become equally or even more important than specific leaf photosynthetic efficiency. Therefore, both laboratory as well as field measurements are needed to provide a realistic picture of performance of a plant community. Unfortunately, field data are scarce.

Jones et al. (1970) showed that at the seedling stage, *Atriplex nummularia* and *A. vesicaria* have high relative growth rates, while very high net assimilation rates were observed for *A. vesicaria*. Photosynthesis measurements of whole plants growing under non-limiting water conditions in the field suggested that *A. nummularia* is highly efficient and compared favourably with *Zea mays*. Unfortunately, transpiration rates were not measured concurrently, and therefore, water use efficiency could not be calculated.

Moore et al. (1972), using laboratory and field measurements, showed that under comparable conditions, transpiration rates of *A. confertifolia* were lower than those of *Eurotia lanata*. However, in the later part of summer (under soil water stress conditions) transpiration of *E. lanata* decreased markedly while *A. conferifolia* maintained active

which could be equal to, or higher than the most productive C_4 plants. De Jong (1978a) compared photosynthetic and transpiration rates of a coastal C_4 species, *Atriplex leucophylla* with those of three C_3 coastal species. The maximum photosynthetic rate of *A. leucophylla* was up to 25% lower than that of the most productive species, *Ambrosia cham-*

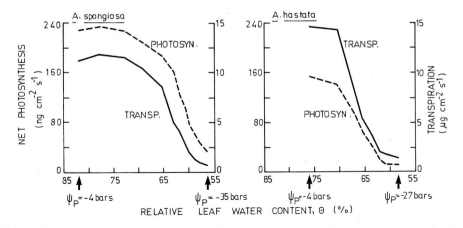

Fig. 9. Effect of imposed water stress on rates of net photosynthesis and transpiration of *Atriplex spongiosa* and *A. hastata*. Water stress was measured in terms of relative leaf water content (θ). Leaf water potentials (ψ_p) corresponding to the highest and lowest θ are shown (redrawn from Slatyer, 1972).

transpiration. A. progressive decrease in transpiration with decreasing soil water or soil water potential was observed, suggesting that there is no 'threshold' value of water potential at which transpiration would decrease abruptly.

Based on glasshouse work, Palmer et al. (1964) concluded that transpiration responses of *A. nummularia* and cotton *(Gossypium hirsutum)* to increasing soil water stress, were similar. The transpiration of both species was reduced over a water potential range of −10 to −60 bars. The fall of evapotranspiration ratio (ratio of actual evapotranspiration of that of potential (measured under unlimited water supply conditions)) with decreasing soil water was very similar for the two species. However, under identical climatic conditions, both the potential as well as actual evapotranspiration rates of cotton were about one and a half times higher than those of *Atriplex*, particularly under high soil water conditions. This difference was ascribed to physiological differences of plants, but no differential physiological attributes were elucidated.

The paucity of field data on water use of chenopods has been stressed (Sharma, 1977). Comparative evapotranspiration of *A. nummularia* and *A. vesicaria*, measured in the field (Sharma, 1976a, 1977) showed that, for both species the evapotranspiration-ratio (ratio of actual evapotranspiration to potential (measured by class A pan) decreased linearly with decreasing soil water or logarithm of soil water potential. Data for *A. vesicaria* (Fig. 10) suggest that there is no 'threshold' value of soil water or water potential at which evapotranspiration decreases abruptly. *A. nummularia* used water more conservatively than *A. vesicaria*, particularly under ample soil water conditions. It appears *A. nummularia* has a better control over water use, and is able to conserve water to sustain itself during dry periods. Measurements showed that average fortnightly evapotranspiration for both species ranged from 0.31 mm d^{-1} to 2.5 mm d^{-1} during the cool season and from 0.09 mm d^{-1} to 4.26 mm d^{-1} during the warm season. The mean fortnightly evapotranspiration ratio for the two communities did not exceed 0.80. The range of this ratio for *A. vesicaria* was 0.72–0.22 for the cool

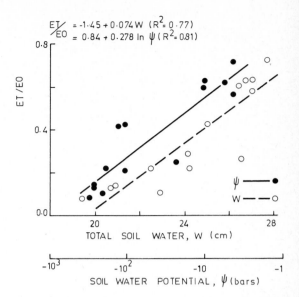

Fig. 10. Relationships between (a) average ET/EO (ratio of evapotranspiration and evaporation from Class A pan) and average soil water store (W), and between (b) ET/EO and average integrated soil water potential in the profile under *A. vesicaria* community (redrawn from Sharma, 1977).

season and 0.42–0.05 for warm season. Similar ranges for *A. nummularia* were 0.69–0.17 and 0.35–0.06 respectively.

Relatively conservative water use by *A. nummularia* was further observed when soil water profiles were measured under several plant communities (*A. nummularia* alone, *A. nummularia* + grassland, grassland). A much higher soil water content in the profile under *A. nummularia* alone, at the end of about two years was founds (Sharma, unpublished). Conservative water use by an *A. vesicaria* community was apparent (Hasick, 1979) by the fact that during spring (when soil water supply was ample) only 50% of the available radiation energy (net radiation − soil heat flux) was utilized in evapotranspiration, and this was further reduced to about 25% during dry autumn period. However, this lower water use may partly be due to lower density of plants.

III. Effects of salinity on water relations

The benefical effects of salinity on the water relations of halophytes have been recognized for

164

some time, however, it has become clearer only recently, how salinity reduces transpiration and leads to higher water use efficiency in chenopods.

Ashby and Beadle (1957) observed that dry matter production and moisture content of leaves of *A. nummularia* increased with increasing salinity of the growing medium. This increase in leaf moisture was explained as a simple hypertrophy resulting from the increased osmotic persuure. Additions of NaCl to the growing medium of *A. nummularia* and *A. inflata* resulted in significantly lower water use per unit dry weight produced. These results are in agreement with data for some other halophytes.

Black (1956) reported that leaves of both *A. vesicaria* and *A. hastata* exhibited greater succulence when NaCl was added to the growing medium. This succulence is a result of increased water content and increased cell size. An accelerated rate of leaf thickening in *A. hastata* was induced by NaCl, the most thickening occurring when a high salt concentration (0.6 M) was used (Black, 1958). Although high salt concentration reduced overall growth, this apparently did not reduce turgor pressures necessary to maintain succulence. Greenway et al. (1966) suggest that this increased succulence for *Atriplex*, by virtue of inducing diluting effects, would be of advantage in the regulation of ion concentration within the plant.

Improvements in the water relations of *A. nummularia* were observed as a result of salinity treatment, in the range of 10–400 m equiv 1^{-1} (Gates, 1972). Salinity resulted in an increased succulence and higher water content of tissues, and in conservation of water through reduction in the transpiration rate. Salinity not only enhanced the capacity of plants to withstand wilting, but also reduced tissue injury during water shortage. Gates suggested that these improvements in water relations would be an ecological advantage for these plants to adapt under drought stress. However, for a North American desert saltbush, *A. polycarpa*, Chatterton and McKell (1969) did not find any correlation between the water percentage of the top growth and the osmotic potential of the culture solution.

In an endeavour to find explanations for the improved water relations of *Atriplex* due to salinity, subsequent studies not only measured leaf succulence, transpiration and photosynthetic rates as a function of increased salinity, but also the resistance to water vapour and CO_2 flow in the leaf (e.g. Gale and Poljakoff-Mayber, 1970; Kaplan and Gale, 1972; De Jong, 1978b). For certain glycophytes, stomatal closure has been reported when grown under salinized conditions, even when plants are osmotically adjusted and leaf turgor is maintained (Gale et al., 1967). A similar situation could exist in chenopods.

Increase in leaf area and succulence of *A. halimus* was observed at relatively low NaCl concentrations (Gale and Poljakoff-Mayber,1970). This was accompanied by an increase in stomatal resistance to water vapour and CO_2. Relatively low salt concentrations reduced mesophyll resistance to CO_2, but high concentrations (osmotic potential < -9 bars) increased mesophyll resistance. It was suggested that the responses to salinity tended to counteract one another in their overall effect on growth, and the dominance of factors will depend on the level and type of salinity and possibly climatic factors.

Later, Kaplan and Gale (1972) observed improved water relations of *A. halimus* with moderate NaCl salinity (< -10 bars osmotic potential), when grown under high evaporative demand. Turgor pressure and percentage saturation of leaves increased with increasing salinity, while transpiration and photosynthetic rates decreased. However, the decrease in transpiration was proportionately much larger than in photosynthesis (Table 3). There was a large mesophyll resistance to water vapour loss in the control (3.4 s cm^{-1}) and this resistance was considerably increased (to 13.9 s cm^{-1}) in the salinized plants. Due to this large increase in resistance the transpiration rate was reduced by 50%, but the photosynthetic rate was reduced by only 20%, thus improving the transpiration ratio by about 40%. The mesophyll resistance was identified as a significant factor, through which improvement in water relations was achieved by reducing transpiration in a manner which only slightly reduced photosynthesis. The greater turgor lead to larger leaf area which was available for increased photosynthesis and growth.

Improvements in water relations at high salinity

165

Table 3. Photosynthesis, transpiration, transpiration ratio and diffusion resistance of leaves of control and NaCl-salinized *Atriplex halimus* plants grown under conditions of high evaporative demand (Kaplan and Gale, 1972).

Parameter	Control culture solution	Salinized culture solution (−10 bars)
Photosynthesis (mg/dm^{-2} h^{-1})	21	17
Transpiration (mg/dm^{-2} h^{-1})	1183	578
Transpiration ratio	56	34
Stomatal resistance to water vapour (s cm^{-1})	2.1	5.8
Mesophyll resistance to water vapour (s cm^{-1})	3.4	13.9

levels (0 to −20 bars) were observed by De Jong (1978b) for a coastal C_4 species, *Atriplex leucophylla*, compared with those of a C_3 coastal species, *Abronia maritima* (Fig. 11). Increasing salinity generally decreased net photosynthesis and leaf conductance (inverse of resistance), but in-

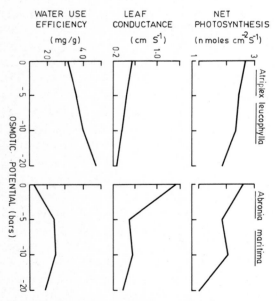

Fig. 11. Salinity responses of *Atriplex leucophylla*, a C_4 species and *Abronia maritima*, a C_3 species measured in terms of net photosynthesis, leaf conductance to water vapour and water use efficiency. The salinity levels were created by artificial sea salt (redrawn from De Jong, 1978b).

creased water use efficiency. *A. leucophylla* had higher mesophyll conductance and water use efficiency at all salinity levels than the C_3 plant. The higher water use efficiency of C_4 relative to C_3 plants suggests that the C_4 photosynthetic pathway would be an advantage in saline situations.

These explanations appear logical on unit leaf area basis, but in the absence of any field data, it is uncertain how the overall transpiration and photosynthesis of a plant community will be modified by salinity, since aerodynamic resistance and community structure play an important role in the performance of a community.

IV. Mechanisms of drought tolerance

There has been a controversy in defining the term 'drought tolerance' (Parker, 1968). Here, it is defined as the ability of plants to withstand drought of soil and their aerial environment without cessation of the living process, and their ability to perform normally when the conditions are favourable. There are various ways in which plants perform these xerophytic functions. Here, some of these features found in chenopods are explored.

Several attempts have been made to compare drought tolerance of various chenopods within this group (e.g. Wood, 1923; Carrodus and Specht, 1965; Slatyer, 1970b, 1972; Sharma 1976a; Williams, 1979) and also with other plants (e.g. Trumble, 1932; Palmer et al., 1964; Moore et al., 1972). All chenopods do not necessarily have the same mechanisms for drought tolerance and there is also a considerable variability with respect to the degree of development of a drought tolerant feature. Important considerations which emerge from various studies are discussed below.

(i) Defoliation under drought
Reduction of surface area for transpiration by defoliation (leaf drop) is the usual mechanism by which xerophytes conserve water under stress conditions. Chenopods do exhibit defoliation, however, not only does the extent of defoliation differ among various members (Carrodus and Specht, 1965; Sharma, 1976a; Williams, 1979), but also the rate of recovery and survival (Leigh and Mulham,

1971). For example, *A. vesicaria* defoliates much more than *A. nummularia*, while *A. nummularia* is capable of recovering from much more severe defoliation. Carrodus and Specht (1965) found that although the rate of defoliation of *A. vesicaria* and *Maireana sedifolia*, under imposed water stress was about the same, the rate of recovery was much higher for *M. sedifolia*. Thus, a marked ability to defoliate in response to drought may be an advantage for *A. vesicaria*, provided droughts are not so severe that plants may not recover.

(ii) Deep and extensive rooting systems
Possession of deep and extensive rooting system by many arid plants is considered an advantage in their drought tolerance, because they can exploit much larger volume of soil. Not all chenopods are deep rooted, for example the most widely spread Australian chenopod, *A. vesicaria* is very shallow rooted, possessing a fibrous rooting system with up to 90% of its roots confined to the top 15 or 30 cm soil layer (Jones and Hodgkinson, 1970; Sharma, 1976a). By contrast, other common chenopods such as *A. nummularia*, *M. sedifolia* and *M. pyramidata* have much deeper rooting systems. Osborn et al. (1935) observed roots of *M. sedifolia* spreading laterally to 5 m with strong secondary roots penetrating to more than 2 m deep. Similar roots were observed for *M. pyramidata* (Williams, 1979). Jones and Hodgkinson (1970) found *A. nummularia* sending its roots down to 3.5 m and displaying an extensive root system. Chenopods such as *A. vesicaria* with shallow rooting system must rely on other drought resistance mechanism.

(iii) Ability of plants to extract water at low water potentials
Like many arid plants, chenopods can extract water from the soil when water potential is much lower than at the conventional permanent wilting point (Fig. 7). This increases the range of water available for plant use. Water extraction is performed by the ability of the plant to maintain a lower leaf water potential compared to that in the soil. The accumulation of solutes in the leaves of most chenopods helps in this process (Sharma et al., 1972) because of regulated salt accumulation in the leaf cells, which

enables these plants to develop low water potential and maintain turgor. In a field study, Sharma (1976a) showed that water extracted by an *A. vesicaria* community amounted to about 41% more than the water available within conventionally accepted limits of water potential (-0.3 to -15 bars).

(iv) Role of vesicles in water economy
The presence of a reflective salt crust on the leaf surface is observed when vesicles collapse, this is likely to reduce heat and radiation load on the plant. However, reduction in water loss is likely to be minimal, since radiation is usually not the limiting factor in arid regions. Chenopod leaves have been shown to absorb considerable amounts of water from an atmosphere of high humidity (Wood, 1925). This water absorption may mainly be by the vesicles since they are extremely rich in solutes and thus develop exceedingly low water potentials. However, it is unlikely that this water enters the mesophyll cells. Carrodus and Specht (1965) observed that final defoliation of *A. vesicaria* and *M. sedifolia* exposed to drought, was not affected when plants were exposed to high humidity. Mozafar and Goodin (1970) could not observe any water absorption by the vesicles of *A. halimus*. Even if we accept that water absorption by leaves does occur in nature, either in the liquid form from dew or in the vapour form from the atmosphere, it is argued that this would be of little practical value. In arid and semiarid regions of Australia, most dew-falls and periods of high humidity are reported to coincide with rainy winter months (Sharma, 1976b), when drought is usually not a problem.

(v) Resistance to desiccation
Relationships between relative leaf water content and plant water potential of two *Atriplex* species are compared with those of three other species (Fig. 12). Both *Atriplex* species show much larger resistance in the reduction of their leaf water content compared with all the other plants, including the highly drought resistant *Acacia aneura*. Slatyer (1960) claimed that this resistance to desiccation is directly related to drought tolerance of plants. However, it is not certain whether such low water

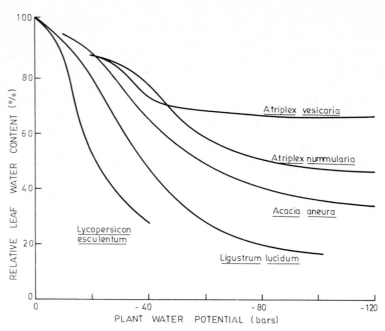

Fig. 12. Desiccation resistance of five species, measured as a drop in relative leaf water content with increasing water stress (decreasing plant water potential). Data of *Lycopersicon esculentum*, *Ligustrum lucidum* and *Acacia aneura* from Slatyer, 1960 and for *Atriplex vesicaria* and *A. nummularia* from Sharma, 1976a.)

potentials are a real advantage to the plant or are merely a result of their survival when water potentials are lowered to such levels (Parker, 1968).

(vi) Resistance of water vapour and CO_2 flow
Generally observed lower transpiration rates for chenopods, even under non-limiting water supply conditions, may be an aid in conserving water which would help these plants during dry periods. This is due to their high leaf resistance to water vapour flow, since leaf resistance is inversely proportional to transpiration rate. The leaf resistance consists of cuticular and stomatal resistance. Slatyer (1970a) showed that under non-stress conditions stomatal resistance of an arid *Atriplex*, *A. spongiosa* (1.15 s cm^{-1}) was higher than a coastal species, *A. hastata* (0.75 s cm^{-1}) and also that this resistance increased to about 2.3 s cm^{-1} in *A. spongiosa* while it remained constant in *A. hastata*. Mature leaves of fully grown *A. nummularia* exhibit even higher resistance of about 5.5 s cm^{-1} (Sharma, unpublished data). Similarly, mature *A. vesicaria* is reported to have a high stomatal resistance (3 to 5 s cm^{-1}) under ample soil water conditions and this

resistance increases by an order of magnitude during dry periods (Hasick, 1979). The cuticular resistance of arid *Atriplex* (90 s cm^{-1} for *A. spongiosa*; 150 s cm^{-1} *A. nummularia*) are also considerably higher than that of *A. hastata* (22 s cm^{-1}). This suggests that water loss from arid chenopods during the night, when stomata are closed, would be negligible. The stomatal resistance of *Atriplex* not only increases with water stress but also with increasing salinity (Gale and Poljakoff-Mayber, 1970; De Jong, 1978b), thus reducing transpiration rates under these conditions.

Slatyer (1970a, 1972) demonstrated that *A. spongiosa* has much lower internal resistance (r_{int}) to CO_2 intake than *A. hastata* and also that these resistances do not change with stress. The 'r_{int}' represents not only a 'resistance' to liquid phase diffusion inside the cell but also 'resistance' associated with the photosynthetic processes. The measured values of r_{int} for *A. spongiosa* and *A. hastata* were 1.1 and 2.8 s cm^{-1}, while for wheat, cotton and maize they were, 3.0, 2.6 and 0.80 s cm^{-1} respectively. This lower r_{int} in *A. spongiosa* gives rise to higher photosynthetic rates which compares favourably with maize. However, with increasing stress it is expected that *A. spongiosa* would have higher photosynthetic efficiency, since r_{int} for maize increases rapidly with increasing water stress while for *A. spongiosa* it remains constant. Slatyer (1970a, 1972) suggested that the combination of relatively high stomatal resistance and relatively low r_{int} values are responsible for substantially higher water use efficiency in most arid chenopods. Similarly, higher water use efficiency observed for a coastal C_4 species, *Atriplex leucophylla* compared with that of three C_3 species, was also associated with relatively high stomatal resistance and low mesophyll resistance (De Jong, 1978a).

It should be appreciated that the resistances measured on unit leaf area basis cannot be directly translated for the functioning of a plant community. Simultaneous field measurements of photosynthetic and evapotranspiration rates are required to see how far these laboratory findings are applicable under real world situations.

4. Summary and conclusions

Most Australian chenopods occur in the arid and semi-arid regions, and are important both from pastoral and land stability viewpoints. These plants are halophytic xerophytes. They can withstand much more severe salinity and water stress at their seedling and mature stages than at germination. While germination is drastically reduced at −9 bars osmotic potential, seedlings can be established at a 1 M NaCl concentration ($\approx$ −45 bars). This higher resistance at seedling and later stages is due to development of salt regulatory mechanisms in the leaves.

The halophytic functions in chenopods are performed by salt accumulation and redistribution of salt into specific parts of the plant. Electrolytes are accumulated in the vesicles against a concentration gradient. This compartmentalization and accumulation of salt in vesicles reduces the excessive salt accumulation in mesophyll cells. Vesicles can be extremely rich in salt and may develop osmotic potentials down to −559 bars in an external 0.2 M solution. It is not fully understood how such a large electrolyte concentration gradient between vesicles and the mesophyll cells is maintained. The growth and replacement of vesicles is also not fully explained. Regulation of solute concentration in the leaves of chenopods may also be achieved by increased succulence through dilution effect.

Sodium is recognized as an essential micronutrient for chenopods, and growth responses to low sodium concentrations are commonly observed. Generally, there is a preferential uptake of Na^+ over K^+, even under well supplied K^+ conditions. There is no satisfactory explanation for this. Moreover, there is usually an excess uptake of $Na^+ + K^+$ over Cl^-, which is balanced by oxalates.

Large accumulation of solutes in chenopod leaves may be a necessary evil as regards their survival under high electrolyte concentration, however, it is not a desirable feature as regards the quality of pasture. High leaf salinity usually renders these plants less palatable to grazing animals and increases their water requirement. This is a problem particularly during summer droughts when animals rely on these plants and during this period the salt concentration of leaves is at its highest. Some of this salt can be leached by rains. The presence of soluble oxalates may also pose problems, as they may be toxic to some animals.

Like many arid plants, chenopods are able to tolerate soil water potentials much lower than − 15 bars. They are reported to transpire and photosynthesize at exceedingly low soil and plant water potentials. No 'threshold' soil or plant water potential can be defined at which an abrupt change in these processes occurs.

In general, chenopods transpire at lower rates compared to mesophytic plants, but photosynthesize at a much higher rate, thus having a considerably higher water use efficiency. Similar to some tropical and arid plant species, most arid chenopods have the 'C_4' photosynthetic pathway, compared with the Calvin cycle 'C_3' pathway found in mesophytic plants. The C_4 plants are equipped with a much more efficient apparatus for CO_2 assimilation. Some *Atriplex* compare favourably with *Zea mays* as regards their photosynthetic rates. High water use efficiency of chenopods is believed to occur due to the combination of relatively high stomatal resistance to water vapour, and low mesophyll resistance to CO_2. In some cases, salinity may increase this efficiency. There is a scarcity of field data on photosynthetic and transpiration rates to explain the behaviour of these plants on a community basis.

Although understanding of the mechanisms of drought tolerance is not complete, it is suggested that all chenopods do not have the same mechanisms and also that they differ considerably with respect to the degree of development of a drought tolerant feature. Various features such as defoliation under stress, deep rooting, ability to develop low plant water potentials, presence of vesicles, resistance to desiccation and consideration of resistances to water vapour and CO_2 transport, are discussed to explain the xerophytic behaviour of some of these plants.

Acknowledgements

I am indebted to several colleagues at the CSIRO, Division of Land Resources Management, and to Dr. H. Greenway, University of Western Australia, for helpful criticism of the manuscript.

Literature cited

Ashby, W.C. and Beadle, N.C.W. 1957. Studies in halophytes. III. Salinity factors in the growth of Australian saltbushes. *Ecology* 38: 344–352.

Beadle, N.C.W. 1952. Studies in halophytes. I. The germination of the seed and establishment of the seedlings of five species of *Atriplex* in Australia. *Ecology* 33: 49–62.

Beadle, N.C.W., Whalley, R.D.B. and Gibson, J.B. 1957. Studies in halophytes. II. Analytical data on the mineral constituents on three species of *Atriplex* and their accompanying soils in Australia. *Ecology* 38: 340–344.

Black, R.F. 1956. Effect of NaCl in water culture on the ion uptake and growth of *Atriplex hastata* L. *Aust. J. Biol. Sci.* 9: 67–80.

Black, R.F. 1958. Effect of sodium chloride on leaf succulence and area of *Atriplex hastata* L. *Aust. J. Bot.* 6: 306–321.

Black, R.F. 1960. Effects of NaCl on the ion uptake and growth of *Atriplex vesicaria* Heward. *Aust. J. Biol. Sci.* 13: 249–266.

Brownell, P.F. 1965. Sodium as an essential micronutrient element for a higher plant (*Atriplex vesicaria*). *Plant Physiol.* 40: 460–468.

Carrodus, B.B. and Specht, R.J. 1965. Factors affecting the relative distribution of *Atriplex vesicaria* and *Kochia sedifolia* (Chenopodiaceae) in the arid zone of South Australia. *Aust. J. Bot.* 13: 419–433.

Charley, J.L. 1959. Soil salinity-vegetation patterns in western New South Wales and their modification by over grazing. Ph.D. Thesis, University of New England.

Chatterton, N.J. and McKell, C.M. 1969. *Atriplex polycarpa*. I. Germination and growth as affected by sodium chloride in water cultures. *Agron. J.* 61: 448–450.

Chatterton, N.J., McKell, C.M., Goodin, J.R. and Bingham, F.T. 1969. *Atriplex polycarpa*. II. Germination and growth in water cultures containing high levels of boron. *Agron. J.* 61: 451–453.

Collander, R. 1941. Selective absorption of cations by higher plants. *Plant Physiol.* 16: 691–720.

De Jong, T.M. 1978a. Comparative gas exchange of four California beach Taxa. *J. Ecol.* (in press).

De Jong, T.M. 1978b. Comparative gas excange and growth responses of C_3 and C_4 beach species grown at different salinities. *J. Ecol.* (in press).

Ehleringer, J. and Björkman, O. 1977. Quantum yields for CO_2 uptake in C_3 and C_4 plants. *Plant Physiol.* 59: 86–90.

Ellern, S.J., Samish, Y.B. and Lachover, D. 1974. Soil and oxalic acid content of leaves of the saltbush *Atriplex halimus* in the Northern Negev. *J. Range Manage.* 27: 267–271.

Gale, J., Kohl, H.C. and Hagan, R.M. 1967. Changes in the water balance and photosynthesis of onion, bean and cotton plants under saline conditions. *Physiol. Pl.* 20: 408–420.

Gale, J. and Poljakoff-Mayber, A. 1970. Interrelations between growth and photosynthesis of salt bush (*Atriplex halimus* L.) grown in saline media. *Aust. J. Biol. Sci.* 23: 937–945.

Gates, C.T. 1972. Ecological response of the Australian native species, *Acacia harpophylla* and *Atriplex nummularia* to soil salinity: effects on water content, leaf area and transpiration rate. *Aust. J. Bot.* 20: 261–272.

Gates, C.T. and Muirhead, W. 1967. Studies of the tolerance of *Atriplex* species. I. Environmental characteristics and plant response of *A. vesicaria*, *A. nummularia* and *A. semibaccata*. *Aust. J. Ex. Agric. Anim. Husb.* 7: 39–49.

Greenway, H. 1965. Plant responses to saline substrates. IV. Chloride uptake by *Hordeum vulgare* as affected by inhibitors, transpiration and nutrients in the medium. *Aust. J. Biol. Sci.* 18: 249–268.

Greenway, H., Gunn, A. and Thomas, D.A. 1966. Plant response to saline substrates. VIII. Regulation of ion concentrations in saltsensitive and halophytic species. *Aust. J. Biol. Sci.* 19: 741–757.

Greenway, H. and Osmond, C.B. 1970. Ion relations, growth and metabolism of *Atriplex* at high external electrolyte concentrations. In *The Biology of Atriplex*, ed. R. Jones, pp. 49–56. CSIRO, Canberra, Australia.

Hasick, D.J. 1979. Heat and water vapour fluxes in an arid zone community. In *Studies of the Australian Arid Zone – 4. Chenopod Shrublands*, ed. R.D. Graetz and K.M.W. Howes. CSIRO, Melbourne (in press).

Hatch, M.D., Osmond, C.B. and Slatyer, R.O. 1972. *Photosynthesis and Photorespiration*. Wiley Interscience, Sydney.

Hayward, H.D. and Wadleigh, C.H. 1949. Plant growth on saline and alkali soils. *Adv. Agron*. 1: 1–38.

Jackson, E.A. 1958. A study of the soils and some aspects of the hydrology at Yudnapinna Station, South Australia. CSIRO Aust. Div. of Soils, Soils and Land Use Series No. 24.

Jones, R.M. 1969. Soil moisture and salinity under bladder saltbush (*Atriplex vesicaria*) pastures in the New South Wales Riverine Plain. *Aust. J. Exp. Agric. Anim. Husb*. 9: 603–609.

Jones, R. and Hodgkinson, K.C. 1970. Root growth of rangeland chenopods: morphology and production of *Atriplex nummularia* and *A. vesicaria*. In *The Biology of Atriplex*, ed. R. Jones, pp. 77–85. CSIRO, Canberra, Australia.

Jones, R., Hodgkinson, K.C. and Rixon, A.J. 1970. Growth and productivity in rangeland species of *Atriplex*. In *The Biology of Atriplex*, ed. R. Jones, pp. 31–42. CSIRO, Canberra, Australia.

Jones, R.J. Seawright, A.A. and Little, D.A. 1970. Oxalate poisoning in animals grazing the tropical grass *Setaria sphacelata*. *J. Aust. Inst. Agr. Sci*. 36: 41–43.

Kaplan, A. and Gale, J. 1972. Effect of sodium chloride salinity on the water balance of *Atriplex halimus*. *Aust. J. Biol. Sci*. 25: 895–903.

Lachover, D. and Tadmor, N. 1965. Qualitative studies of *Atriplex halimus* growing as plant fodder in the semi-arid conditions of Israel. 1. Seasonal variations in chloride content, in essential minerals and the presence of oxalates in the different parts of the plant. *Agron. Trop. Paris* 20: 309–322.

Larcher, W. 1975. *Physiological Plant Ecology*. Springer-Verlag, Berlin.

Leigh, J.H. 1972. Saltbush and other chenopod browse shrubs. In *The Use of Trees and Shrubs in the Dry Country of Australia*, ed. N. Hall, pp. 284–298. Australian Government Publishing Service, Canberra.

Leigh, J.H. and Mulham, W.E. 1971. The effect of defoliation on the persistence of *Atriplex vesicaria*. *Aust. J. Agric. Res*. 22: 239–244.

Malcolm, C.V. 1969. Use of halophytes for forage production on saline wastelands. *J. Aust. Inst. Agric. Sci*. 35: 38–49.

Moore, R.T., White, R.S. and Caldwell, M.M. 1972. Transpiration of *Atriplex confertifolia* and *Eurotia lanata* in relation to soil, plant and atmospheric moisture stress. *Can. J. Bot*. 50: 1411–1418.

Mooney, H.A., Ehleringer, J. and Berry, J.A. 1976. High photosynthetic capacity of a winter annual Death Valley. *Science* 194: 322–324.

Mozafar, A. and Goodin, J.R. 1970. Vesiculated hairs: a mechanism for salt tolerance in *Atriplex halimus* L. *Plant Physiol. Lancaster* 45: 62–65.

Mozafar, A., Goodin, J.R. and Oertli, J.J. 1970. Na and K interactions in increasing the salt tolerance of *Atriplex halimus* L. I. Yield characteristics and osmotic potential. *Agron. J*. 62: 478–481.

Oertli, J.J. 1968. Extracellular salt accumulation, a possible mechanism of salt injury in plants. *Agrochima* 12: 461–469.

Osborn, T.G.B., Wood, J.G. and Paltridge, T.B. 1935. On the climate and vegetation of the Koonamore Vegetation Reserve to 1931. *Proc. Linn. Soc. N.S.W*. 60: 392–427.

Osmond, C.B. 1966. Divalent cation absorption and interaction in *Atriplex*. *Aust. J. Biol. Sci*. 19: 37–48.

Osmond, C.B. 1967. Arid metabolism in *Atriplex*. I. Regulation of oxalate synthesis by the apparent excess cation absorption in leaf tissue. *Aust. J. Biol. Sci*. 29: 575–587.

Osmond, C.B. 1970. Carbon metabolism in *Atriplex* leaves. In *The Biology of Atriplex*, ed. R. Jones, pp. 17–21. CSIRO, Canberra, Australia.

Osmond, C.B., Lüttge, U., West, K.R., Pallaghy, C.K. and Shacher-Hill, B. 1969. Ion absorption in *Atriplex* leaf tissue. II. Secretion of ions to epidermal bladders. *Aust. J. Biol. Sci*. 22: 797–814.

Pallaghy, C.K. 1970. Salt relations of *Atriplex* leaves. In *The Biology of Atriplex*, ed. R. Jones, pp. 57–62. CSIRO, Canberra, Australia.

Palmer, J.H., Trickett, E.S. and Linacre, E.T. 1964. Transpiration response of *Atriplex nummularia* Lindl. and upland cotton vegetation to soil-moisture stress. *Agr. Meteorol*. 1: 282–293.

Parker, J. 1968. Drought-resistance mechanisms. In *Water Deficits and Plant Growth*, ed. T.T. Kozlowski, Vol. 1, pp. 195–235. Academic Press, New York.

Richardson, A.E.V. and Trumble, H.C. 1928. The transpiration ratio of farm crops and pasture plants in the Adelaide district. *J. Agr. South Aust*. 32: 224–244.

Sankary, M.N. and Barbour, M.G. 1972. Autecology of *Atriplex polycarpa* from California. *Ecology* 53: 1154–1162.

Sharma, M.L. 1973. Simulation of drought and its effect on germination of five pasture species. *Agron. J*. 65: 982–987.

Sharma, M.L. 1976a. Soil water regimes and water extraction patterns under two semi-arid shrub (*Atriplex* spp.) communities. *Aust. J. Ecol*. 1: 249–258.

Sharma, M.L. 1976b. Contribution of dew in the hydrologic balance of a semi-arid grassland. *Agr. Meteorol*. 17: 321–331.

Sharma, M.L. 1977. Water use by chenopod shrublands. In *Studies of the Australian Arid Zone. III. Water in Rangelands*, ed. K.M.W. Howes, pp. 139–149. CSIRO, Melbourne, Australia.

Sharma, M.L. and Tongway, D.J. 1973. Plant induced soil salinity patterns in two saltbush (*Atriplex* spp.) communities. *J. Range. Manage*. 26: 121–125.

Sharma, M.L. Tunny, J. and Tongway, D.J. 1972. Seasonal changes in sodium and chloride concentration of saltbush (*Atriplex* spp.) leaves as related to soil and plant water potential. *Aust. J. Agric. Res*. 23: 1007–1019.

Slatyer, R.O. 1960. Aspects of the tissue of water relationships of an important arid zone species (*Acacia aneura* F. Muell.) in comparison with two mesophytes. *Bull. Res. Counc. of Israel* 8: 159–168.

Slatyer, R.O. 1970a. Comparative photosynthesis, growth and transpiration of two species of *Atriplex*. *Planta* (*Berl.*). 93: 175–189.

Slatyer, R.O. 1970b. Carbon dioxide and water vapour exchange in *Atriplex* leaves. In *The Biology of Atriplex*, ed. R. Jones, pp. 23–29. CSIRO, Canberra, Australia.

Slatyer, R.O. 1972. Effects of short periods of water stress on leaf photosynthesis. In *Plant Response to Climatic Factors*, ed. R.O. Slatyer, pp. 271–276. Proceedings of the Uppsala Symposium, UNESCO, Paris.

Thompson, W.W. and Liu, L.L. 1967. Ultrastructural features of the salt gland of *Tamarix aphylla* L. *Planta* (*Berl.*) 73: 207–220.

Trumble, H.C. 1932. Preliminary investigations on the cultivation of indigenous saltbush (*Atriplex* spp.) in an area of winter rainfall and summer drought. *J. Counc. Sci. and Indust. Res.* 5: 152–161.

Wallace, A., Mueller, R.T. and Romney, E.M. 1971. Sodium relations in dsesert plants. 2. Distribution of cations in plant parts of three different species of *Atriplex*. *Soil Sci.* 115: 390–394.

West, K.R. 1970. The anatomy of *Atriplex* leaves. In: *The Biology of Atriplex*, ed. R. Jones, pp. 11–15. CSIRO, Canberra, Australia.

Williams, D.G. 1979. The comparative ecology of two perennial Chenopods. In *Studies of the Australian Arid Zone – Chenopod Shrublands*, ed. R.D. Graetz and K.M.W. Howes. CSIRO, Melbourne (in press).

Wilson, A.D. 1966. The intake and excretion of sodium by sheep fed on species of *Atriplex* (saltbush) and *Kochia* (Bluebush). *Aust. J. Agric. Res.* 17: 155–163.

Wilson, P.G. 1975. Taxanomic revision of the genus *Maireana* (Chenopodiaceae). *Nuytsia* 2: 12–82.

Wood, J.G. 1923. On transpiration in the field of some plants from the arid portions of South Australia, with notes on their physiological anatomy. *R. Soc. South Aust.* 47: 259–279.

Wood, J.G. 1925. The selective absorption of chlorine ions and absorption of water by the leaves of the genus *Atriplex*. *Aust. J. Exp. Biol. Med. Sci.* 2: 45–56.

CHAPTER 5

Senescence in mangroves

S.M. KARMARKAR

Department of Biology, Ramniranjan Jhunjhunwala College,
Ghatkopar, Bombay 400 086, India

1. Introduction

The shedding of plant parts is a universal phenomenon, whereby plant parts are lost through the process of abscission or, alternatively, through death and/or withering. Perennials are capable of shedding several of their vegetative parts (branches, leaves, roots, etc.) or reproductive parts (flowers, inflorescences, fruits, etc.) through abscission. During the last two decades a great deal of time and effort has been expended in attempts to unravel the mysteries underlying the process of abscission. By and large, these researches have been fruitful and today, we do have a sizeable information regarding the physiology and ecology of senescence and abscission. It must be admitted, however, that in view of the enormity of the problem, there are several facets which still must be probed into in greater detail. The information available on senescence and abscission till 1973 has been documented by Kozlowski (1973) in the 'shedding of plant parts', while certain aspects of senescence in relation to post-harvest physiology, have been reviewed by Sacher (1973). Surprisingly, if not all, the literature on senescence and abscission pertains to the leaves of non-halophytic plants, and as far as the author is aware, information dealing with senescence in halophytes is very scanty.

Abscission has great ecological significance, since its effect on the environment can be beneficial or harmful. It is useful to the extent that, through the loss of parts, especially the leaves, the plant be-

comes adapted to overcome drought. Abscission is a mechanism through which the plant brings about self-pruning, whereby injured or diseased parts are removed; at the same time, enormous quantities of litter are added to the soil. This litter forms the major component of soil organic matter and therefore contributes towards keeping the soil in a dynamic state through the intense biological activity which accompanies its decomposition.

From their studies on litter production in the mangrove forests of Southern Florida and Puerto Rico, Pool et al. (1975) have observed that mangroves have developed a leaf fall strategy whereby, leaves are dropped continuously, throughout the year, with higher rates of leaf fall during the wet season and lower rates during the dry season, coinciding with the changes in temperature. This pattern also coincides with growth conditions, as demonstrated by Gill and Tomlinson (1971), in the red mangrove, *Rhizophora mangle.* These workers (Pool et al., 1975) have further observed that salinity stress has a direct influence on leaf fall i.e. whenever salinity crosses a certain threshold, it is metabolically less costly to drop the leaves than to overcome the stress, and therefore, leaf fall rates increase above normal. On the other hand, under normal conditions, as new leaves are produced, older ones fall off, enabling the plants to maintain constant photosynthetic rates. The annual leaf fall in the Rookery Bay mangrove forest in Southern Florida was estimated to be of the order of 485 g/m^2; of this 200 g/m^2 in the form of organic matter

Tasks for vegetation science, Vol. 2 ed. by D.N. Sen and K.S. Rajpurohit

was washed into the estuary. The leaf litter thus becomes useful, not only as a source of carbon, but also because it helps to replenish the soils with mineral elements when it undergoes decomposition. The leaf litter contributes to a very large extent to the detritus in the marine ecosystem. On the other hand in certain mangrove communities in Westernport Bay, Australia, comprising *Avicennia marina*, there is low leaf litter production and most of the soil nutrients are recycled through the loss of roots.

The harmful effects of abscission have become evident in situations where the human element has brought about induced defoliation through the use of herbicides, particularly in South Vietnam. In this region, mangroves have been defoliated for military purposes, resulting in a deterioration of the environment. Such a defoliation has resulted in soil erosion, death of soil micro-organisms, loss of nutrients from the soil and even a disturbance of the food chain (Kozlowski, 1973). Teas and Kelly (1975) studied the effect of herbicides on the mangroves of South Vietnam and noted that 80 percent of the mangrove vegetation is sensitive to herbicidal action. While *Rhizophora, Ceriops, Bruguiera* and *Kandelia* showed sensitivity to the action of Agent White (2,4-dichlorophenoxyacetic acid and picloram, both as triisopropanolamine salts) and Agent Orange (2,4,5-trichlorophenoxyacetic acid and 2,4-dichlorophenoxyacetic acid, both as n-butyl esters), these herbicides had no effect on *Avicennia*. The sensitivity in the former group of plants was found to be due to accumulation of the herbicide in small buds, due to which the buds are killed. The genus *Excoecaria* was also found to be resistant like *Avicennia*. At about the same time, the effects of mangrove defoliation on the estuarine ecology of South Vietnam were studied by de Sylva and Michel (1975). Their studies revealed that higher siltation, resultant turbidity and possibly lower oxygen concentration in the estuaries of the defoliated mangrove community reduced the spawning habitat for fishes, thus partly destroying their nursery ground, thereby causing an ecological imbalance.

Although in most halophytes the leaves are shed throughout the year, an interesting observation has been made by Amonkar (1977) that in *Salvadora*

persica, a halophyte common along the west coast of India, there are marked periods in which senescence occurs; consequently, leaf fall is generally restricted from December to May. On the other hand, in *S. persica* growing in a non-saline environment, the period of leaf fall is limited to late summer only. Obviously, the number of leaves shed in *S. persica* (saline) is much greater than that of *S. persica* (non-saline). As will be discussed at greater length later, the mature senescent leaves act as storehouses for sodium and chloride ions, and in the case of *S. persica*, they also store calcium ions. Since the extent of accumulation is greater in the plants from the saline environment, the leaves from these plants are shed more frequently, thereby getting rid of excess salts periodically. In *S. persica*, from the non-saline environment, since leaf fall is less frequent, these plants remove larger quantities of salts through their senescent leaves; this enables the plants to regulate their internal salt levels. Proof for this is provided by the fact that the ash content of *S. persica* (saline) increases by 29 percent in senescent leaves as compared to *S. persica* (non-saline) where the increase is only 20 percent. Stated differently, in *S. persica* (non-saline), although the number of leaves which senesce and are ultimately shed is markedly lesser than in *S. persica* (saline), the quantity of salt removed each time in the former plant is far in excess of that of the latter (Table 1).

As stated earlier, complete information is lacking on the physiological changes which precede the abscission of leaves. It is a common knowledge that, as plant parts begin to age, variations occur in the structure of the plant, form and function are thereby altered. This deteriorative phase represents senescence. In some plants the internal, degradative processes find external expression in the entire plant which therefore senesces at one and the same time; in others the process is gradual, spread over a period of time. This is externally manifested by the loss of one part or another. The latter process is referred to as organ senescence. Since most halophytes are perennials, senescence in these plants is characterized by a shedding of leaves.

As leaves reach maturity, metabolic activity is at its peak; thereafter, metabolic functions begin to decline. Such changes foreshadow the onset of

Table 1. Ash content in young, mature and senescent leaves of some halophytes.

Species	Saline			Non-saline			Author
	young	mature	senescent	young	mature	senescent	
Salvadora persica	22.88	23.39	28.14	15.48	17.21	22.24	Amonkar (1977)
Clerodendrum inerme	—	23.12	26.14	—	—	—	Mishra (1967)
Acanthus ilicifolius	—	11.95	12.55	—	—	—	Siddhanti (1977)

Values expressed in g per 100 g fresh tissue.

senescence. Externally this becomes evident either by a loss of leaf colour, or sometimes, by a change in leaf colour due to the development of carotenoids. Internally, there is a translocation of mineral elements from the leaf and a diminuition in its auxin supply (Leopold, 1961), increased tannin production (Scott et al., 1948) and reduced photosynthetic activity (Meyer, 1918; Leopold, 1961). Leopold (1961) believes that senescence has two primary assets: (i) it brings about recovery through retranslocation of the bulk of the nutrition from the leaf; and (ii) it results in a shedding of the ineffective leaves from the plant. Osborne (1973) is of the opinion that the metabolic changes which occur during senescence are certainly deleterious, but these changes play a positive role in senescence in as much as they enable the plant to undergo an ecological adaptation. Changes which accompany senescence, therefore, precede and, in fact, act as a signal for leaf abscission.

While in deciduous species the stimulus for senescence is provided by changes in day length, in evergreens such changes are ineffective. In the former plants decrease in day length affects photosynthetic activity; less metabolites are therefore available for retranslocation. In the latter, competition from new leaf growth acts as a stimulus for the shedding of the older leaves (Kozlowski, 1973). Wildman and Bonner (1947) noted that in ageing spinach leaves, more than 70 percent of the soluble protein disappeared during the life of the leaf. This fraction of protein was an enzymatic protein (Ribulose biphosphate carboxylase) suggesting that the loss of this major carboxylating component

affects the rate of photosynthesis; this could be an important factor responsible for initiating senescence. Due to a loss of metabolites cellular activity is impaired. Later cytolysis occurs in the cells of the separation layer, formed in the abscission zone (at the base of the petiole). Ultimately, due to cellular disorganization, the leaves are abscissed.

2. Environmental factors on senescence

Several external environmental factors influence the process of senescence and the consequent shedding of the leaves. As a result of these climatic (light, temperature, rainfall, wind, etc.), edaphic (soil moisture deficiency and toxicity of minerals, soil salinity) and other factors (soil-borne diseases, pollutants, etc.) several changes occur in the plants. To counteract the effects of such external factors, plants have evolved physiological mechanisms which permit them to overcome adverse conditions. Such mechanisms find expression through senescence and abscission (Dostal, 1967; Kozlowski, 1973) i.e. the environmental factors exert their influence on the internal factors, which then find expression through correlation and competition. An example of correlation can be witnessed in the physiological changes which accompany senescence – loss of proteins, carbohydrates and degradation of chlorophyll. These changes occur in one part of the plant (leaf blade) but they cause changes in another part (base of petiole), resulting in abscission. On account of the differences in nutritional status, hormone levels, metabolic activity, etc. be-

175

tween different leaves, there is a severe competition among them. As a consequence, weaker leaves senesce and undergo abscission (Kozlowski, 1973). 1973).

3. Physical properties

Several workers have studied the changes in physical properties of the leaves of halophytes, during senescence. Jamale and Joshi (1976) studied such changes in *Sonneratia alba*, and *Excoecaria agallocha*, collected from the rich mangrove vegetation of Ratnagiri, along the west coast of India. In Bombay, similar studies have been conducted in our laboratory on *Sonneratia apetala*, and *Sonneratia alba* (Rao, 1974) *Acanthus ilicifolius* (Siddhanti, 1977) and *Salvadora persica* (Amonkar, 1977). Earlier, Mishra (1967) carried out similar studies in *Clerodendrum inerme*, Gaertn. From these results, summarized in Table 2, it is observed that, in general, there is an increase in the water content of the leaves during senescence. Figures available for *A. ilicifolius* and *E. agallocha* indicate that there is a loss of organic material as tissues age. Further, the senescent leaves of all halophytes investigated, possess a greater density and they also display a greater degree of succulence than the mature green leaves.

In mangroves, as the leaves age, there is an increase in their sodium and chloride content (discussed later in this chapter). In *Atriplex hastata* L. it has been observed that increased salination causes increased succulence (Mendoza, 1971). Thus the increase in the degree of succulence of halophytic leaves (Table 2) may be associated with their increased sodium and chloride content. According to Mendoza (1971) the increase in the succulence is seen in the total plant volume/surface ratio and not in the water content of the plant in terms of g/fresh weight or g/cm³; in other words, there is an increase in the cell water content per surface area. In a subsequent review of the physiology of desert halophytes, Caldwell (1974) has remarked that succulence is usually taken to mean increased thickness of plant organs. He further observes that there may not be a change in the total plant water content and therefore, increase in plant succulence can also be associated with increase in cellular volume/surface ratio and cell water content.

4. Elemental composition

Walsh (1974) has remarked that there is a paucity of literature data concerning the elemental composition of mangroves. While taking note of the work done by Sokoloff et al. (1950), Sidhu (1963a,b) and

Table 2. Physical properties of the green and senescent leaves of some halophtes.

Species	Leaf	Av. leaf wt (g) (M)	Av. leaf area (cm²) (A)	Av. leaf thickness (mm) (d)	M/A (g/cm²)	Density (20M/A.d) g/cm³	Moisture % *	Dry matter	Degree of succulence **	Author
Sonneratia apetala	green	3.305	35.66	1.67	0.09	0.115	71.37	—	6.61	Rao (1974)
	senescent	5.716	36.12	1.18	0.11	0.26	82.13	—	13.00	
Sonneratia alba	green	0.978	16.26	0.61	0.06	0.19	60.95	—	4.17	Rao (1974)
	senescent	1.003	12.96	0.47	0.07	0.33	81.56	—	6.31	
Acanthus ilicifolius	green	1.73	36.5	0.68	0.05	1.39	77.43	10.62	3.65	Siddhanti (1977)
	senescent	0.95	30.7	0.43	0.63	1.43	80.09	7.36	2.5	
Excoecaria agallocha	green	1.49	27.6	0.054	0.05	1.00	74.1	25.9	—	Jamale and Joshi (1976)
	senescent	1.81	18.1	0.11	0.1	0.91	84.5	15.5	—	

* Values expressed in 100 g of fresh tissue.
** Degree of succulence: water content in g per dm. sq.

Golley (1969), Walsh (1974) further states that little information is available in relation to environmental factors and age of the trees. Subsequently, Joshi et al. (1975), while studying ion regulation in the mangroves of the west coast of India, reported changes in ionic content in the leaves of *Sonneratia alba*, *Lumnitzera racemosa*, and *Excoecaria agallocha* during senescence. Meanwhile, Rao (1974), Jamale and Joshi (1976) Siddhanti (1977) and Amonkar (1977) have also studied ion regulation, during senescence, in various other halophytes. Their results have been discussed in detail later in this chapter. In view of Walsh's (1974) remarks the data for changes in the mineral content, during senescence, of the leaves of mangroves (Table 3) assume very great significance. Presumably, these collective data are being presented for the first time.

The increase in the ash content of the senescent leaves (Table 1) is due to the overall increase in the mineral content of their tissues (Table 3). As compared to the green mature leaves, the senescent leaves have a higher content of ions like sodium, calcium and chlorides. On the other hand, potassium and magnesium contents decline with age.

Keller (1925), working on *Salicornia*, noted that increments in salt brought about a corresponding increase in succulence. Walter (1955) is of the opinion that succulence is due to the presence of salt which causes a hydration of proteins. Karmarkar and Joshi (1969) studied the effect of salt concentration on metabolic activity in *Kalanchoe pinnata* (*Bryophyllum pinnatum*). They observed that this garden succulent is rich in chlorides and these values compare favourably with chloride values for halophytes. Karmarkar and Joshi (1969) concluded that possibly, chlorides along with sodium help in developing succulence in this plant. On the basis of these observations, Joshi and Mishra (1970) have attributed the increase in the water content of the senescent leaves of *Clerodendrum inerme* to the hydration of proteins, brought about by an increase in chloride ions. Further, these workers are of the

Table 3. Inorganic constituents in green and senescent leaves of some halophytes.

Species	Leaf	Sodium	Potassium	Calcium	Magnesium	Phosphorus	Chloride	Author
Acanthus	green	163.5	62.4	80.0	108.3	28.1	136.6	Sidhanti
ilicifolius	senescent	177.1	65.3	98.0	75.0	42.3	146.4	(1977)
Excoecaria	green	109.11	77.21	164.09	—	1.8	156.6	Jamale and
agallocha	senescent	272.08	29.77	93.54	—	0.41	334.16	Joshi (1976)
Clerodendrum	green	56.52	77.44	124.00	58.33	—	176.85	Mishra
inerme	senescent	304.78	46.66	127.00	45.00	—	230.00	(1967)
Salvadora	green	310.09	99.05	293.5	145.0	25.00	215.2	Amonkar
persica	senescent	339.65	55.5	300.0	88.17	19.50	278.64	(1977)
(saline)								
Salvadora	green	188.63	104.83	268.05	155.66	33.00	140.26	Amonkar
persica	senescent	202.86	62.74	326.34	117.11	21.00	150.36	(1977)
(non-saline)								
Sonneratia	green	124.44	42.77	42.62	—	—	—	Jamale and
alba	senescent	463.19	42.62	—	—	—	—	Joshi (1978)
Avicennia	green	150.20	31.19	26.79	—	—	103.58	
alba	senescent	171.15	24.08	31.77	—	—	123.12	
Thespesia	green	113.04	28.20	90.00	68.33	35.00	93.71	Kotmire
populnea	senescent	216.52	25.12	130.00	85.83	30.00	190.57	and Bhosale
(saline)								(1978)
Thespesia	green	16.52	39.48	170.00	58.33	46.66	52.28	Kotmire
populnea	senescent	26.95	29.23	232.00	80.83	33.3	58.00	and Bhosale
(non-saline)								(1978)

Values expressed as meq per 100 g dry tissue.

opinion that senescence is a process which brings about a partial desalination of the plants; i.e. excess of salts accumulate in the leaves, causing them to senesce and as abscission follows, it helps to regulate the internal levels of salts in the plant. This is apparently true for all halophytes, since there is, in general, an increase in the chloride content with age (Table 3). In a later review, Walter and Stadelmann (1974) have observed that since chlorides accumulate in high concentrations in the cell sap of salt resistant plants, they presumably cannot have a toxic effect on the protoplasm. On the other hand, the effect of chloride ions on protoplasm appears to be specific, resulting in the increase in the succulence of the leaves. Thus according to these authors, a high chloride concentration is essential for causing succulence.

The increase in the ash content of the senescent leaves, accompanied by an increase in the water content, is of great significance, since, according to Jennings (1968), it represents a mechanism by which toxicity of salts, if any, can be counteracted due to the dilution of the cell sap. Jennings (1968) also believes that there is a correlation between the sodium content and water content in halophytes, whereas potassium content does not show any appreciable relationship with succulence. Gale and Poljakoff-Mayber (1970) have also made a similar observation.

Walter and Stadelman (1974) are of the opinion that the presence of ions in the protoplasm modifies the capacity of the protoplasm for hydration and that the kind of ion is of great significance, e.g. swelling is increased more by univalent ions than by bivalent and trivalent ions. Further, the addition of monovalent ions enhances hydration. Atkinson et al. (1967) also noted a correlation between succulence and the salt content in the leaves, relative to the age of the leaf. In their studies on the ionic content in successive leaf pairs of *Rhizophora mucronata* and *Aegialitis annulata*, these workers noted that in the former plant sodium and chloride content increased with age; however, when these results are computed in terms of fresh weight, there is no change in the chloride content while potassium content declines and sodium content increases. The accumulation of ions was ascribed to the absence of

an excretory mechanism in *R. mucronata*. On the other hand, *A. annulata* was capable of regulating its sodium and chloride content through salt glands, and thus, in this plant these ions do not accumulate with age (Table 4).

Table 4. Concentrations of sodium, potassium and chlorides in leaf samples (A, B & C) of *Rhizophora mucronata* and *Aegialitis annulata* (adapted after Atkinson et al., 1967).

	R. nucronata			*A. annulata*		
	A	B	C	A	B	C
Dry weight (g)	0.5	0.61	0.63	0.45	0.5	0.44
Water (% fresh wt.)	65	65	69	60	59	60
Na (μ-equiv/leaf)	290	480	645	325	280	235
Na (μ-equiv/ml H_2O)	313	435	461	480	388	356
K (μ-equiv/leaf)	81	48	45	106	93	70
K (μ-equiv/ml H_2O)	88	44	32	157	129	106
Cl (μ-equiv/leaf)	520	585	730	290	361	386
Cl (μ-equiv/ml H_2O)	562	530	522	429	361	386

Leaf samples A, B & C in sequence of increasing age.

Although in *Salvadora persica* there is a tendency to accumulate sodium and chloride ions, during senescence, osmotic adjustments are made by the leaves becoming more succulent (Table 5); thereby the internal concentrations of these ions are reduced (Amonkar, 1977). Similar results have been obtained with *Acanthus ilicifolius* (Siddhanti, 1977). In *Sonneratia alba*, *Excoecaria agallocha*, *Lumnitzera racemosa* and *Clerodendrum inerme*, however, there is a definite increase in the sodium and chloride content on a fresh weight basis, while the potassium content declines during senescence (Jamale and Joshi, 1976). In *Thespesia populnea*, an associate mangrove, there is no appreciable change in the ionic content on a fresh weight basis (Kotmire and Bhosale, 1978). The above data apparently support Jennings' (1968) contention that halophytes possess a mechanism for avoiding salt toxicity through an increase in succulence.

Table 5. Changes in mineral content in green and senescent leaves of *Acanthus ilicifolius* and *Salvadora persica*.

Species	Leaf	% wt.	Moisture content	Sodium	Potassium	Magnesium	Chlorides	Author
Acanthus	green	dry		163.5	62.4	108.3	136.6	
ilicifolius		fresh	77.45	36.9	14.08	24.44	30.83	Siddhanti
	senescent	dry		177.1	65.3	75.0	146.4	(1977)
		fresh	80.09	35.26	13.0	14.92	29.15	
Salvadora	green	dry		310.09	99.05	145.0	215.2	
persica		fresh	68.31	98.27	31.79	—	68.19	
(saline)	senescent	dry		331.65	55.5	88.17	278.64	
		fresh	86.10	46.10	7.71	—	38.37	Amonkar
Salvadora	green	dry		188.63	104.83	155.66	140.26	(1977)
persica		fresh	60.04	75.25	41.79	—	56.05	
(non-saline)	senescent	dry		202.86	62.74	117.11	150.36	
		fresh	73.14	54.48	13.79	—	40.39	

Mineral content values expressed as meq per 100 g leaf tissue.

The decrease in the potassium content, as tissues age, has been correlated with the retranslocation of potassium from ageing cells (Nimbalkar and Joshi, 1975; Jamale and Joshi, 1976; Amonkar, 1977). The change in the K:Na ratio, according to these workers, is not due to retention of sodium but due to retranslocation of potassium (Table 6). With an increase in the age of the leaf there is an increase in the calcium content in all plants investigated (Table 3) except in *E. agallocha* where it shows an appreciable decline. Earlier, Leopold (1961) and Varner (1961) have independently recorded calcium accumulation in older tissues. It is now believed that changes in calcium content bring about changes in cell permeability. This loss of membrane permeability is suspected to be a causative factor of senescence.

There are conflicting reports regarding the role of magnesium during senescence. Joshi and Mishra (1970) noted that in *Clerodendrum inerme*, magnesium is withdrawn from the senescent leaves so as to facilitate the accumulation of sodium, calcium and chloride ions. Nimbalkar and Joshi (1974) observed that in sugar cane var. Co. 740, there is an

Table 6. Potassium : sodium ratios in leaves of some halophytes.

Species	Leaf	Sodium	Potassium	K:Na	Author
Acanthus	green	163.5	62.4	0.38	Siddhanti (1977)
ilicifolius	senescent	177.1	65.3	0.36	
Excoecaria	green	109.11	78.21	0.71	Jamale and Joshi (1976)
agallocha	senescent	272.08	29.77	0.11	
Clerodendrum	green	56.52	77.44	1.37	Mishra (1967)
inerme	senescent	304.78	46.66	0.15	
Salvadora	green	310.09	99.05	0.31	Amonkar (1977)
persica	senescent	339.65	55.5	0.16	
Sonneratia	green	124.44	42.77	0.34	Rao (1974)
alba	senescent	463.19	42.62	0.09	
Avicennia	green	150.2	31.19	0.21	Jamale and (Joshi (1976)
alba	senescent	171.15	24.08	0.14	
Thespesia	green	113.04	28.2	0.25	Kotmire andBhosale (1978)
populnea	senescent	216.52	25.12	0.12	

Values for sodium and potassium expressed in meq per 100 g dry tissue.

accumulation of silica, calcium and magnesium ions, accompanied by a withdrawal of potassium and phosphorus. Amonkar (1977) has reported that in *Salvadora persica*, while there is an accumulation of sodium, calcium and chloride ions, there is a decline in the levels of potassium, phosphorus and magnesium. Kotmire and Bhosale (1978) have also observed that in *Thespesia populnea*, the magnesium content of the leaf increases with age, while the phosphorus content shows a decline.

The decline in the phosphorus content in senescent tissues has been linked by several workers to the decline in metabolic activity. Williams (1938), Link and Swanson (1960), McCollum and Skok (1960), Oata (1964), and Seth and Wareing (1967) have observed that phosphorus compounds are transported with other metabolites, from the mature organs to the developing vegetative or reproductive structures. More recently, Clough and Attiwill (1975) studied nutrient recycling in a mangrove community in Australia. The life span of the leaves of *Avicennia marina*, the main constituent of this community, was found to be about three years. Analysis of the dead leaves in the litter revealed that more than one third of nitrogen and phosphorus is withdrawn from the leaves before death. There is a loss of metabolites when tissues begin to age. It is thus likely that with the loss of metabolites, phosphorus compounds are transported from the senescent leaves. While concurring with the above view, Jamale and Joshi (1976) have further opined that a diminishing phosphorus pool could be a prime factor in initiating senescence. Amonkar's (1977) results also lend support to this view (Table 3).

5. Organic acid metabolism

Joshi and Mishra (1970), Amonkar (1977) and Kotmire and Bhosale (1978) have reported an increase in titratable acidity in the senescent tissues of *Clerodendrum inerme*, *Salvadora persica* and *Thespesia populnea*, respectively (Table 7). Their results lend support to the views expressed by Wiesner (1905) and Facey (1950). While the former has noted that the increase in acidity in senescent leaves accelerates abscission, the latter believes that

Table 7. Titratable acid number (TAN) in green and senescent leaves of *Clerodendrum inerme* and *Thespesia populnea*.

Species		Green	Senescent	Author
Clerodendrum inerme		48.21	150.06	Joshi and Mishra (1970)
Thespesia populnea	saline	29.07	42.5	Kotmire and
	non-saline	57.69	76.5	Bhosale (1978)

Values expressed in terms of ml of decinormal NaOH required to neutralize the acid in an extract obtained from 100 g fresh tissue.

changes in cellular pH could possibly affect auxin level and this might initiate abscission.

During abscission there is an increase in respiration and according to Addicot and Lynch (1955) it may account for a higher rate of synthesis of TCA cycle acids. With the use of $^{14}CO_2$, Joshi and Mishra (1970) showed that in *Clerodendrum inerme*, fatty acid synthesis is vigorous in the senescent leaves. On account of this there is a greater synthesis of malonic acid. It is likely that the accumulation of malonic acid could cause an accumulation of succinic acid due to inhibition of succinate dehydrogenase activity by malonate. In addition, the heavy labelling in leucine, according to Joshi and Mishra (1970) could be due to either its non-utilization in protein synthesis or alternatively, its high rate of synthesis (Table 8). Their results corroborate the observations of Ben-Zioni et al. (1966) who believe that a similar accumulation of leucine in the leaves of *Nicotiana* resulted in inducing senescence. Joshi and Mishra (1970) believe that the accumulation of palmitic acid, stearic acid and oleic acid in senescent leaves, due to an enhancement of fatty acid synthesis, represents a mechanism to withstand desiccation.

6. Photosynthesis

Photosynthetic $^{14}CO_2$ fixation in the green and senescent leaves of *Clerodendrum inerme* has been studied by Joshi and Mishra (1970). They noted a considerable loss in the photosynthetic activity of senescent leaves, which according to them, could be attributed to the disruption of chloroplasts. Joshi

Table 8. Distribution of radioactivity in ethanol soluble compounds following 1 h $^{14}CO_2$ fixation in green and senescent leaves of *Clerodendrum inerme*.

Compound	Green	Senescent	Compound	Green	Senescent
Sugar phosphates			*Amino acids*		
Phosphoenol pyruvate + phosphoglyceric acid	6.92	9.51	Glutamate	9.08	6.9
			Aspartate	21.67	10.76
Sugar monophosphate	1.79	1.45	Cystine	2.02	2.29
Sugar diphosphate	3.27	3.26	Glycine-serine	6.35	5.86
Dihydroxy acetone phosphate	1.65	—	Amino butyric acid	0.98	0.38
			Threonine	4.07	1.5
Sugars			Histidine	6.18	1.65
Glucose	4.68	3.9	Proline	2.01	—
Fructose	3.82	6.01	Alanine	1.65	1.05
Sucrose	8.24	4.21	Leucines	1.95	3.6
Maltose	1.07	3.7			
Rhamnose	1.69	8.47	*Organic acids*		
			Citrate	1.95	5.92
			Malate	5.57	3.46
			Succinate	1.97	6.01
			Fumarate	1.56	0.96
			Malonate	—	7.12
			Rate of fixation	41.8	7.6

Values of incorporation of radioactivity in individual compounds are expressed as percentage of total activity counted on chromatograms while the rate of fixation is expressed as counts/min/mg fresh tissue.

and Mishra (1970) further noted a difference in the carbohydrate metabolism of senescent leaves (Table 8). In the aged tissue there was a greater synthesis of maltose, fructose and rhamnose, while glucose and sucrose were synthesized in appreciably lesser quantities. Winter (1975) noted that in the halophytes, *Mesembryanthemum nodiflorum* and *Suaeda maritima*, when net photosynthetic CO_2 assimilation is calculated on a fresh weight or dry weight basis, there is apparently a decrease in net CO_2 uptake with increased salinity. However, if net CO_2 assimilation is expressed per mg chlorophyll there is no such decrease. In fact, net CO_2 uptake is correlated with plant productivity at different levels of salinity.

More recently, Jamale and Joshi (1976) have reported that the senescent leaves of *Sonneratia alba* and *Excoecaria agallocha* contain much less chlorophyll than the green, mature leaves and consequently the rate of $^{14}CO_2$ fixation, per mg of tissue is higher in the latter (Table 9). In *E. agallocha* CO_2 fixation pattern in the senescent leaves indicates that there is no loss of activity in chlorophyll, but the decline in CO_2 fixation is due to degradation of chlorophyll.

7. Changes in chlorophyll, carbohydrates and protein

In *Sonneratia apetala* and *Sonneratia alba* (Rao, 1974), *Acanthus ilicifolius* (Siddhanti, 1977) and *Lumnitzera racemosa* (Jamale and Joshi, 1976) there is a loss of chlorophyll and a reduction in the reducing and total sugars, as the leaves senesce (Table 10). According to Lewington and Simon (1969), the decline in photosynthesis and carbohydrate is associated with a rapid translocation of products of hydrolysis out of the senescent leaves.

Boucaud and Ungar (1976) have shown that in the halophyte, *Suaeda*, the chlorophyll content tends to decline as salt concentrations increase. Similarly, in *Thespesia populnea*, the plants from the saline habitat have a lower chlorophyll content than the same species occurring in a non saline environment (Kotmire and Bhosale, 1978). Comparatively the loss of chlorophyll *a* is much greater than the loss of chlorophyll *b*, in *Thespesia populnea* (Table 10).

The protein content declines with age in *Son-*

Table 9. Total chlorophylls and $^{14}CO_2$ incorporation in green and senescent leaves of *Sonneratia alba* and *Excoecaria agallocha*.

Species	Leaf	Chlorophyll (a + b) mg/ 100 g fresh tissue	Ethanol soluble fraction		Ethanol insoluble fraction		Author
			cpm/mg chlorophyll cpm × 10⁴	cpm/mg tissue	cpm/mg chlorophyll cpm × 10⁴	cpm/mg tissue	
Sonneratia	green	97.46	49.74	485	25.64	25	Jamale
alba	senescent	24.31	15.1	37	8.16	2	and Joshi (1976)
Excoecaria	green	119.24	53.45	636	78.99	94	Jamale
agallocha	senescent	21.77	70.91	156	36.36	8	and Joshi (1976)

Table 10. Chlorophylls, carbohydrates, protein and polyphenol content in green and senescent leaves of some halophytes.

Species	Leaf	Total chlorophylls	Reducing sugars	Total sugars	Starch	Protein	Total polyphenols	Author
Sonneratia	green	97.46	0.23	0.28	2.1	—	2.31	Jamale and
alba	senescent	23.31	0.08	0.09	1.9	—	1.19	Joshi (1976)
Sonneratia	green	23.26	0.02	0.50	0.48	2.38	—	Rao (1974)
apetala	senescent	2.09	0.01	0.29	0.13	2.23	—	
Sonneratia	green	29.84	0.03	0.66	0.61	1.09	—	Rao (1974)
alba	senescent	4.10	0.01	0.22	0.05	0.83	—	
Acanthus	green	22.48	1.34	1.56	3.55	0.55	—	Siddhanti
ilicifolius	senescent	3.00	0.25	0.23	0.32	0.21	—	(1977)
Lumnitzera	green	65.4	0.19	0.29	0.9	—	2.39	Jamale and
racemosa	senescent	15.21	0.04	0.05	0.6	—	1.43	Joshi (1976)
Thespesia	green	210.21	0.55	0.68	2.78	—	5.62	Kotmire and
populnea (saline)	senescent	16.9	0.05	0.28	2.60	—	3.78	Bhosale (1978)
Thespesia	green	348.37	0.58	0.59	3.43	—	5.31	Kotmire, and
populnea (non-saline)	senescent	12.09	0.19	0.39	2.80	—	4.02	Bhosale (1978)

Values expressed as mg per 100 g fresh tissue.

neratia apetala, *Sonneratia alba* (Rao, 1974) and *Acanthus ilicifolius* (Siddhanti, 1977). Spencer and Titus (1972) and Tetley and Thimann (1974) have also noted a decrease in nucleic acids, proteins, chlorophyll and sugars, during senescence.

8. Enzymes

In *Clerodendrum inerme* the activity of malate dehydrogenase increases while that of succinate dehydrogenase decreases, as tissues begin to age (Joshi and Mishra, 1970). The decrease in the activity of succinate dehydrogenase is believed to be due to the inhibition of the enzyme by malonate, which accumulates in the senescent leaves. Sodium and chloride ions are believed to stimulate malate dehydrogenase activity.

Since halophytes grow in a medium rich in sodium and chloride ions, these plants have developed various mechanisms by means of which salt uptake/exclusion is regulated. On account of this the internal salt levels in the various parts, particularly the leaves, show wide variations. In order to understand the intricate mechanism of salt tolerance at the cellular level; several studies have been carried out, using halophytic and non-halophytic salt tolerant species; in most of these studies, spread over more than two decades, the effect of salt on the activity of different enzymes has been determined (Hiatt and Evans, 1960; Joshi et al., 1962; Craigie, 1963; Hason-Porath and Poljakoff-Mayber, 1969; Weimberg, 1970, 1975; Karmarkar and Ranganathan, 1971; Greenway and Osmond, 1972; Karmarkar and Amonkar, 1974; Kalir and Poljakoff-Mayber, 1975, 1976; Amonkar, 1977; Siddhanti, 1977). While Weimberg (1970, 1975) and Greenway and Osmond (1972) believe that enzymes do not vary in their response to the presence of salt in the cellular medium, the other school of thought, mainly led by Poljakoff-Mayber, subscribes to the view that enzymes show variable responses to the presence of salt. Experiments conducted in our laboratory on several enzymes from halophytic species indicate that the presence of salt does affect the activity of the enzyme, however more detailed studies are still necessary to conclusively prove whether enzyme activity is altered by salts which accumulate in senescent leaves.

The effect of age on polyphenols, polyphenol oxidase and peroxidase in *Sonneratia alba*, *Avicennia alba* and *Excoecaria agallocha* has been studied by Jamale and Joshi (1978). The decrease in polyphenols in the senescent leaves, according to them, could be due to either a translocation of these substances to parts like bark, for storage, or their utilization in other metabolic activities. Jamale and

Joshi (1978) further noted that the enzymes, monophenol-oxidase and diphenol-oxidase showed a low activity in young leaves, but gradually the activity increased as the leaves developed, reaching a steady state in the mature leaves (Table 11). With the onset of senescence, there was a sudden spurt in the activity of both the enzymes; but as the leaves senesced fully, monophenolase activity showed a slight increase while diphenolase activity was reduced. On the other hand, peroxidase activity in the senescent leaves showed a 50–60 fold increase over that of mature leaves in *S. alba*.

9. Effect of amino acids, gibberellic acid and calcium in inducing senescence

It is now well known that certain amino acids and growth promoting substances do play a positive role in inducing or retarding senescence. Martin and Thimann (1972) have observed that in oat leaves, L-serine is capable of inducing senescence but its effect can be counteracted by L-arginine. Thus they are of the opinion that L-arginine is an anti-senescence substance and that gibberellic acid retards senescence by promoting protein synthesis. In similar experiments conducted in our laboratory on *Acanthus ilicifolius* (Siddhanti, 1977) and *Sonneratia apetala* and *Sonneratia alba* (Rao, 1974) it was observed that, during senescence, chloroplastic proteins are degraded earlier than cytoplasmic proteins (Fig. 1). Further, in *A. ilicifolius*, L-serine (0.5 mM to 50 mM) is capable of inducing senescence,

Table 11. Distribution of polyphenols as well as polyphenol-oxidase and peroxidase enzyme activity in progressively older leaves of *Sonneratia alba*.

Stage of leaf	Polyphenols*	Polyphenol-oxidase**		Peroxidase**	Author
		monophenolase	diphenolase		
young	3.77	0.28	0.18	Trace	Jamale and
premature	3.55	1.67	2.17	0.01	Joshi (1978)
mature	2.36	1.40	2.25	0.06	
onset of senescence	2.18	6.80	26.80	0.12	
senescent	1.80	9.20	20.80	0.58	

* Values in g/100 g fresh tissue; ** Enzyme activity expressed as change in O.D./ mg protein.

183

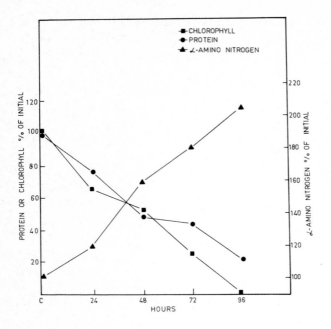

Fig. 1. Changes in the chlorophyll, protein and α-amino-nitrogen content with time after darkening, in *A. ilicifolius*.

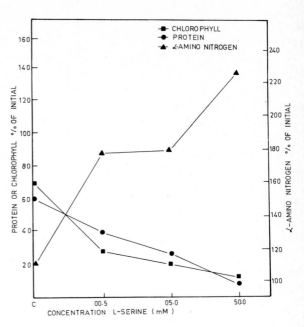

Fig. 2. Effect of L-serine on the chlorophyll, protein and α-amino-nitrogen content in *A. ilicifolius*, 96 h after darkening.

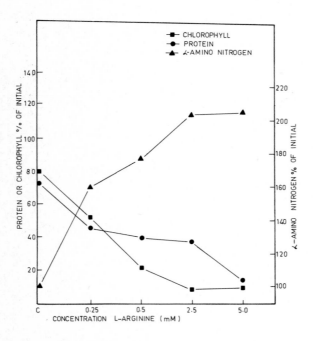

Fig. 3. Effect of L-arginine on the chlorophyll, protein and α-amino-nitrogen content in *A. ilicifolius*, 96 h after darkening.

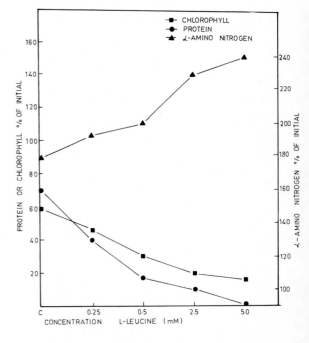

Fig. 4. Effect of L-leucine of the chlorophyll protein and α-amino-nitrogen content in *A. ilicifolius* 96 h after darkening.

184

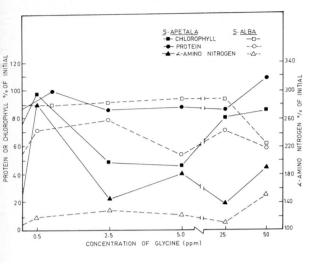

Fig. 5. Effect of glycine on the chlorophyll, protein and α-amino-nitrogen content in *S. apetala* and *S. alba*, 72 h after darkening.

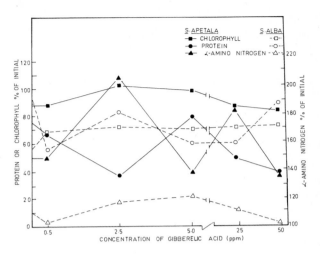

Fig. 6. Effect of gibberillic acid on the chlorophyll, protein and α-amino-nitrogen content in *S. apetala* and *S. alba*, 72 h after darkening.

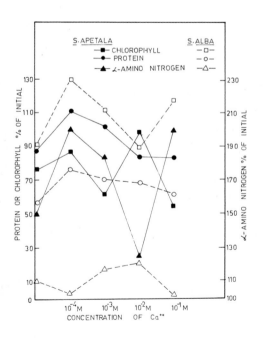

Fig. 7. Effect of calcium chloride on the chlorophyll, protein and α-amino-nitrogen content in *S. apetala* and *S. alba*, 72 h after darkening.

while L-arginine (0.25 mM to 5 mM) is equally effective (Figs. 2 & 3). Thus the antagonistic nature of these two amino acids, reported by Martin and Thimann (1972) could not be observed in our experiments with the leaves of mangroves. L-leucine is also capable of inducing senescence in *A. ilicifolius* (Fig. 4).

In *S. apetala*, L-glycine, at a concentration of 2.5 ppm to 5 ppm, is capable of causing chlorophyll and protein breakdown (Fig. 5), while in *S. alba* the same effect is produced by L-glycine at 50 ppm concentration only. Gibberellic acid is capable of retaining chlorophyll in both the species of *Sonneratia* (Fig. 6); however, in *S. alba* senescence can be deferred only if the GA concentration is above 50 ppm while in *S. apetala* the same effect can be produced by 5 ppm GA.

It was formerly believed that the accumulation of calcium ions impaired cell wall permeability and therefore it was believed to hasten senescence. Pooviah and Leopold (1973) demonstrated that in *Rumex*, senescence could be deferred by an increase in the calcium content. In our experiments it was observed that in *S. apetala*, the protein content could be maintained above that of the control with the application of 10^{-4} to 10^{-3} M calcium chloride

(Fig. 7). In *S. alba*, even higher concentrations of calcium chloride (10^{-1} M) brought about a similar effect, indicating that acculumulation of calcium in mature leaves could be of use in retarding senescence.

Acknowledgements

The author thanks Professor G.V. Joshi, Shivaji University, Kolhapur, India and Professor Howard J. Teas, University of Miami, Coral Gables, Florida, U.S.A. for offering useful criticism in the preparation of this manuscript.

Literature cited

Addicot, F.T. and Lynch, R.S., 1955. Physiology of abscission. *Ann. Rev. Plant Physiol.* 6: 211–238.

Amonkar, D.V. 1977. Physio-ecological studies in halophytes – Studies in *Salvadora persica* Linn. Ph.D. Thesis, University of Bombay.

Atkinson, M.R., Findlay, G.P., Hope, A.P., Pitman, M.G., Saddler, H.D.W. and West, K.R. 1967. Salt regulation in the mangroves, *Rhizophora mucronata* Lam. and *Aegialitis annulata* R. Br. *Aust. J. Bil. Sci.* 20: 589–599.

Ben-Zioni, A., Itai, C. and Waadia, Y. 1966. *Plant Physiol.* 47: 361. as cited by Joshi, G.V. and Mishra, S.D. 1970. Photosynthesis and mineral metabolism in senescent leaves of *Clerodendron inerme* Gaertn. *Indian J. Exp. Biol.* 8: 41–43.

Boucaud, J. and Ungar, I.A. 1976. Influence of hormonal treatments on the growth of two halophytic species of *Suaeda*. *Am. J. Bot.* 63: 694–699.

Caldwell, M.M. 1974. Physiology of Desert Halophytes. In *Ecology of Halophytes*, eds. .J. Reimold and W.H. Queen, pp. 355–378 Academic Press, New York.

Clough, B.F. and Attiwill, P.M. 1975. Nutrient cycling in a community of *Avicennia marina* in a temperate region of Australia. In *Proc. Int. Symp. Biol.* and *Management of Mangroves*, Vol. I, eds. G.E. Walsh, S.C. Snedaker and H.J. Teas, pp. 137–146. Int. Food Agric. Sci., Univ. Florida, Gainesville.

Craigie, J.S. 1963. Dark fixation of C-14 carbonate by marine algae. *Can. J. Bot.* 41: 317–325.

De Sylva, D.P. and Michel, H.B. 1975. Effect of mangrove defoliation on the estuarine ecology of South Vietnam. In *Proc. Int. Symp. Biol.* and *Management of Mangroves*, Vol. II, eds. G.E. Walsh, S.C. Snedaker and H.J. Teas, pp. 710–718. Int. Food Agric. Sci., Univ. Florida, Gainesville.

Dostal, R. 1967. *On Integration in Plants.* Harvard University Press, Cambridge, Massachusetts.

Facey, V. 1950. Abscission in leaves of *Fraxinus americana* L. *New Phytol.* 49: 103–116.

Gale, J. and Poljakoff-Mayber, A. 1970. Interrelations between growth and photosynthesis of saltbush (*Atriplex halimus* L.) grown in saline media. *Aust. J. Biol. Sci.* 23: 937–945.

Gill, A.M. and Tomlinson, P.B., 1971. Studies on the growth of red mangrove (*Rhizophora mangle* L.). III. Phenology of the shoot. *Biotropica* 3: 109–124.

Golley, F.B. 1969. Caloric value of wet tropical forest vegetation. *Ecology* 50: 517–519.

Greenway, H. and Osmond, C.B. 1972. Salt responses of enzymes from species differing in salt tolerance. *Plant Physiol.* 49: 256–259.

Hason-Porath, E. and Poljakoff-Mayber, A. 1969. The effect of salinity on the malic dehydrogenase of pea roots. *Plant Physiol.* 44: 1031–1034.

Hiatt, A.J. and Evans, H.J. 1960. Influence of salts on activity of malic dehydrogenase from spinach leaves. *Plant Physiol.* 35: 662–672.

Jamale, B.B. and Joshi, G.V. 1976. Physiological studies in senescent leaves of mangroves. *Indian J. Exp. Biol.* 14: 697–699.

Jamale, B.B. and Joshi, G.V. 1978. Effect of age on mineral constituents, polyphenols, polyphenol-oxidase and peroxidase in leaves of mangroves. *Indian J. Exp. Biol.* 16: 117–119.

Jennings, D.H. 1968. Halophytes, succulence and sodium in plants – a unified theory. *New Phytol.* 67: 899–911.

Joshi, G.V., Dolan, T., Gee, R. and Saltman, P. 1962. Sodium chloride effect on dark fixation of CO_2 by marine and terrestrial plants. *Plant Physiol.* 37: 446–449.

Joshi, G.V., Jamale, B.B. and Bhosale, L.J., 1975. Ion regulation in Mangroves. *Proc. Int. Symp. Biol.* and *Management of Mangroves*, Vol. II, eds. G.E. Walsh, S.C. Snedaker and H.J. Teas, pp. 595–607. Int. Food Agric. Sci., Univ. Florida, Gainesville.

Joshi, G.V. and Mishra, S.D. 1970. Photosynthesis and mineral metabolism in senesçent leaves of *Clerodendron inerme* Gaertn. *Indian J. Exp. Biol.* 8: 41–43.

Kalir, A and Poljakoff-Mayber, A. 1975. Malic dehydrogenase from *Tamarix* roots. Effect of sodium chloride *in vivo* and *in vitro*. *Plant Physiol.* 55: 155–162.

Kalir, A. and Poljakoff-Mayber, A. 1976. Effect of salinity on respiratory pathways in root tips of *Tamarix tetragyna*. *Plant Physiol.* 57: 167–170.

Karmarkar, S.M. and Amonkar, D.V. 1974. Effect of univalent and divalent ions on enzyme activity in *Salvadora persica* Linn. and *Cressa cretica* Linn. *M.V.M. Patrika* 9: 77–82.

Karmarkar, S.M. and Joshi, G.V. 1969. Mineral-constitutents of *Bryophyllum pinnatum* and their correlation with Crassulacean Acid Metabolism. *J. Uni. Bombay* 65: 73–78.

Karmarkar, S.M. and Ranganathan, T.P. 1971. Effects of sodium chloride on dehydrogenases and transaminases in *Bryophyllum pinnatum* and *Aloe vera*. *Indian J. Exp. Biol.* 9: 123–124.

Keller, B. 1925. Halophyten und xerophyten studien. *J. Ecol.* 13: 224.

Kotmire, S.Y. and Bhosale, L.J. 1978. Physiological studies in *Thespesia populnea* Corr. under saline and non saline habitats.

Jour. Biol. Sci. (in press).

Kozlowski, T.T. 1973 Extent and Significance of Shedding. In *Shedding of Plant Parts*, ed. T.T. Kozlowski, pp. 1–44. Academic Press, New York and London.

Lewington, R.J. and Simon, E.W. 1969. The effects of light on the senescence of detached cucumber cotyledons. *J. Exp. Bot.* 20: 138–144.

Link, A.J. and Swanson, C.A. 1960. A study of several factors affecting the dstribution of phosphorus-32 from the leaves of *Pisum sativum*. *Plant and Soil* 12: 57–68.

Leopold, A.C. 1961. Senescence in plant development. *Science* 134: 1727–1732.

McCollum, J.P. and Skok, J. 1960. Radiocarbon studies on the translocation of organic constituents into ripening tomato fruits. *Proc. Amer. Soc. Hort. Sci.* 75: 611–616.

Martin, C. and Thimann, K.V.1972. Role of protein synthesis in the senescence of leaves. I. The formation of protease. *Plant Physiol.* 49: 64–71.

Mendoza, M.M. 1971. The effects of NaCl on anatomical and physiological processes in *Atriplex hastata* L. M.S. Thesis. Univ. Utah, Salt Lake City.

Meyer, A. 1918. Eiweisstoffwechsel und Vergilben der Laubblatter von *Tropaeolum majus*. *Floa* (Jena) 11–12: 85–127.

Mishra, S.D. 1967. Physiological studies in mangroves of Bombay. Studies in *Clerodendron inerme* Gaertn. Ph.D. Thesis, University of Bombay.

Nimbalkar, J.D. and Joshi, G.V. 1974. Photosynthesis and mineral metabolism in senescent leaves of sugar cane Var. Co. 740. *Indian Agric.* 18: 257–264.

Nimbalkar, J.D. and Joshi, G.V. 1975. Physiological studies in senescent leaves of sugar cane Var. 740. *Indian J. Exp. Biol.* 13: 384–386.

Oata, Y. 1964. RNA in developing plant cells. *Ann. Rev. Plant Physiol* 15: 17–36.

Osborne, D.J. 1973. Internal factors regulating abscission. In *Shedding of Plant Parts*, ed. T.T. Kozlowski, pp. 125–147. Academic Press, New York and London.

Pool, D.J., Lugo, A.E. and Snedaker, S.C. 1975. Litter production in mangrove forests of Southern Florida and Puerto Rico. In *Proc. Int. Symp. Biol. and Management of Mangroves*, Vol. I, eds. G.E. Walsh, S.C. Snedaker and H.J. Teas, pp. 213–237 Int. Food Agric. Sci., Univ. Florida, Gainesville.

Pooviah, B.W. and Leopold, A.C. 1973. Defferal of Leaf Senescence with calcium. *Plant Physiol.* 52: 236–239.

Rao, U. 1974. Physiological studies in halophytes – Studies in *Sonneratia apetala* Buch-Ham. and *Sonneratia acida* Linn. M.Sc. Thesis, University of Bombay.

Sacher, J.A. 1973. Senescence and post harvest physiology. *Ann. Rev. Plant Physiol.* 24: 197–224.

Scott, F.M. Schroeder, M.R. and Turell, F.M. 1948. Development, cell shape, suberisation of internal surface and abscission in the leaf of the Valencia orange, *Citrus sinensis*. *Bot. Gaz.* (Chicago) 109: 381–411.

Seth, A.K. and Wareing, P.F. 1967. Hormone directed transport of metabolites and its possible role in plant senescence. *J. Exp. Bot.* 18: 65–77.

Siddhanti, L. 1977. Physiological studies in halophytes – Studies in *Acanthus ilicifolius* L. M.Sc. Thesis, University of Bombay.

Sidhu, S.S. 1963a. Studies on the mangroves of India. I. East Godavari region. *Indian For.* 89: 337–351.

Sidhu, S.S. 1963b. Studies on mangroves. In symposium on Ecological Problems in the Tropics. *Proc. Nat. Acad. Sci.*, India 33: 129–136.

Sokoloff, B., Redd, J.B. and Dutscher, R. 1950. Nutritive value of mangrove leaves (*Rhizophora mangle* L.). *Q. J. Fla. Acad. Sci.* 12: 191–194.

Spencer, P.W. and Titus, J.S. 1972. Biochemical and enzymatic changes in apple leaf tissue during autumnal senescence. *Plant Physiol.* 49: 746–750.

Teas, H.J. and Kelly, J. 1975. Effects of herbicides on mangroves of Southern Vietnam and Florida. *In Proc. Int. Symp. Biol. and Management of Mangroves*, Vol. II, eds. G.E. Walsh, S.C. Snedaker and H.J. Teas, pp. 719–728. Int. Food Agric. Sci., Univ. Florida, Gainesville.

Tetley, R.M. and Thimann, K.V. 1974. The metabolism of oat leaves during senescence. I. Respiration, carbohydrate metabolism and action of cytokinins. *Plant. Physiol.* 53: 294–303.

Varner, J.E. 1961. Biochemistry of senescence. *Ann. Rev. Plant Physiol.* 12: 245–264.

Walsh, G.E., 1974. Mangroves: a review. In *Ecology of Halophytes*, eds. R.J. Reimold and W.H. Queen, pp. 51–174. Academic Press, New York.

Walter, H. 1955. The water economy and the hydrature of plants. *Ann. Rev. Plant. Physiol.* 6: 239–253.

Walter, H. and Stadelmann, E. 1974. A new approach to the water relations of desert plants. In *Desert Biology*, Vol. II, ed. G.W. Brown, Jr., pp. 212–310. Academic Press, New York.

Weimberg, R. 1970. Enzyme levels in pea seedlings grown on highly salinised media. *Plant Physiol.* 46: 466–470.

Weimberg, R. 1975. Effect of growth in highly salinised media on the enzymes of the photosynthetic apparatus in pea seedlings. *Plant Physiol.* 56: 8–12.

Wiesner, J. 1905. *Ber deut botan Ges.* 23: 49. as cited by Joshi, G.V. and Mishra, S.D. 1970. Photosynthesis and mineral metabolism in senescent leaves of *Clerodendron inerme* Gaertn. *Indian J. Exp. Biol.* 8: 41–43.

Wildman, H.G. and Bonner, J. 1947. As cited by Woolhouse, H.W. 1972. *Ageing Processes in Higher Plants*. Oxford Univ. Press London.

Williams, R.F. 1938. Physiological ontogeny in plants and its relation to nutrition. *Aust. J. Exp. Biol. Med. Sci.* 16: 65–83.

Winter, K. 1975. The effect of salinity on growth and photosynthesis in the halophytes, *Mesembryanthemum nodiflorum* L. *and Suaeda maritima* L. Dumort. *Oecologia* 17: 317–324.

CHAPTER 6

Ecophysiological aspects of some tropical salt marsh halophytes

A.J. JOSHI

Department of Botany , Sir P.P. Institute of Science,
Bhavnagar University, Bhavnagar 364 002, India

1. Introduction

Although the oldest reference to halophytes dates back to 1563, when Dodoens described *Plantago maritima* for the first time (cf. Uphof, 1941), the term 'halophyte' was introduced much later by Pallas, a Russian botanist in 1803 (cf. Strogonov, 1973). Subsequent literature has been discussed periodically in many reviews and books (Uphof, 1941; Adriani, 1958; Chapman, 1960; Waisel, 1972; Reimold and Queen, 1974; Poljakoff-Mayber and Gale, 1975; Flowers, 1975; Flowers et al., 1977; Ungar, 1978). Strogonov (1964, 1973) has contributed two excellent reviews of the work done in the U.S.S.R. Navalkar (1973) and Joshi (1976) discussed the results of Indian workers mainly on mangroves. Moreover, recent information on halophytes as well as on mangroves has been discussed fully in other chapters of this book. Only additional data on Indian halophytes will therefore be discussed in the following pages.

2. Soils

About 3,56,500 hectares of coastal area in India has been occupied by halophytes, including mangroves (Blasco et al., 1975). In spite of having such a large area, limited attempts have been made to know the mechanical composition of such soils (Table 1). It is evident that halophytes occur on soils having a sandy loam to silty loam texture.

Raheja (1966, 1968) gave an exhaustive account of saline soils and their problems in India, but few efforts have been made to carry out detailed investigations in this direction. Analysis of soil samples collected during different seasons from three habitats inhabited by different halophytes near Bhavnagar (21°45′ N, 72° 14′ E) showed that pH of saturation extracts varies from 7.4 to 8.3 (Table 2). The salinity of the habitat was at a minimum during monsoon but increased during the dry period i.e. summer through winter. Exchangeable sodium percentage (ESP) and sodium adsorption ratio (SAR), important factors affecting swelling and dispersion of clay particles, were found to be extremely high in all the cases. These observations suggest that the soils have saline-alkaline characteristics.

The soils inhabited by halophytes have a wide range of texture. Indian halophytes occur on sandy loam to silty loam soils. Ungar (1965, 1968) and Ungar et al. (1969) found *Suaeda depressa* and other halohytes growing mostly on sandy loam soils in different parts of the U.S.A. The difference in the soil structure observed by them distinguishes the inland halophytic habitats. However, Chapman (1947) reported *Suaeda maritima* growing on silty clay soil. Wide range in soil texture of halophytic habitats in general and of a single habitat (e.g. *Suaeda nudiflora*) in particular suggest that distribution of halophytes has little relation with soil texture.

Rao and Aggarwal (1964, 1966) and Rao et al. (1963a, b, 1966) have studied pH, total soluble salts

Table 1. Texture percentage of saline soils in India.

Species	Locality	Coarse sand	Fine sand	Silt	Clay	Reference
Suaeda nudiflora	Old Port, Bhavnagar	0.33	23.82	61.85	14.00	Rao and Mukherjee (1967)
	Jafrabad	9.34	30.06	47.55	13.05	Rao and Mukherjee (1967)
	Victor Albert Port	6.42	58.33	17.50	17.75	Rao and Mukherjee (1967)
	Gogla (Diu)	0.90	58.60	37.50	3.00	Rao and Aggarwal (1964)
	Bhavnagar	—	12.70	37.50	27.50	Joshi and Iyengar (unpubl.)
Suaeda maritima	Cauvery Delta	1.50	44.06	11.00	36.00	Blasco (1975)
Salicornia brachiata	Bhavnagar	—	27.50	38.80	31.80	Joshi and Iyengar (unpubl.)
Salicornia sp.	Rameshwaram	67.40	11.70	15.60	5.30	Rao et al. (1963a)
Aeluropus lagopoides	Okha	0.40	33.30	51.40	14.90	Rao and Aggarwal (1964)

Table 2. Salinity characteristics of salt marsh soils.[a]

Season	Species	pH	EC mmhos/cm 25°C	ESP	SAR
Monsoon	*Suaeda nudiflora*	7.5	18.32	36.21	39.34
	Salicornia brachiata	7.9	43.82	53.86	79.99
Winter	*Suaeda nudiflora*	8.0	53.16	50.25	69.33
	Salicornia brachiata	7.4	156.71	79.97	271.47
	Sesuvium portulacastrum	—	94.87	69.87	157.09
Summer	*Suaeda nudiflora*	8.2	130.33	76.06	216.30
	Salicornia brachiata	8.3	216.59	83.77	350.77
	Sesuvium portulacastrum	—	109.19	68.21	146.34

ESP = Exchangeable sodium percentage; SAR = Sodium adsorption ratio; [a] Joshi and Tyengat (unpubl.).

(TSS) and other characteristics of soil samples from different coastal localities in India. Blasco (1975) also gave information on habitat of *Suaeda monoica*. These workers reported pH varying from 7.1 to 7.8 and TSS from 1.4 to 3.7 % in 1:5 soil:water extracts. Similarly, Rajpurohit et al. (1979) reported less salinity in *Suaeda fruticosa* habitat. However, our observations suggest that more salinity prevails in *Suaeda nudiflora* and *Salicornia brachiata* habitats.

It has been suggested that pH may be an important factor in plant distribution after a certain salinity has been reached (Keith, 1958; other investigators however do not agree with this (Evans, 1953; Kurz and Wagner, 1957; Ungar, 1974). The small differences in pH of the soil extracts found during present study indicate an absence of any relation between pH and distribution of halophytes.

3. Seed germination

Germination is a critical stage in the life cycle of halophytes since the salinity of the saline habitats affects this stage. Most of the work done on germination of halophytes has been dealt with in detail by Waisel (1972) and Ungar (1978). Thus, only important results on Indian species are presented in Table 3.

Seed coat in *S. nudiflora* was fonund to be hard, and after scarification with sulphuric acid, increase in germination was observed. Scarified seeds germinated well up to 10,000 ppm seawater. 63% seeds of *S. brachiata* germinated in 36,000 ppm seawater although more germination was seen in less saline conditions. This indicates that seeds of this species can germinate even in seawater per se.

The seeds of *Suaeda nudiflora* germinated in seawater up to 10,000 ppm. Similarity in salt tolerance by other species of *Suaeda* has been reported earlier. For example, 18.6% of the seeds of *S. novae-zealandiae* (Chapman, 1960) and 23% of the seeds of *S. linearis* (Ungar, 1962) germinated in 1% NaCl. Germination of *S. monoica* (Waisel and Ovadia, 1972) and *S. depressa* (Williams and Ungar, 1972) occurred in 0.6 M (3.08 %) and 4% NaCl, respectively. Thus *S. nudiflora* is lesser salt tolerant than the last two species.

Many species of *Salicornia* showed a high degree of salt tolerance at germination stage. For instance, 38, 36 and 12% seeds of *S. stricta* germinated in seawater, 5% and 10% NaCl, respectively (Chapman, 1960); 12 and 8% seeds of *S. europaea* in 3 and 5% NaCl (Ungar, 1967); 87% seeds of *S. bigelovii* in 4.04% sea salt solution (Riwers and Weber, 1971). Seeds of *S. europaea* (Ungar, 1974) and *S. herbacea* (Walter, 1968 cited by Waisel, 1972) germinated even in 10% NaCl; those of *S. rubra* in 5% NaCl

Table 3. Effects of salinity on seed germination.

Treatment	*Suaeda nudiflora*[a]		*Salicornia brachiata*[b]
	Unscarified	Scarified	
Dist. water	38	56	89
Seawater (ppm)			
1,000	42	58	88
5,000	40	46	84
10,000	6	20	84
20,000	—	—	63
36,000	—	—	63

[a] Joshi and Iyengar (1977); [b] Joshi (1981a).

(Ungar, 1974). A 63% germination of *S. brachiata* in seawater per se found during the present study is higher than reported so far for any species of *Salicornia*, except *S. bigelovii* (Riwers and Weber, 1971), in seawater or equal NaCl salinity.

4. Mineral composition

Halophytes are able to survive in considerably high concentration of salts in the habitats. In doing so the salts are accumulated either in the plant until it dies or in the deciduous leaves which fall from the plants; the excessive salts are even excreted thorough the glands, etc. Moreover, the changing salinity of habitats have direct effects on mineral uptake and their accumulation in the plants. Observations on such changes in two halophytes, one in which the salts are accumulated in the entire plant (*Salicornia brachiata*) and the other in which the accumulation takes place only in leaves (*Suaeda nudiflora*) are presented in Table 4. Salt content in the latter species increases during winter and summer when the salinity of the habitat is also increased (Table 2) and the leaves become succulent. However, in *S. brachiata* maximum salt content was found during monsoon, when all plants were green and succulent. Less ash content during winter and summer may be due to drying of lower parts of the plants. It was observed that lower part of the main stem and branches dried earlier and so that only half of the distal apical parts which showed senescence were analyzed. Sodium (Na) and chloride (Cl) were found to be the main constituents of the ash and demonstrated a direct relationship with increased salt accumulation. Potassium (K) content showed an inverse relation with that of Na. With regard to divalent cations, magnesium (Mg) accumulation was always more than calcium (Ca). It was not possible to observe any seasonal change in the sulphate (SO_4) content in these halophytes.

A considerably high ash content was found in *Suaeda nudiflora* and *Salicornia brachiata* as a result of excessive accumulation of salts. Waisel and Ovadia (1972) reported 34 to 43% ash in *Suaeda monoica*; while Chaudhri et al. (1964) found 41.2% ash in *Suaeda fruticosa*. This is also true for

Table 4. Seasonal variations in mineral composition of halophytes.[a]

Season	Species	Ash %	Meq/100 g dry wt					
			Na	K	Ca	Mg	Cl	SO$_4$
Monsoon	*Suaeda nudiflora*	44.08	677.0	23.8	29.9	80.6	498.6	46.2
	Salicornia brachiata	57.83	735.2	17.9	25.5	159.6	983.3	38.5
Winter	*Suaeda nudiflora*	50.52	689.6	23.8	47.2	110.2	588.5	71.2
	Salicornia brachiata	47.53	576.5	32.7	32.9	90.5	685.5	23.1
Summer	*Suaeda nudiflora*	54.44	837.0	20.0	37.9	69.1	571.6	47.1
	Salicornia brachiata	44.09	552.2	33.3	26.0	89.6	826.3	26.4

[a] Joshi and Iyengar (unpubl.).

many species of *Salicornia* (Kabanov and Otegenov, 1973; Gorham and Gorham, 1955; Keller, 1951 cited by Kabanov and Otegenov, 1973).

Sodium accumulation, in *Suaeda nudiflora* and *Salicornia brachiata* was found to be higher than any other cation. These observations are further strengthened by similar reports on different species of *Suaeda* (Chaudhri et al., 1964; Waisel, 1972; Wiebe and Walter, 1972; Albert and Popp, 1977) and *Salicornia* (El-Gabaly, 1961; Harward and McNulty, 1965; Hansen and Weber, 1975; Albert and Popp, 1977). However, Zellner (1926 cited by Chapman, 1960) reported maximum (26.69% of dry wt) Na in *Suaeda salsa*. Less Na in red plants of *Salicornia rubra* due to senescence than green ones was observed by Tiku (1975). This also supports the present observation on *Salicornia brachiata*. Excessive Na accumulation can be due to preferential uptake of this ion by halophytes (Flowers, 1975). Jennings (1968) believed that Na ions have a positive role in increased succulence in halophytes.

Potassium (K) content in *Suaeda nudiflora* and *Salicornia brachiata* indicates that K is sufficiently absorbed by them since the adequate level of this cation in plants is considered to be 1% of dry wt (cf. Epstein, 1965). K uptake in halophytes is not fully understood. Rains and Epstein (1967) explained Dual Carrier Mechanism' in *Avicennia marina*. Experiments on *Suaeda maritima* showed that increasing Na concentrations suppressed K uptake from a 5 meq solutions, whereas K uptake from 0.05 meq solution was uneffected until the concentration of Na was greater than 10 meq (Flowers, 1975). These reports call for further information on K uptake from Na rich environments.

Calcium (Ca) in *Suaeda nudiflora* and *Salicornia brachiata* varies from 26 to 47.2 meq /100 g dry wt. Similar values of Ca accumulation in other species of *Suaeda* have been reported (Chaudhri et al., 1964; Waisel, 1972), while in some, less Ca was observed (Albert and Popp, 1977; Wiebe and Walter, 1972). Interestingly, in other species of *Salicornia* reported so far (Wiebe and Walter, 1972; Albert and Popp, 1977) less Ca was found than in *Salicornia brachiata*. NaCl controlled Ca uptake has been elucidated by Waisel (1972), who suggested that Ca uptake decreased when NaCl concentration was increased upto 100 mM, but above this level no significant change in Ca uptake was observed. As Ca has been known to inhibit the growth of halophytes (Van Eijk, 1939; Osmond, 1966; Greenway, 1968), its limited uptake can be considered favourable for halophytes.

The present results on Mg content in *Suaeda nudiflora* and *Salicornia brachiata* show similarity with other species of these genera (Chaudhri et al., 1964; Wiebe and Walter, 1972; Albert and Popp, 1977). The last authors believed that plants growing in saline conditions have more Mg than K. Though the values (Table 4) agree with this observation, it was not always applicable (Joshi, 1979).

Chloride (Cl) content in *Suaeda nudiflora* and *Salicornia brachiata* compensates for cationic balance. However, these values are more than Cl content found in many other species of *Suaeda* (Chaudhri et al., 1964; Waisel, 1972; Wiebe and Walter, 1972; Albert and Popp, 1977) and *Salicornia* (El-Gabaly, 1961; Harward and McNulty, 1965; Hansen and Weber, 1975; Albert and Popp, 1977). Joshi (1976) suggested that Cl uptake in plants subjected to saline conditions may be of 'passive' nature.

Sulphate (SO_4) content is less than Cl in *Suaeda nudiflora* and *Salicornia brachiata*. Many Cl accumulating halophytes generally have less sulphate content (Waisel, 1972), which may be due to its low uptake being divalent anion or its utilization in assimilation of organic compounds (Walter, 1961).

5. Amino acids

Salinity alters many biochemical reactions in plants as a result of which many intermediate products are formed and free amino acids represent a group of such compounds in the plants subjected to salinity (cf. Strogonov, 1973). The investigations on free amino acids in halophytes which occur in extremely saline milieus suggest that proline always occurs in large quantities (Goas, 1965, 1967; Larther, 1970; Stewart and Lee, 1974). However, much remains to be investigated on correlation of free amino acids in halophytes with changing salinity of the habitat.

The results mentioned in Table 5 represent winter values in *S. nudiflora*, *S. brachiata* and *S. portulacastrum*. Proline, aspartic acid, glutamic acid, alanine, glycine, serine are the main amino acids, whereas concentration of other amino acids varies from traces to 32 μg/g fr wt. Some of the acids of the latter group found in one species were not found in others. Proline accumulation was maximum during winter in these halophytes (Joshi, 1979) accompanied by maximum succulence of salt accumulating parts.

Strogonov (1973) assumed that survival of plants in saline environments depends upon the altered biochemical relations and on the quantitative ratio between toxic and protective compounds. According to him proline is one of the protective substances. Interestingly, the present study revealed markedly high accumulation of proline accounting for 20 to 83% of the total amino acids. Stewart and Lee (1974) suggested that proline functions as a source of solute for intracellular osmotic adjustment. Aspartic acid and glutamic acid which also occur in great quantities in *Suaeda nudiflora* and *Salicornia brachiata*, are involved in synthesis of

Table 5. Free amino acids in halophytes.

Amino acid	S. nudiflora	S. brachiata	S. portulacastrum[a]
Proline	98.85[b]	190.90	416.24
Aspartic acid	56.85	73.19	33.33
Glutamic acid	46.85	34.16	20.30
Alanine	122.36	38.65	10.01
Glycine	16.07	20.24	2.04
Serine	25.25	47.26	15.01
γ aminobutyric acid	20.07	11.78	Tr.
Phenylalanine	32.46	—	2.88
Methionine	11.83	6.91	0.36
Valine	6.57	4.76	0.41
Threonine	7.17	10.12	2.29
Leucine	8.52	4.79	Tr.
Isoleucine	4.06	5.05	Tr.
Tyrosine	—	—	Tr.
Glutamine	6.66	—	—
Asparagine	7.88	—	—
Ornithine	9.97	—	—
Total	481.42	447.81	502.87
Proline % of total amino acids	20.54	42.63	82.87

[a] Joshi (1981 b); [b] μg/g fr wt; Tr. = traces.

amides that in turn function as protective substances (Strogonov, 1973). However, alanine, phenylalanine, methionine, leucine, isoleucine, and tyrosine are believed to have toxic effects in plants. The fact that they do not accumulate as much in these halophytes indicates that they are endowed with an adaptative physiology that leads to minimum synthesis of these free amino acids. It is difficult to explain alanine content which was also found in great quantity in other samples also (Joshi, 1979). Thus the present study on halophytes partly supports Strogonov's assumption (1973). More detailed study is called for in this direction.

Acknowledgements

I am greatly indebted to Dr. D.J. Mehta and Dr. E.R.R. Iyengar, Director and Assistant Director, respectively, at the Central Salt and Marine Chemicals Research Institute, Bhavnagar, for the facilities I was allowed to use, and for the encouragement and help at various stages of this research on halophytes.

Literature cited

Adriani, M.J. 1958. Halophyten. *Encycl. Plant Physiol.* 4: 709–736.

Albert, R. and Popp, M. 1977. Chemical composition of halophytes from the Neusiedler Lake region in Austria. *Oecologia* 27: 157–170.

Blasco, F. 1975. The Mangroves of India. Institut Français de Pondicherry Publ., India.

Blasco, F. Caratini, C., Chanda, S. and Thanikaimoni, G. 1975. Main characteristics of Indian Mangroves. In *Proc. Int. Symp. Biol. and Management of Mangroves.*, Vol. I, eds. G.E. Walsh, S.C. Snedaker and H.J. Teas, pp. 71–87. Int. Food Agric. Sci., Univ. Florida, Gainesville.

Chapman, V.J. 1947. *Suaeda maritima* (L.) Dum. *J. Ecol.* 35: 293–302.

Chapman, V.J. 1960. *Salt Marshes and Salt Deserts of the World.* Leonard Hill Book Ltd., London.

Chaudhri, I.I., Shah, B.H., Nagri, N., and Mallik, I.A. 1964. Investigations on the role of *Suaeda fruticosa* in the reclamation of saline and alkaline soils of West Pakistan. *Plant & Soil* 21: 1–7.

El-Gabaly, M.M. 1961. Studies on salt tolerance and specific ion effects on plants. UNESCO, *Arid Zone Res. Salinity Problems in the Arid Zone, Proc. Tehran Symp.* 14: 169–174.

Epstein, E. 1965. Mineral metabolism. In *Plant Biochemistry*, eds. J. Bonner and J.E. Varner, pp. 438–466. Academic Press, New York and London.

Evans, L.T. 1953. The ecology of halophytic vegetation at Lake Ellesmere, New Zealand. *J. Ecol.* 41: 106–122.

Flowers, T.J. 1975. Halophytes. In *Ion Transport in Plant Cells and Issues*, eds. D.A. Baker and J.L. Hall, pp. 309–333. North-Holland Publ. Co., Amsterdam.

Flowers, T.J., Troke, P.F. and Yeo, A.R. 1977. The mechanism of salt tolerance in halophytes. *Ann. Rev. Plant Physiol.* 28: 89–121.

Goas, M. 1965. Sur Le metabolisme azote des halophytes: Etude des acides amines et amides libres. *Soc. Fr. Physiol. Veg. Bull.* 11: 309–316.

Goas, M. 1967. Contribution a l'etude-du metabolisme azote des halophytes: Acides amines et amides libres d'*Aster tripolium* L. in aquiculture. *C.R. Acad. Sci. Paris* 265: 1049–1052.

Gorham, A.V., and Gorham, V. 1955. Iron, manganese, ash and nitrogen in some plants from salt marsh and shingle habitats. *Ann. Bot* 19: 571–577.

Greenway, H. 1968. Growth stimulation by high chloride concentration in halophytes. *Israel J. Bot.* 17: 169–177.

Hansen, D.J., and Weber, D.J. 1975. Environmental factors in relation to the salt content of *Salicornia pacifica* var. *Utahensis. Great Basin Natur.* 35: 86–96.

Harward, M.R., and McNulty, I. 1965. Seasonal changes in ionic balance in *Salicornia rubra. Utah Acad. Proc.* 42: 65–69.

Jennings, D.H. 1968. Halophytes, succulence and sodium in plants – a unified theory. *New Phytol.* 67: 899–911.

Joshi, A.J. 1979. Physiological studies on some halophytes. Ph.D. thesis, Saurashtra Univ., Rajkot.

Joshi, A.J. 1981a. Germination of *Salicornia brachiata* Roxb., a salt marsh halophyte. *Indian J. Plant Physiol.* (in press).

Joshi, A.J. 1981b. Amino acids and mineral constituents in *Sesuvium portulacastrum* L., a salt marsh halophyte. *Aquatic Botany* 10: 69–74.

Joshi, A.J. and Iyengar, E.R.R. 1977. Germination of *Suaeda nudiflora* Moq. *Geobios* 4: 267–268.

Joshi, G.V. 1976. *Studies in Photosynthesis under Saline Conditions.* Report P.L. 480 Project No. A7SWC-95, Shivaji Univ., Kolhapur, India.

Kabanov, V.V., and Otegenov, Zh. 1973. Effects of sodium chlorides and sulphates on plant composition. *Sov. Plant Physiol.* 20: 682–689.

Keith, L.B. 1958. Some effects of increasing soil salinity on plant communities.*Can. J. Bot.* 36: 79–89.

Keller, B.A. 1951. Izbrannye sochineniya, Izdatel'stvo Akademii Nauk SSSR, cited by Kabanov, V.V. and Otegenov, Zh. 1973. Effects of sodium chlorides and sulphates on plant composition. *Sov. Plant Physiol.* 20: 682–689.

Kurz, H., and Wagner, K. 1957. Tidal marshes of the Gulf and Atlantic coasts of North Florida and Charleston, South Carolina. *Fl. St. Univ. St.* 24: 1–168.

Larther, F. 1970. Contribution a l'etude du metabolisme azote de *Limonium vulgare* Mill. Acides amines et amides libres. *Bull. Soc. Sci. de Bretagne* 45: 27–32.

Navalkar, B.S. 1973. Ecology of mangroves. In *Progress of Plant Ecology in India* Vol. I, eds. R. Misra, B. Gopal, K.P. Singh and J.S. Singh, pp. 101–109. Today & Tomorrow's Printers and Publishers, New Delhi, India.

Osmond, C.B. 1966. Divalent cations absorption and interaction in *Atriplex. Aust. J. Biol. Sci.* 19: 37–48.

Poljakoff-Mayber, A., and Gale, J. 1975. *Plants in Saline Environments.* Springer Verlag, Berlin.

Raheja, P.C. 1966. Aridity and salinity – A survey of soils and land use. In *Salinity* and *Aridity*, ed. H. Boyko, pp. 43–127. Dr. W. Junk Publishers, The Hague.

Raheja, P.C. 1968. Saline soil problems with particular reference to irrigation with saline water in India. In *Saline Irrigation for Agriculture and Forestry*, ed. H. Boyko, pp. 217–233. Dr. W. Junk Publishers, The Hague.

Rains, D.W. and Epstein, E. 1967. Preferential absorption of potassium by leaf tissue of the mangrove, *Avicennia marina*: an aspect of halophytic competence in coping with salt. *Aust. J. Biol. Sci.* 20: 847–857.

Rajpurohit, K.S., Charan, A.K. and Sen, D.N. 1979. Micro distribution of plants in an abandoned salt pit at Pachpadra salt basin. *Ann. Arid Zone* 18: 122–126.

Rao, Ananda, T. and Aggarwal, K.R. 1964. Ecological studies of Saurashtra coast and neighbouring islands. III – Okhamandal point to Diu coastal areas. *UNESCO, Proc. Symp. Prob. Indian Arid Zone*, Jodhpur. pp. 31–42.

Rao, Ananda, T. and Aggarwal, K.R. 1966. Ecological studies of Saurashtra coast and neighbouring islands. II – Beyt Island. *Bull. Bot. Surv. Indian* 8: 16–24.

Rao, Ananda T. and Mukherjee, A.K. 1967. Ecological studies of Saurashtra coast and neighbouring islands. V – Jafrabad to

Bhavnagar coastal area. *Ibid.* 9: 79–87.

Rao, Ananda T., Aggarwal K.R. and Mukherjee, A.K. 1963 a. An ecological account of the vegetation of Rameswaram Island. *Ibid.* 5: 301–323.

Rao, Ananda T., Aggarwal, K.R. and Mukherjee, A.K. 1963 b. Ecological studies on the soil and vegetation of Krusadi group of islands in the Gulf of Manaar. *Ibid.* 5: 141–148.

Rao, Ananda T., Aggarwal, K.R. and Mukherjee, A.K. 1966. Ecological studies of Saurashtra coast and neighbouring islands. IV – Piram Island. *Ibid.* 8: 60–68.

Reimold, R.J. and Queen, W.H. eds. 1974. *Ecology of Halophytes.* Academic Press, New York.

Riwers, W.G. and Weber, D.J. 1971. The influence of salinity and temperature on seed germination in *Salicornia bigelovii. Physiol. Plant.* 24: 73–75.

Stewart, G.R. and Lee, J.A. 1974. The role of proline accumulation in halophytes. *Planta* 120: 279–289.

Strogonov, B.P. 1964. *Physiological Basis of Salt Tolerance of Plants.* IPST, Jerusalem.

Strogonov, B.P. 1973. *Structure and Function of Plant Cells in Saline Habitats.* IPST, Jerusalem.

Tiku, B.L. 1975. Ecophysiological aspects of halophyte zonation in saline sloughs. *Plant & Soil* 43: 355–369.

Ungar, I.A. 1962. Influence of salinity on seed germination in succulent halophytes. *Ecology* 43: 763–764.

Ungar, I.A. 1965. An ecological study of the vegetation of the big salt marsh. Stafford county, Kansas. *The Univ. Kansas Sci. Bull.* 46: 1–99.

Ungar, I.A. 1967. Influence of salinity and temperature on seed germination. *Ohio J. Sci.* 67: 120–123.

Ungar, I.A. 1968. Species-soil relationships on the Great Salt Plains of Northern Oklahoma. *Amer. Midl. Nat.* 80: 392–406.

Ungar, I.A. 1974. Inland halophytes of the United States. In *Ecology of Halophytes,* eds. R.J. Reimold and W.H. Queen, pp. 235–305. Academic Press, New York.

Ungar, I.A. 1978. Halophyte seed germination. *Bot. Rev.* 44: 233–264.

Ungar, I.A., Hogan, W. and McClelland, M. 1969. Plant communities of saline soils at Licoln, Nebraska. *Amer. Midl. Nat.* 82: 564–577.

Uphof, J.C.T. 1941. Halophytes. *Bot. Rev.* 7: 1–58.

Van Eijk, M. 1939. Analyse der Wirkung des NaCl auf die Entwicklung sukkulenze und transpiration bei *Salicornia herbacea,* sowie Untersuchungen über den Einfluss der Salzaufnahme, auf die Würzelatmung bei *Aster tripolium. Rec. Trav. Bot. Neerl.* 36: 559–657.

Waisel, Y. 1972. *Biology of Halophytes.* Academic Press, New York.

Waisel, Y. and Ovadia, S. 1972. Biological flora of Israel. *Suaeda monoica* Forssk. ex J.F. Gmel. *Israel J. Bot.* 21: 42–52.

Walter, H. 1961. The adaptation of plants to saline soils. UNESCO, *Arid Zone Res. Salinity Problems in the Arid Zones, Proc. Tehran Symp.* 14: 129–134.

Walter, H. 1968. *Die Vegetation der Erde in Okologischer Betrachtung,* Band II. Fischer Verlag, Stuttgart, cited by Waisel, Y. 1972. *Biology of halophytes,* Academic Press, New York.

Wiebe, H.H. and Walter, H. 1972. Mineral ion composition of halophytic species from Northern Utah. *Amer. Midl. Nat.* 87: 241–245.

Williams, M.D. and Ungar, I.A. 1972. The effect of environmental parameters on the germination, growth, and development of *Suaeda depressa* (Pursh) Wats. *Amer. J. Bot.* 59: 912–918.

Zellner, J. 1926. Zur Chemie der Halophyten. *Ber. Acad. Wiss. Wien. Math-Naturwiss. Kl. Abt.* IIb 135: 585–592, cited by Chapman, V.J. 1960. *Salt Marshes and Salt Deserts of the World.* Leonard Hill Books Ltd., London.

CHAPTER 7

Adaptation of plants to saline environments: salt excretion and glandular structure

NILI LIPHSCHITZ and YOAV WAISEL

Department of Botany, Tel-Aviv University, Tel-Aviv, Israel

1. Introduction

Excretion of ions by special salt glands is a well known mechanism for regulating the mineral content of many halophytic plants.

Such glands were discovered and described for various plant species already since the second half of the nineteenth century (Volkens, 1884; Marloth, 1887). The first salt glands were described in the Tamaricaceae (Marloth, 1887) and in the Plumbaginaceae (Volkens, 1884; Schtscherback, 1910; Ruhland, 1915; DeFraine, 1916). Since then an increasing attention was given to the structure and function of salt glands, to the mechanisms of salt excretion, and to its ecological significance.

Salt glands are common in many plant species; most of them appear in present-day halophytes. Some glands appear in species that today occupy rather non-saline environments. Excretion occurs in such plants only when they are transferred from the glycophytic to the semi-halophytic or halophytic habitat (e.g. Gramineae family - c.f. Liphschitz and Waisel, 1974, 1978).

Till now, salt glands were discovered in 11 families including 17 genera of the Dicotyledons and in one family of the Monocotyledons, i.e. the Gramineae. The latter includes 16 genera all of the Chloridoideae subfamily, with active secreting salt glands and 17 genera of the Panicoideae subfamily with non-functioning or slightly functioning glands (Table 1).

2. Structure of salt glands

The structure of salt glands varies greatly in different plant species but it is usually similar in plants within one family.

The simplest type of gland is found in the Gramineae. Glands are present on both sides of the leaves in longitudinal rows parallel to the veins. Each gland consists of two cells: a basal cell and a cap cell. The basal cell is the collecting cell whereas the upper cell is the excreting one. The cells contain dense cytoplasm and a prominent nucleus but lack a central vacuole. The gland is characterized by cutinized or suberized cell walls. Cutinization is heaviest on the outer walls of the upper excreting cell and on those walls of the lower cell which border the neighbouring epidermal cells. The walls of the basal cell are lignified in their upper part, i.e. around the 'bottle-neck' region of the gland (Plate I).

The basic structure of the glands is similar in all genera of the two subfamilies, i.e. Chloridoideae and Panicoideae. The Chloridoideae includes three tribes and the Panicoideae includes two tribes. Nevertheless, in spite of the basic uniform structure, variations in form and function of the salt gland occur. Variations in form can be observed in the basal as well as in the cap cell (Plate II).

The glands of the Chloridoideae are sunken into the epidermis in some species (e.g. *Spartina*), or located above it (e.g. *Bouteloua*). Transition forms which are semi-sunken are also seen in some species

Table 1. Occurrence of salt glands in the various families, subfamilies, and genera of plants.

Family (subfamily)	Genus at present occupying saline habitat	at present occupying non-saline habitat
Acanthaceae	*Acanthus* (Walter and Steiner, 1936)	
Avicenniaceae	*Avicennia* (Biebel and Kinzel, 1965)	
Combretaceae	*Laguncularia* (Biebel and Kinzel, 1965)	
Convolvulaceae	*Cressa* (Volkens, 1887)	
Frankeniaceae	*Frankenia* (Volkens, 1887)	
Gramineae Chloridoideae	*Spartina (Skelding and Winterbotham, 1939)*	*Chloris* (Liphschitz et al., 1974)
	Aeluropus (Pollak and Waisel, 1970)	*Cynodon* (Liphschitz and Waisel, 1974)
	Distichlis (Wiebe and Walter, 1972)	*Bouteloua* (Liphschitz and Waisel, 1974)
		Buchloe (Liphschitz and Waisel, 1974)
		Coelachyrum (Liphschitz and Waisel, 1974)
		Crypsis (Liphschitz and Waisel, 1974)
		Dactyloctenium (Liphschitz and Waisel, 1974)
		Dinebra (Liphschitz and Waisel, 1974)
		Eleusine (Liphschitz and Waisel, 1974)
		Enteropogon (Liphschitz and Waisel, 1974)
		Sporobolus (Liphschitz and Waisel, 1974)
		Tetrachne (Liphschitz and Waisel, 1974)
		Tetrapogon (Liphschitz and Waisel, 1974)
Panicoideae		*Andropogon* (Liphschitz and Waisel, 1978)
		Brachiaria (Liphschitz and Waisel, 1978)
		Cenchrus (Liphschitz and Waisel, 1978)
		Chrysopogon (Liphschitz and Waisel, 1978)
		Coix (Liphschitz and Waisel, 1978)
		Dichanthium (Liphschitz and Waisel, 978)
		Digitaria (Liphschitz and Waisel, 1978)
		Echinochloa (Liphschitz and Waisel, 1978)
		Erianthus (Liphschitz and Waisel, 1978)
		Hyparrhenia (Liphschitz and Waisel, 1978)
		Panicum (Liphschitz and Waisel, 1978)
		Paspalum (Liphschitz and Waisel, 1978)
		Paspalidium (Liphschitz and Waisel, 1978)
		Saccharum (Liphschitz and Waisel, 1978)
		Setaria (Liphschitz and Waisel, 1978)
		Sorghum (Liphschitz and Waisel, 1978)
		Tricholaena (Liphschitz and Waisel, 1978)
Myrsinaceae	*Aegiceras* (Scholander et al., 1962)	
Plumbaginaceae	*Aegialitis* (Atkinson et al., 1967)	
	Armeria (Berry, 1970)	
	Limonium (Arisz et al., 1955)	
	Statice (Volkens, 1884)	
Primulaceae	*Glaux* (Rozema, 1975)	
Rhizophoraceae	*Rhizophora* (Atkinson et al., 1967)	
	Ceriops (Walter and Steiner, 1936)	
	Bruguiera (Walter and Steiner, 1936)	
Sonneratiaceae	*Sonneratia* (Walter and Steiner, 1936)	
Tamaricaceae	*Tamarix* (Marloth, 1887)	
	Reaumuria (Volkens, 1887)	

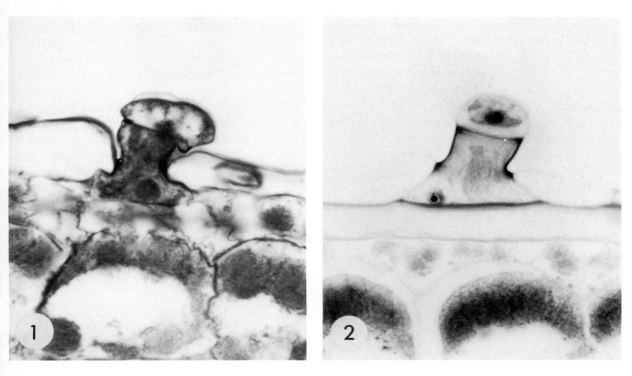

Plate I. Salt gland of *Chloris gayana*. Lower collecting cell and upper secreting cell. 1. × 1330; 2. × 1330.

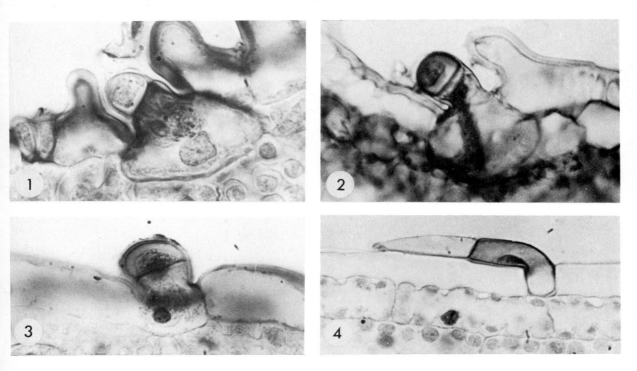

Plate II. Salt glands of various Gramineae species. 1. *Sporobolus arenarius*, × 997.5; 2. *Aeluropus litoralis*, × 997.5; 3. *Dactyloctenium aegyptium*, × 997.5; 4. *Bouteloua curtipendula*, × 997.5.

(e.g. *Dinebra*, *Tetrapogon*, etc.). In some species the gland resembles a trichome, with a narrow and elongated cap cell, growing out of a narrow base cell (e.g. *Bouteloua*, etc.) (Plate III). Various types of glands are also distinguishable in the Sporoboleae. Sunken glands appear in *Sporobolus* whereas in *Crypsis* the trichome-like gland appears. In the Aeluropodeae only semi-sunken glands were observed. Variations in the form of the basal as well as of the cap cell occur also within the Panicoideae. Two main types can be distinguished: the semi-sunken gland (e.g. *Paspalum*, *Coix*) and an emerging trichome-like gland with a narrow elongated cap cell growing out of a narrow base (e.g. *Panicum*) (Plate IV).

Investigation into the ultrastructure of the salt gland of *Spartina foliosa* (Levering and Thomson, 1971) shows that there is no cuticular layer separating between the mesophyll and the salt gland. The dense cytoplasm of the basal cell contains a high number of mitochondria, rod-like wall protuberances and infolding plasmalemma extending into the basal cell. The protuberances originate on the wall between the cap and basal cells and are isolated from the cytoplasm of the basal cell by infoldings of the plasmalemma. The basal cell contains also small

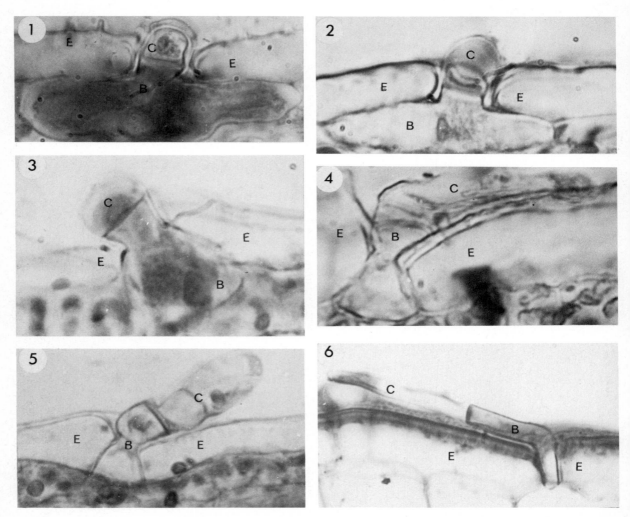

Plate III. Salt glands of various Gramineae species, E = epidermis, C = cap cell, B = basal cell. 1. *Spartina alterniflora*, ×997.5; 2. *S. patens*, ×997.5; 3. *Distichlis spicata*, ×997.5; 4. *Paspalum distichum*, ×997.5; 5. *Coix lacryma*, ×633.75; 6. *Panicum repens*, ×997.5.

vesicles, multivesicular bodies, free ribosomes, endoplasmic reticulum, dictyosomes, and a few plastids. The nucleus is quite large. In the cap cell there is a dense cytoplasm and numerous mitochondria but there are no partitioning membranes or wall protuberances. The portion of the cuticle above the cap cell is thicker than that along the sides of the cap cell. The dome-shape cuticular cap has numerous apparent pores. It contains also free ribosomes, a few dictyosomes, and plastids. Plasmodesmata occur in the wall between the basal and mesophyll cells.

The glands of the Avicenniaceae are of a different type (Plate V). The salt gland of *Avicennia marina* consists of several, mostly two to four collecting cells, one disc-like stalk cell and eight (sometimes twelve) excretory cells. The excretory cells, contain dense cytoplasm with few small vacuoles, large nuclei, many golgi bodies and mitochondria, numerous vesicles and membraneous strands. At the top of the gland the cuticle contains narrow pores.

The protoplast of the stalk cell is similar to that of the secretory cells, with a large nucleus and cutinized side wall. The uncutinized upper and lower transverse walls of the cell contain numerous plasmodesmata. The collecting cells resemble the epidermal cells, though their nucleus is larger (Shimony et al., 1973; Fahn and Shimony, 1977).

The Tamaricaceae have a more complicated gland (Plate VI), composed of eight cells arranged in four pairs. The salt gland of *Tamarix aphylla* has been frequently investigated. It consists of two collecting cells and six small excretory cells. The excretory cells are coated with a sheath of cuticle, except for few loci, which connect them with the collecting cells. These cells have dense cytoplasm, numerous mitochondria and plastids, especially in the upper two pairs. Plants grown under saline conditions contain more ribosomes and polysomes than plants grown under non-saline conditions (Thomson and Liu, 1967; Shimony and Fahn, 1968). The collecting cells are highly vacuolated.

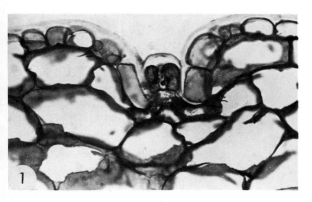

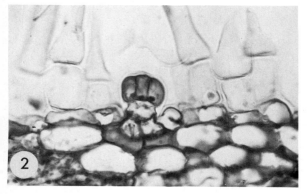

Plate IV. Salt glands of *Avicennia marina*. 1. Adaxial side of the leaf, ×633.75; 2. Abaxial side of the leaf, ×633.75.

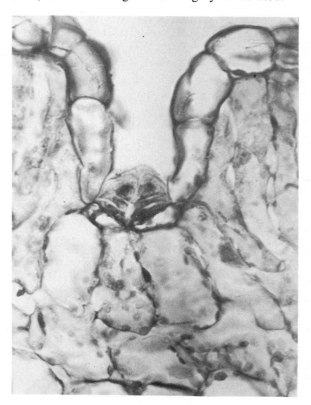

Plate V. Salt gland of *Tamarix aphylla*, ×845.

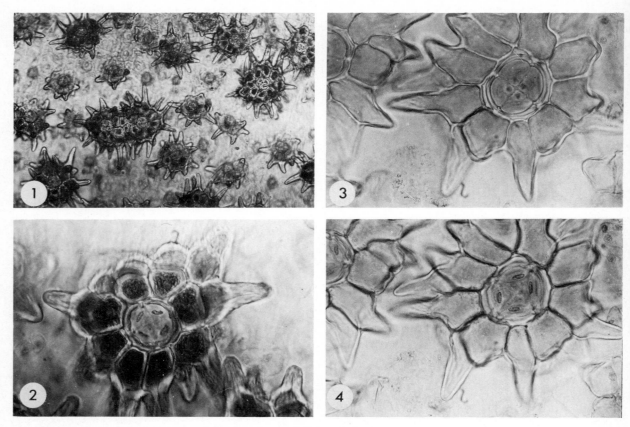

Plate VI. Salt glands of *Statice pruinosa*. 1. Surface view; 2–4. Enlarged glands.

In the Primulaceae the salt glands of *Glaux maritima* was investigated (Rozema, 1975; Rozema and Riphagen, 1977; Rozema et al., 1977). The glands are sunken in depressions of the epidermis and appear to be surrounded by six regular arranged epidermal cells. Three main cell types can be distinguished in the gland: a large collecting basal cell with a central vacuole and a narrow layer of peripheral cytoplasm containing an elongated nucleus and some chloroplasts. Next to the collecting cell there is a relatively large flat stalk-cell with dense cytoplasm, a large nucleus and numerous free ribosomes. Its lateral walls are suberized and cutinized. The outer part of the gland consists of four to eight excretory cells; the bases of these cells are all contiguous on the top of the slightly conic stalk cells. The excretory cells have large nuclei and large mitochondria and dense cytoplasm with well distinguished many tubular outgrowths of the cell walls covered by large plasmamembrane. They are

covered by a cuticle cap which is in continuation from the lateral cell walls of the stalk cell. The outer wall of the gland is ten times thicker than the adjacent epidermal wall. Only some times a pore was found in the cuticle cap, thus the existence of pores is still unclear. The boundaries between the mesophyll and collecting cell contain few plasmodesmata. The junction of the collecting cells with the stalk cell contain many plasmodesmata. Numerous plasmodesmata occur between the stalk cell and the excretory cell (Rozema et al., 1977).

Multicellular glands comprised of sixteen cells are found in the Plumbaginaceae. In some species (e.g. *Statice pruinosa*) the stem glands are located on the top of special structures (Plate VII).

The gland of *Limonium* consists of four excretory cells arranged in a circle. Each cell is accompanied at its outer side by a smaller adjoining cell. Both, the excretory and the adjoining cells, are surrounded by two cup-shaped cell layers, each comprised of

202

four flat cells, arranged similarly to the excretory cells (Ruhland, 1915; De Fraine, 1916; Ziegler and Lüttge, 1966) (Plate VIII).

The upper side of the gland and the neighbouring epidermal cells are covered by a thick cuticle. The outer walls of the outermost cup-shaped layer of cells are also heavily cutinized. The cuticle extends to a slight degree to the adjacent continuous walls, thus forming a rigid structure below the epidermis in which the gland is located (Helder, 1956).

One small single pore (with a diameter of approximately 1μ) is located in the center of the cuticle covering the top of each excretory cell (Plate VIII). Large pits are also found in the walls of gland cells adjacent to the assimilating tissue of the leaves. Through these pits there is a contact with four large cells – the collecting cells. The gland cells contain dense cytoplasm, large nucleus, small vacuoles and thin membranes.

The salt gland in *Aegiceras* of the Myrsinaceae is larger and composed of more cells than those of *Tamarix*, *Limonium* or *Aegialitis*, (cf. Cardale and Field, 1971). The salt gland of *Aegiceras corniculatum* consists of a large number of excretory cells and a single large basal cell. The excretory cells and the basal cell are joined by plasmodesmata. The cuticle covering the top of the gland differs from the cuticle covering its sides by the type or quantity of wax deposits. The excretory cells resemble those of *Tamarix*, in their ultrastructure except for the cell-wall protuberance in *Tamarix*. The basal cell is analogous to the collecting cells of *Tamarix* and *Limonium* by being connected to the excretory cells

via plasmodesmata and by being vacuolated. In contrast to *Limonium* and *Tamarix* the basal cell in *Aegiceras* is surrounded by cuticle (Cardale and Field, 1971).

3. Location and development of salt glands

Salt glands are found most abundantly on leaves. Only in *Statice pruinosa* their number on the stem exceeds that of the leaves (Waisel, 1972). Usually their number is fewer than stomata (cf. Ruhland, 1915). In *Glaux maritima* glands occur only on leaf surfaces, but no glands are found on the stem (Rozema et al., 1977).

Development of salt glands was followed in *Chloris gayana* of the Gramineae (Liphschitz et al., 1974). Salt glands are initiated already early in the development of the leaves. They are differentiated earlier than many other leaf tissues such as the mesophyll, but later than the initiation of the trichomes. Glands appear on both sides of the leaves in longitudinal rows, parallel to the veins. Their number per unit leaf area or per number of epidermal cells is equal for the adaxial and the abaxial sides of the leaves. Fewer glands than stomata appear on the leaves but the number of glands equals that of trichomes (cf. Liphschitz et al., 1974) (Table 2). The number of glands per unit area and per epidermal cell are only very slightly affected by addition of salt to the growth medium or by application of growth substances such as 2,4–D, Kinetin and GA. Only Benzyl Adenine almost doubled the number of glands and trichomes per unit area, while the density of epidermis cells and stomata did not increase as much. Thus, the ratio of glands to epidermis cells was increased by B.A. (Liphschitz et al., 1974).

In other plant species addition of salt to the growth medium affected the number of glands. Leaves of *Glaux maritima* which were taken from plants grown on relatively high salinity contain more salt glands than those taken from non-salty sites (Rozema et al., 1977).

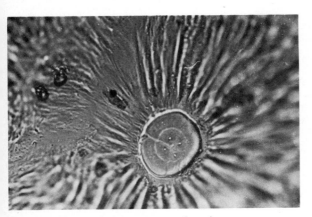

Plate VII. Salt gland of *Limonium*, surface view.

Plate VIII. *Dactyloctenium aegyptium* grown under saline and non-saline conditions for 10 days. 1. General view of a plant; 2. Salt crystals on leaf surface; 3. Enlarged section of 2. Salt whiskers are well distinguished; 4. Plants grown on (1) Hoagland, and on (2) 50 mM NaCl in Hoagland solution.

Table 2. Number of glands, epidermis cells, stomata and trichomes per unit leaf-area (0.5 mm^2) in Rhodes grass plants grown under controlled conditions (mean and standard deviations of 100 measurements). (Liphschitz et al., 1974).

Treatment	No. of glands	No. of epidermis cells	No. of stomata	No. of trichomes	No. of glands/ 1000 epidermis cells
H plants ($\frac{1}{2}$ Hoagland's nutrient solution)	9.6± 1.1	255.2±41.4	52.1±11.6	9.6±3.0	37.6
H plants sprayed with 100 ppm BA	18.2±10.5	310.5±45.2	53.3±11.0	20.2±7.4	58.0
H plants sprayed with 100 ppm Kinetin	11.4± 2.7	281.2±23.3	60.4± 4.7	12.5±2.2	40.7
H plants sprayed with 100 ppm 2,4-D	9.0± 1.2	291.2±24.9	47.2±12.6	11.4±2.7	31.0
H plants sprayed with 100 ppm GA	8.9± 0.9	278.7±22.7	54.1± 2.5	10.2±1.4	32.1
H + NaCl 0.2 M	10.5± 1.7	258.8±24.8	54.0±11.7	10.7±4.2	42.1

4. Correlation between salt gland location and structure and efficiency of salt excretion in the Gramineae

Variations in excretion rates between various species of the Gramineae i.e. Chloridoideae and Panicoideae were observed (cf. Pollak and Waisel, 1970; Liphschitz and Waisel, 1974, 1978). Excretion efficiency values, i.e. salt excretion related to the leaf salt content were compared with the gland's structure and location, in order to reveal whether a correlation exists between salt gland's structure and excretion efficiency.

As seen from the data presented in Tables 3 and 4, the more sunken the gland, the higher its excretion efficiency.

Species with sunken glands (e.g. Spartina) have a large oval-shaped basal cell and excrete salts more efficiently than species having trichome-like glands with a narrow basal cell and an elongated cap cell (e.g. Bouteloua). Such a correlation was observed in species of all other tribes investigated. In the Panicoideae, Paspalum distichum with a semi-sunken gland type excretes more efficiently than Panicum repens with a trichome-like gland. In the Chloridoideae, Spartina alterniflora, S. patens and Distichlis spicata have the most efficient glands among the species examined of the Chloridoideae and Aeluropodeae, and their glands are sunken.

Tight relationships can also be found between excretion efficiency and basal cell dimensions. A large sunken basal cell seems to transport outwards larger quantities of ions as compared with narrow, small and elongated basal cells. It is also possible that the distance from the basal cell through the cap outwards also affects ion movement and excretion efficiency. Elongated basal and cap cells comprise a longer path for ion movement and are thus inferior. On the other hand, it is possible that not the distance from the basal cell affects excretion efficiency, but the excretory mechanism itself.

Furthermore, it seems so far, that grasses of the Chloridoideae and Panicoideae, are also salt tolerant and can withstand a relatively high salinity in their growth media. Such tolerance can be achieved in two alternative ways: by salt exclusion or by salt endurance. The first type is usually found in plants with sunken glands whereas salt endurance or protoplasmic salt resistance is found in species that lost their excretion efficiency, i.e. in those that have elevated and trichome-like glands and cannot get rid of the excess of salt by excretion mechanism. Of course, one can find intermediate types of salt gland structure in salt secretion efficiency and in salt endurance.

5. Ecological factors affecting salt excretion

Various factors affect the rate of salt excretion. One of the main factors which dominates this process is the salt concentration of the growth medium. Secretion efficiency is also dependent upon the ions which are present in the growth medium. Other factors which affect salt excretion are light, temperature, oxygen pressure and the presence of metabolic inhibitors.

Table 3. Effect of NaCl concentration in the culture solution, and of time exposure, on secretion and internal content of Na^+, K^+ and Cl^- in leaves of various species of the Chloridoideae and Panicoideae.

Tribe	Species	Treatment solution	Secretion			Internal content		
			Na	K	Cl	Na	K	Cl
Chloridoideae	Bouteloua curtipendula	H. + 50 mM NaCl	0.0694	0.0153	0.4337	0.3398	0.7170	0.8114
			±0.0333	±0.0101	±0.2436	±0.2034	±0.3850	±0.3405
		H. + 100 mM NaCl	0.0149	0.0036	1.0192	0.7736	0.7737	2.1930
			±0.0218	±0.0024	±0.6786	±0.2969	±0.5188	±0.8470
	Chloris gayana	H. + 100 mM NaCl	0.2639	0.0173	0.2822	0.4156	0.2458	0.5906
			±0.1316	±0.0082	±0.0978	±0.1198	±0.0992	±0.1211
		H. + 150 mM NaCl	1.3666	0.1061	1.4893	1.2849	0.3842	1.1809
			±0.4665	±0.0720	±0.4349	±0.7972	±0.1532	±0.4891
	Cynodon dactylon	H. + 100 mM NaCl	0.0263	0.0336	0.1073	0.2012	0.2788	0.5215
			±0.0124	±0.0322	±0.0617	±0.0903	±0.1044	±0.2052
		H. + 150 mM NaCl	0.0341	0.0179	0.1907	0.3212	0.5273	0.7655
			±0.0214	±0.0122	±0.0963	±0.1100	±0.3297	±0.4627
	Dactyloctenium aegyptium	H. + 50 mM NaCl	1.4400	0.5086	0.8128	1.5684	1.6591	0.8948
			±0.4457	±0.2458	±0.3922	±0.7346	±0.4808	±0.3489
		H. + 100 mM NaCl	1.6209	0.4471	1.1104	1.2476	1.0405	1.1907
			±0.5071	±0.2185	±0.4501	±0.5076	±0.5811	±0.6917
	Dinebra retroflexa	H. + 50 mM NaCl	0.1344	0.0460	0.1975	0.1448	0.1217	0.2720
			±0.1090	±0.0372	±0.1186	±0.085	±0.0395	±0.1133
	Eleusine indica	H. + 50 mM NaCl	0.0239	0.0114	0.0563	0.2442	0.6533	0.6564
			±0.0115	±0.0079	±0.0246	±0.2143	±0.1956	±0.3466
		H. + 100 mM NaCl	0.0614	0.0406	0.1696	1.4110	1.7084	2.1023
			±0.0351	±0.0205	±0.1011	±0.6320	±0.6138	±1.0624
	Sporobolus arenarius	H. + 50 mM NaCl	0.035	0.0038	0.0471	0.3445	0.2237	±0.4738
			±0.0211	±0.0029	±0.0323	±0.2284	±0.1920	±0.2925
		H. + 100 mM NaCl	0.0467	0.0046	0.1023	0.6951	0.2897	0.7812
			±0.0195	±0.0028	±0.0529	±0.3594	±0.2496	±0.4813
		H. + 150 mM NaCl	0.1265	0.0261	0.4176	3.7684	0.4161	5.1528
			±0.1058	±0.0170	±0.2557	±1.5473	±0.1563	±1.9527
Panicoideae	Digitaria sanguinalis	H. + 100 mM NaCl	0.075	0.010	0.153	1.603	0.543	2.306
			±0.024	±0.004	±0.072	±0.433	±0.128	±0.124
		H. + 150 mM NaCl	0.010	0.005	0.102	1.360	0.473	1.697
			±0.004	±0.003	±0.036	±0.552	±0.163	±0.720
	Panicum repens	H. + 100 mM NaCl	0.028	0.007	0.066	0.707	0.888	1.032
			±0.020	±0.004	±0.016	±0.235	±0.119	±0.452
		H. + 150 mM NaCl	0.020	0.003	0.039	0.513	0.767	0.706
			±0.004	±0.005	±0.012	±0.142	±0.246	±0.152
	Paspalum distichum	H. + 100 mM NaCl	0.020	0.003	0.021	0.365	0.371	—
			±0.007	±0.001	±0.011	±0.063	±0.194	—
		H. + 150 mM NaCl	0.032	0.004	0.019	0.692	0.330	0.893
			±0.021	±0.002	±0.010	±0.216	±0.190	±0.279

Values are the means of ten replicates and are given as μ-equivalents per mg dry weight leaf. Secretion was measured 5 days after an equilibration period, and internal content values are thus for 12 days.

Growth conditions: LD; $27\pm2°C$.

Treatment solution: full strength Hoagland nutrient solution + NaCl in concentrations shown.

Table 4. Excretion efficiency (salt excretion/internal salt content) of various species of the Chlorideae, Paniceae and Aeluropodeae treated with 100 mM NaCl for 5 days.

Tribe	Species	Excretion efficiency of different gland type		
		Sunken	Semi-sunken	Trichome-like
Chlorideae	*Spartina patens*	2.12		
	Dactyloctenium aegyptium	1.30		
	Chloris gayana		0.63	
	Cynodon dactylon		0.13	
	Eleusine indica		0.04	
	Bouteloua curtipendula			0.02
Paniceae	*Paspalum distichum*		0.06	
	Digitaria decumbens			0.046
	Panicum repens			0.040
Aeluropodeae	*Distichlis spicata*	5.46		
	Aeluropus litoralis		2.52	

I. Concentration effects

Salt glands are assumed to transport and excrete ions against a concentration gradient. This phenomenon was first shown by Arisz et al. (1955) for *Limonium latifolium*. The concentration of the excreted fluid of leaf discs of *Limonium latifolium* exceeded that of the external salt solution only at lower values of the latter (Arisz et al., 1955). In various mangrove species (e.g. *Aegialitis*, *Aegiceras* and *Avicennia*) NaCl concentration in the excreted solution exceeded the NaCl concentration of seawater (Scholander et al., 1962). A similar phenomenon was observed for *Aeluropus litoralis* where the concentrations of the excreted solution up to 1 M NaCl were always higher than the concentration of the treatment solution (Pollak and Waisel, 1970). The same holds true for *Tamarix aphylla* where the excreted brine's concentration was higher than that of the external medium (Berry, 1970).

Under dry conditions the excreted salt crystalizes and covers the leaf either in a form of scales (*Limonium*, *Spartina*) or in the form of whiskers (*Tamarix*, *Aeluropus*) (cf. Decker, 1961; Pollak and Waisel, 1970). The growing whiskers dissolve eventually and recrystalize into salt cubes.

The amount of excreted salt may be rather high. In *Limonium latifolium* leaf discs excreted brine up to half their weight (70 mg fluid from a 150 mg leaf discs) in 24 hours (Arisz et al., 1955). *Statice* also excreted high amounts of brine (Ruhland, 1915). High values for chloride excretion (90 peq.cm^{-2}.sec^{-1}) were obtained by Atkinson et al. (1967) for *Aegialitis* leaves. High amounts of fluid were excreted also by *Tamarix aphylla* (Waisel, 1961; Berry, 1970), *Aeluropus litoralis* (Pollak and Waisel, 1970), *Chloris gayana* and *Sporobolus arenarius* and other Gramineae (Liphschitz and Waisel, 1974), depending on environmental conditions.

Increasing salinity stimulates excretion up to a certain level, above which salt excretion is adversely affected. Results obtained for chloride excretion by *Limonium latifolium* leaf discs subjected to a NaCl solution showed that increasing the concentration of the external medium up to a certain level (up to 0.3 M NaCl) increased the brine excretion (Arisz et al., 1955). In *Tamarix aphylla* reduction in NaCl excretion started at 400 mM NaCl (Waisel, 1961). In *Aeluropus litoralis* sodium excretion and sodium content were positively correlated with the increase of NaCl concentration of the external solution up to 0.2 M to 0.4 M NaCl. In other grasses like *Distichlis spicata*, *Spartina patens* and *Spartina alterniflora* maximal excretion was distinguished at 200 mM NaCl (Fig. 1). Concentrations of 100 mM to 150 mM NaCl of the external medium were shown to be optimal for sodium and chloride excretion of most species of the Chloridoideae (e.g. *Cynodon dactylon*, *Eleusine indica*, *Chloris gayana*, *Dactyloctenium aegyptium*, *Sporobolus arenarius*) (cf. Liphschitz and Waisel, 1974). It is interesting to note that in all these species, better growth was observed when at least 50 mM NaCl was added to the basic nutrient solution. In some of them growth ceased altogether after a short exposure to a salt-free nutrient solution (Liphschitz and Waisel, 1974).

In *Glaux maritima* increasing salinity up to 150 mM NaCl stimulated salt excretion but at 300 mM a reduction of excretion was observed (Rozema, 1975).

II. Specificity of ions

The ionic excrete of the salt glands is constituted mainly of sodium and chloride. Nevertheless, some other ions, e.g. potassium, calcium, magnesium, sulfate, nitrate and phosphate, as well as various organic substances are present in the brine (Waisel, 1972). Excretion of phosphate and sulfate was distinguished in *Statice gmelini* (Schtscherback, 1910). Excretion of nitrate, sulfate and phosphate in *Tamarix* was detected when plants were amply

supplied with those ions. Minor amounts of magnesium were also excreted in this species (Waisel, 1961).

Brine excreted by various plant species was found to contain sugars, amino acids, amines and proteins (Waisel, 1972).

When the excretion process is selective, i.e. the specific ion sodium, chloride or any other ion, may be excreted in larger quantities though it is present in the medium in equal or in lower amounts than the other ions.

For instance, in *Aeluropus litoralis* and in *Tamarix aphylla*, sodium is preferentially excreted even though its concentration in the growth me-

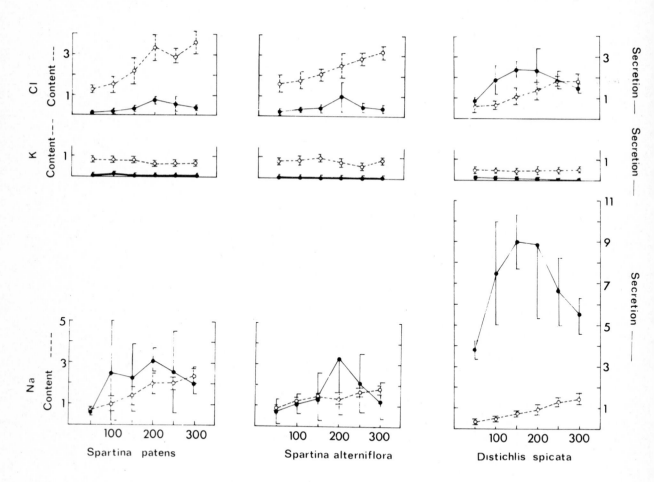

Fig. 1. Effects of NaCl concentration in the treatment solution on secretion and internal content of Na^+, K^+ and Cl^- in detached leaves of *Spartina patens*, *S. alterniflora* and *Distichlis spicata*. Measurements were made 24 h after leaf subjection to the treatment solutions. Values are means of 6 replications and are given as μmoles/mg dw/24 h for secretion and as μmoles/mg dw for salt content. Growth conditions: LD, $27° \pm 2° C$.

dium may equal that of potassium (Pollak and Waisel, 1970; Waisel, 1961). In both plant species inter-relationships between the secreted ions do exist. In *Aeluropus litoralis* the preference of excretion is in the following order: Na > K > Ca whereas in *Tamarix aphylla* this order is: Na > Ca > K. Presence of K, Mg and Ca in the irrigation solution may depress the excretion of Na by *Aeluropus* as well as by *Tamarix*. Excess of $CaCl_2$ and $MgCl_2$ decreases the amount of K excreted by *Tamarix*.

The amount of Ca excreted is not affected by the concentration of the external solution or its composition in both plant species (Waisel, 1961; Pollak and Waisel, 1970).

When excretion is not selective, the excreted brine composition depends on the ions present in the growth medium. In *Aegiceras corniculatum* ions were excreted in accordance with their relative concentrations in the external solution, i.e. in seawater (Scholander et al., 1962). Increasing the sodium concentration in the growth medium of *Tamarix* was followed by an increase in the excretion of sodium and concomitantly by a decrease in potassium excretion (Berry and Thomson, 1967; Thomson et al., 1969). Moreover, addition of Cs or of Rb to the growth medium brought about their excretion by *Tamarix* (Thomson et al., 1969), and by *Limonium* (Hill, 1967) though naturally those elements are not present in the plant.

III. Effect of time exposure on excreted fluid composition

The relationships between the excreted sodium and chloride varies with the time of exposure of the plants to the salt solution. In various Chloridoideae examined, the amount of chloride excreted during the first days of exposure was higher than that of sodium, but after 10 days of exposure to the same solution the amount of the excreted sodium turned to be 2 to 3 times higher than that of chloride (Liphschitz and Waisel, 1974).

IV. Temperature

Under high temperature conditions (in the range of

25° to 30° C) excretion rates are usually accelerated. Rise in temperature from 5° to 25° accelerated excretion rates of *Limonium* by a factor of 8 (Arisz et al., 1955). Transfer of leaves of *Aeluropas litoralis* from 20° C to 30° C or from 10° C to 20° C accelerated their excretion rates (Pollak, 1975).

V. Metabolic inhibitors

Low concentrations of metabolic inhibitors (KCN, Na azide, NaF, Arsenite, DNP) stimulated slightly the chloride excretion, whereas marked inhibition of excretion processes takes place when their concentration is above $5 \times 10^{-3}M$ (Arisz et al., 1955). The effect of the inhibitors was mostly on the volume of the excrete and not on its composition. C CCP was reported to inhibit the excretion process in *Aegialitis* (Atkinson et al., 1967).

VI. Light

Excretion process in *Limonium* was reported to be light dependent (Arisz et al., 1955) and in the dark much lower rates of excretion were distinguished. In *Aeluropus litoralis* (Pollak and Waisel, 1970, 1978) as well as in *Aegialitis* (Atkinson et al., 1967) excretion rates were higher in the light only in intact plants, while detached leaves did not respond to light conditions.

The differences in excretion rates which were observed in the light and in the dark treatments resulted from different rates, probably, of the transpiration stream in light and dark (Pollak and Waisel, 1978).

Besides the prominent effect of light on the absolute amount of sodium excreted from intact leaves of *Aeluropus litoralis*, light also enhanced sodium transport through the leaves.

VII. Diurnal variations in excretion

Diurnal variations in salt excretion were reported for *Aegialitis* and *Aegiceras* with highest rates at noontime (Scholander et al., 1962). Diurnal variation in the excretion rates of *Avicennia marina* were also distinguished (Waisel P.C.) but with two peaks, one at noontime and the other at midnight.

A diurnal rhythm of salt excretion (sodium, potassium and chloride) was also reported for *Distichlis spicata* (Hansen et al., 1976).

VIII. Ion location and site of excretion

In spite of the fact that the structure and the ultrastructure of many salt glands were studied, only scanty data are available regarding ion location within salt glands and the exact site of the excretion. Moreover, the role played by each of the gland's cells is also still obscure.

It is generally accepted that the excretion of salts by salt glands is an active process (Waisel, 1972). Two main hypotheses regarding the site of excretion have been suggested. According to one hypothesis only one site of active transport exists inside the gland. Arisz et al. (1955), Sutcliffe (1962), Levering and Thomson (1971), Bostrom and Field (1972), and Pollak (1975), all assumed that the site of excretion is located only inside the gland cells, but on the other hand these investigators disagree as to its precise position. Arisz et al. (1955) assumed that the active step takes place when ions are absorbed into the vacuoles of the gland cells, whereas according to Sutcliffe (1962) ions are actively transported outwards across the outer membranes of the cap cells of the glands, followed by an osmotic withdrawal of water from the tissue to the pore region and the formation of brine droplets. Levering and Thomson (1971) suggested that salts are actively transported from the basal cell to the cap cell of the bi-cellular gland of *Spartina foliosa*. However, from there onwards, the solution leaks out passively through pores in the cuticle. Other investigators also located the active mechanism in the gland cells, but give no evidence for its exact site (cf. Bostrom and Field, 1972; Pollak, 1975).

The other hypothesis suggests two sites of the excretion process. According to Shimony et al. (1973) two metabolic pumps seem to be involved. One pump seems to be located in the xylem parenchyma cells, Thus forming a downhill gradient from the xylem outwards the gland. A second pump may be located somewhere inside the excretory cells, causing the excretion if ions from those cells outwards.

Attempts have also been made to investigate the site and nature of the excretion mechanism (Ziegler and Lüttge, 1967; Shimony et al., 1973). However, these attempts have been unsuccessful, because of the fixation methods used in these investigations. The leaves of *Limonium* (Ziegler and Lüttge, 1967) or of *Avicennia marina* (Shimony et al., 1973) used in these investigations were fixed slowly, movement of ions could be possible and therefore the exact location and paths of ion transport remained uncertain.

Using a new technique of immediate fixation Ramati and Liphschitz (1975) enabled to locate soluble ions within the various cell compartments and thus provided a better picture of the situation in situ. Investigations made on *Sporobolus arenarius*, using the new technique, showed that higher quantities of ions occurred in the cell walls than in the cytoplasm (Table 5).

High and similar quantities of sodium and chloride ions were found in the cell walls of the mesophyll and basal cells. A drastic reduction in the content of these ions (60%) was observed in the walls of the cap cells. These results suggest that the process of excretion seems to include the existence of selective barriers between the various cells. Indeed, cell walls are part of the free space and the entrance of ions into the walls was therefore passive. However, beyond the walls, ions must have passed through a selective membrane. Three steps separated by selective barriers can thus be observed: (1) A barrier between the mesophyll cells and the basal gland cell, resulting in accumulation of ions within the basal cell; (2) A barrier between the basal cell and the cap cell, resulting in differences in accumulation of sodium and chloride; (3) A selective barrier between the cap cell and the leaf surface. This last selective step is implied by the high ratio Na/K and Cl/K (Table 5).

Early investigations on salt excretion (Levering and Thomson, 1971; Liphschitz and Waisel, 1974) suggested that excretion is mainly conducted in the basal gland cell. Nevertheless, later on it was observed in *Sporobulus arenarius* that both gland cells participate in the excretion process (Ramati et al., 1976).

Table 5. Ion content and ion ratios in leaf-cells of *Sporobolus arenarius*. Leaves were given 150 mM NaCl in Hoagland's solution for 24 h (values are means of ten replications, and expressed as 10^{-6} μmol/excitation volume) (Ramati et al., 1976).

	Ion contents	Mesophyll cell	Basal cell	Cap cell	
Na	cell wall	687.30±139.36	683.53±189.54	251.24± 65.93	
	cytoplasm	41.07± 11.01	70.18± 25.71	39.18± 11.80	
K	cell wall	1629.70±730.60	1344.46± 71.92	499.82±171.15	
	cytoplasm	184.01±119.76	294.52±147.64	67.48± 46.68	
Cl	cell wall	3360.44±324.30	2990.56±396.06	1685.37± 30.97	
	cytoplasm	40.10± 6.20	65.90± 11.90	30.70± 7.10	

	Ion ratios	Mesophyll cell	Basal cell	Cap cell	Secreted brine
Na/K	cell wall	0.42	0.50	0.56	
	cytoplasm	0.23	0.24	0.58	23.08
Na/Cl	cell wall	0.19	0.23	0.15	
	cytoplasm	1.02	1.09	1.27	0.83
Cl/K	cell wall	2.21	2.22	3.75	
	cytoplasm	0.20	0.20	0.40	31.40

6. Salt excretion, salt resistance and C_4 syndrome in the Gramineae

Two characteristics were known up to 1976 to be common to most species of the Panicoideae and the Chloridoideae: presence of Kranz type anatomy (cf. Smith and Brown, 1973) and the possession of the C_4 syndrome (cf. Smith and Brown, 1973; Downton and Tregunna, 1968; Gutirrez et al., 1974; Raghavendra and Ramadas, 1976). Kranz type anatomy was found by Brown (1958) in the Panicoideae which are considered of tropical origin.

Many C_4 plants were shown not only to tolerate but also to require sodium as micronutrient (Brownell and Crossland, 1972). Moreover, the primary adaptation of C_4 plants was probably to saline environment (Leatch, 1974). C_4 plants (as well as CAM plants) seem to be frequently either halophytes or of halophytic origin (Table 6) (Liphschitz and Waisel, 1974, 1978). Evolution of structural and functional adaptations to such environment were rather a prerequisite for efficient assimilation and growth processes (Leatch, 1974).

All plants belonging to the Panicoideae and the Chloridoideae subfamilies are salt resistant and withstand a relatively high salinity in their growth medium. Such tolerance is achieved in two alternative ways; by salt excretion or by salt endurance. The first type is usually found in plants with sunken glands whereas salt endurance or protoplasmic salt resistance is found in species that lost their excretion efficiency and have trichome-like glands.

In view of these findings it seems so far that species belonging to the Panicoideae and the Chloridoideae have evolved from closely related ancestors which occupied saline (coastal?) habitats. Though some species remained in saline habitats, most species of these groups immigrated later from saline to non-saline habitats. Such migration probably occurred not too long ago, as those plants still retain many characteristics of their halophytic ancestors.

Table 6. Correlation between presence of salt glands, C_4 products, high values of the C^{13}/C^{12} ratio, low compensation point and Kranz anatomy in various genera of the Chloridoideae and Panicoideae.

Species	Tribe	Presence of salt glands[1] or trichomes	C_4 first products and known enzymatic systems	C^{13}/C^{12} below-16[5]	Low CO_2 Comp.[5]	Kranz anatomy[7]
	Chlorideae					
Astrebla pectinata		−	−	−	+	+
A. lappacea		+	−	−	−	+
Bouteloua curtipendula		+	NAD-ME[2]+PEP-CK[2]	+	+	+
Buchloe dactyloides		+	NAD-ME[2]	+	+	+
Chloris gayana		+	NAD-ME+PEP-CK[2]	+	+	+
Coelachyrum brevifolium		+	−	−	−	+
Cynodon dactylon		+	+[3]	+	+	+
Dinebra aegyptiaca		+	−	−	−	+
Eleusine indica		+	NAD-ME[2]	−	+	+
Dactyloctenium aegyptium		+	−	−	+	+
Enteropogon sp.		+	+[3]	+	−	+
Tetrachne dregei		+	−	−	−	+
Spartina alterniflora		+	−	+	−	+
Spartina anglica		+	+[6]	−	−	+
	Sporoboleae					
Sporobolus poiretii		−	NAD-ME[2] PEP-CK[2]	+	+	+
Sporobolus arenarius		+	−	−	−	+
Crypsis faktorovskyi		+	−	+[8]	−	+
	Aeluropodeae					
Aeluropus litoralis		+	+[4]	−	−	+
Distichlis spicata		+	−	+	−	+
	Paniceae					
Brachiaria mutica		+	−	−	+	+
Cenchrus ciliaris		+	−	+	+	+
C. pauciflorus		−	NADP-ME[2]	−	−	+
Digitaria sanguinalis		+	NADP-ME[2]	+	+	+
Echinochloa colonum		+	NADP-ME[2]	+	+	+
E. crus-galli		+	NADP-ME[2]	+	+	+
Panicum maximum		+	NAD+ME[2]+NADP-ME[2]+ PEP-CK[2]	+	+	+
P. miliaceum		+	NAD-ME[2]+NADP-ME[2]+ PEP-CK[2]	+	+	+
P. repens		+	−	+	+	+
Paspalum dilatatum		+	NADP-ME[2]	+	+	+
Paspalidium geminatum		+	−	+[8]	−	+
Pennisetum purpureum		−	NADP-ME[2]	+	+	+
P. asperifolium		+	−	−	−	+
Saccharum officinarum		−	NADP-ME[2]	+	+	+
S. biflorum		+	−	−	−	+
Setaria italica		+	NADP-ME[2]	+	+	+
S. viridis		+	NADP-ME[2]	−	+	+
Tricholaena teneriffae		+	−	−	−	+

Table 6 *(continued)*.

Species	Tribe	Presence of salt glands[1] or trichomes	C_4 first products and known enzymatic systems	C^{13}/C^{12} below-16[5]	Low CO_2 Comp.[5]	Kranz anatomy[7]
	Andropogoneae					
Andropogon scoparius		−	NADP-ME[2]	+	+	+
A. distachyus		+	−	−	−	+
Chrysopogon gryllus		+	−	−	+	+
Coix lacryma jobi		+	+[6]	−	+	+
Dichanthium aristatum		−	−	−	+	+
Dichanthium annulatum		+	−	−	−	+
Erianthus sp.		+	+[3]	−	−	+
Hyparrhenia hirta		+	−	−	−	+
Sorghum halepense		+	NADP-ME[2]	+	+	+
Zea mays		+	NADP-ME[2]	+	+	+

1. Data based on the present work, and on Liphschitz and Waisel (1974).
2. Data based on Gutirrez et al. (1974).
3. Data based on Raghavendra and Ramadas (1976).
4. Data based on Shomer-Ilan and Waisel (1973).
5. Data based on Downton and Tregunna (1968), Smith and Brown (1973), Raghavendra and Ramadas (1976).
6. Data based on Byott (1976).
7. Data based on all mentioned authors in 2, 4, 5 and self examinations.
8. Measurements were made by Dr. A. Nissabaum of the Weizman Inst., Rehovot.
− Data inavailable.

Literature cited

Arisz, W.H., Camphuis, I.J., Heikens, H. and van Tooren, A.J. 1955. The secretion of the salt gland of *Limonium latifolium* KTZE. *Acta Bot. Neerl.* 4: 322–338.

Atkinson, M.R., Findlay, G.P., Hope, A.B., Pitman, M.G., Saddler, H.D.M. and West, K.R. 1967. Salt regulation in the mangroves *Rhizophora mucronata* Lam. and *Aegialitis annulata*. R. Br. *Aust. J. Biol. Sci.* 20: 589–599.

Berry, W.L. 1970. Characteristics of salts secreted by *Tamarix aphylla*. *Amer. J. Bot.* 57: 1226–1230.

Berry, W.L. and Thomson, W.W. 1967. Composition of salt secreted by salt glands of *Tamarix aphylla*. *Can. J. Bot.* 45: 1774–1775.

Biebel, R. and Kinzel, H. 1965. Blattban und Salzhaushalt von *Laguncularia racemosa* (L.) Gaertn. und andere Mangrovebaume auf Puerto Rico. *Österr. Bot. Z.* 112: 56–93.

Bostrom, T.E. and Field, C.D. 1972. Electrical potential in the salt gland of *Aegiceras*. In *Ion Transport in Glands*: pp. 385–392. Academic Press, N.Y.

Brown, W.V. 1958. Leaf anatomy in grass systematics. *Bot. Gaz.* 119: 170–178.

Brownell, F.P. and Crossland, C.J. 1972. The requirement for sodium as micronutrient by species having the C_4 dicarboxylic photosynthetic pathway. *Plant Physiol.* 49: 794–797.

Byott, G.S. 1976. Leaf air space system in C_3 and C_4 species. *New Phytol.* 76: 295–299.

Cardale, S. and Field, C.D. 1971. The structure of the salt gland of *Aegiceras corniculata*. *Planta* 99: 183.

Decker, J.P. 1961. Salt secretion by *Tamarix pentandra* Pall. *Forest Sci.* 7: 214–217.

Downton, W.J.S. and Tregunna, E.B., 1968. Carbon dioxide compensation – its relation to photosynthetic carboxylation reaction, systematics of the Gramineae and leaf anatomy. *Can. J. Bot.* 46: 207–215.

de Fraine, E. 1916. The morphology and anatomy of the genus *Statice* as represented at Blakenely Point. I. *Statice binervosa* G.E. Smith and *Statice bellidifolia* D.C. (= *S. reticulara*). *Ann. Bot.* 30: 239–282.

Fahn, A. and Shimony, C. 1977. Development of the glandular and nonglandular leaf hairs of *Avicennia marina* (Forsskal) Vierh. *Jour. Linn. Soc.* 74: 37–46.

Gutirrez, M., Gracen, V.E. and Edwards, G.E. 1974. Biochemical and cytological relationships in C_4 plants. *Planta* 119: 279–300.

Hansen, D.J., Dayananden, P., Kaufman, P.B. and Brotherson, J.D. 1976. Ecological adaptations of salt marsh grass *Distichlis spicata* (Gramineae), and environmental factors affecting its growth and distribution. *Amer. J. Bot.* 63: 635–650.

Helder, R.J. 1956. The loss of substances by cells and tissues (salt glands). In: *Encyclopedia of Plant Physiol.* II. 469–486.

Hill, A.E. 1967. Ion and water transport in *Limonium*. I. Active transport by the leaf gland cells. *Biochem. Biophys. Acta* 135:

454–460.

Hill, A.E. 1967. Ion and water transport in *Limonium*. II. Short circuit analysis. *Ibid.* 135: 461–465.

Hill, A.E. 1967. Ion and water transport in *Limonium*. III. Time constant of the transport system. *Ibid.* 136: 66–72.

Hill, A.E. 1967. Ion and water transport in *Limonium*. IV. Delay effects in transport process. *Ibid.* 136: 73–79.

Leatch, W.M. 1974. The C_4 syndrome: a structural analysis. *Ann. Rev. Plant Physiol.* 25: 27–52.

Levering, C.A. and Thomson, W.W. 1971. The ultrastructure of the salt gland of *Spartina foliosa*. *Planta* 97: 183–196.

Liphschitz, N., Shomer-Ilan, A., Eshel, A. and Waisel, Y. 1974. Salt glands on leaves of Rhodes grass *Chloris gayana* Kth. *Ann. Bot.* 38: 459–462.

Liphschitz, N. and Waisel, Y. 1974. Existence of salt glands in various genera of the Gramineae. *New Phytol.* 73: 507–513.

Liphschitz, N. and Waisel, Y. 1978. Salt glands in two subfamilies of the Gramineae: Panicoideae and Chloridoideae. *Intecol.* 3: 9.

Marloth, R. 1887. Zur bedeutung der Salz abscheidenen Drüsen der Tamariscineen. *Ber. d. Bot. Fes.* 5: 319–324.

Pollak, G. 1975. Physiological and ecological aspects of salt excretion in *Aeluropus litoralis*. Ph.D. Thesis, Tel-Aviv University, Tel-Aviv.

Pollak, G. and Waisel, Y. 1970. Salt secretation in *Aeluropus litoralis* (Willd.) Parl. *Ann. Bot.* 34: 879–888.

Pollak, G. and Waisel, Y. 1979. Ecophysiological aspects of salt excretion in *Aeluropus litoralis*. *Physiol. Plant.* 47: 177–184.

Raghavendra, A.S. and Ramadas, V.S. 1976. Distribution of the C_4 dicarboxylic acid pathway of the photosynthesis in local monocotyledonous plants and its taxonomic significance. *New Phytol.* 76: 301–305.

Ramati, A. and Liphschitz, N. 1975. Preparation of plant material for microautoradiography and electron probe microanalysis: the xylene technique. *Experientia* 37: 1108.

Ramati, A., Liphschitz, N. and Waisel, Y. 1976. Ion localization and salt secretion in *Sporobolus arenarius*, (Gou.) Duv. Jour. *New Phytol.* 76: 289–294.

Rozema, J. 1975. An eco-physiological investigation into the salt tolerance of *Glaux maritima* L. *Acta Bot. Neerl.* 24: 407–417.

Rozema, J. and Riphagen, I. 1977. Physiological and ecological relevance of salt secretion by the salt gland of *Glaux maritima* L. *Oecologia* 29: 349–357.

Rozema, J., Riphagen, I. and Sminia, T. 1977. A light and electron microscopical study on the structure and function of the salt gland of *Glaux maritima* L. *New Phytol.* 79: 665–671.

Ruhland, W. 1915. Untersuchungen über die Hautdrüsen der Plumbaginacean. Ein Beitrage zur der Halophyten. *Jb. Wiss. Bot.* 55: 409–498.

Schtscherback, J. 1910. Über die Salzausscheidung durch die Blätter von *Statice gmelini*. *Ber. d. Bot. Ges.* 28: 30–34.

Scholander, P.F., Hammel, H.T., Hemmingsen, E.A. and Garey, W. 1962. Salt balance in mangroves. *Plant Physiol.* 37: 722–729.

Shimony, C. and Fahn, A. 1968. Light and electron microscopical studies on the structure of salt glands in *Tamarix aphylla* L. *J. Linn. Soc. Bot.* 60: 283–288.

Shimony, C., Fahn, A. and Reinhold, L. 1973. Ultrastructure and ion gradients in the salt glands of *Avicennia marina* (Forssk) Vierh. *New Phytol.* 72: 27–36.

Shomer-Ilan, A. and Waisel, Y. 1973. The effect of sodium chloride on the balance between the C_3 and C_4 carbon fixation pathway. *Physiol. Plantarum* 29: 190–193.

Skelding, A.D. and Winterbotham, J. 1939. The structure and development of the hydathodes of *Spartina townsendii* Groves. *New Phytol.* 38: 69–79.

Smith, B.N. and Brown, W.V. 1973. The Kranz syndrome in the Gramineae as indicated by carbon isotopic ratios. *Amer. J. Bot* 60: 505–513.

Sutcliffe, J.F. 1962. *Mineral Salt Absorption in Plants*. Pergamon Press, London.

Thomson. W.W. and Liu, L.L. 1967. Ultrastructural features of the salt gland of *Tamarix aphylla* L. *Planta* 73: 201–220.

Thomson, W.W., Berry, W.L. and Liu, L.L. 1969. Localization and secretion of salt by the salt glands of *Tamarix aphylla*. *Proc. Nat. Acad. Sci.* (Wash.) 63: 310–317.

Volkens, G. 1884. Die Kalkdrusen der Plumbaginean. *Ber. d'. Bot. Ges.* 2: 334–342.

Volkens, G. 1887. *Die Flora der Aegyptisch-Arabischen Wüste auf grundlage Anatomische-Physiologischer Forschungen dargestellt*. Gebrunder Borntraeger, Berlin.

Waisel, Y. 1961. Ecological studies on *Tamarix aphylla* (L.) Karst. III. The salt economy. *Plant and Soil.* 4: 356–364.

Waisel, Y. 1972. *Biology of Halophytes*. Academic Press, New York.

Walter, H. and Steiner, M. 1936. Die Okologie der Ostafrikanischen Mangroven. *Z. Bot.* 30: 65–193.

Wiebe, H.H. and Walter, H. 1972. Mineral ion composition of halophytic species from northern Utah. *Amer. Mid. Naturalist* 67: 241–245.

Ziegler, H. and Lüttge, U. 1966. Die Saltzdrüsen von *Limonium vulgare*. I. Die Feinstruktur. *Planta* 70: 193–206.

Ziegler, H. and Lüttge, U. 1967. Die Saltzdrüsen von *Limonium vulgare*. II. Die lokalisierung des Chlorides. *Planta* 74: 1–17.

214

CHAPTER 8

The role of bladders for salt removal in some Chenopodiaceae (mainly *Atriplex* species)

UTE SCHIRMER and SIEGMAR-W. BRECKLE
Department of Ecology, University of Bielefeld,
P.O. Box 8640, D-4800 Bielefeld, F.R.G.

1. Introduction

Though there are numerous articles on halophytes there is still no conformity what a halophyte really is and how the difference to glycophytes may best be defined. According to Schimper (1935) halophytes are plants which complete their whole life cycle in saline habitats; usually there, soils are characterized by high NaCl levels and low osmotic potentials.

According to Repp (1961) salt tolerance of plasma is a direct measure for the salt tolerance of a species. She defines salt tolerance as the capability of a plant cell to survive 24 h in a distinct salt solution.

Other authors look upon salt tolerance on enzymatic levels and restrict the term to some bacteriae whose enzymes can withstand higher external salt concentrations than enzymes of non-salt-tolerant organisms do (Baxter and Gibson, 1954). If this criterion is strictly applied salt tolerance has not been detected in higher plants. According to Greenway and Osmond (1970) isolated enzymes of halophytic *Atriplex* species display the same salt tolerance as those from salt sensitive *Phaseolus* plants. In this context one must point to the necessary distinction between physiologically (Hedenström and Breckle, 1974) and ecologically defined salt tolerance, discussed by Albert (1980/81).

Salt resistance of phanerogams is understood more likely by the importance of avoidance mechanisms (Levitt, 1956), that means, plants have developed special mechanisms to withstand sali-

nities by avoiding that their sensitive cell structures contact high salt concentrations. According to these mechanisms halophytes can be divided into two principal groups: salt excluders and salt absorbers. Salt excluders have developed high ion selectivity in their roots, thus only minor amounts of ions (Na^+, Cl^-) enter the plant body at all; salt absorbers have high ion levels in their cells and struggle with toxic effects and low osmotic potential. One mechanism against toxic levels of ions is found in succulence. High salt levels are partially compensated with high water storage in plant organs, thus diluting concentrations. This mechanism is found typically in the stem-succulent *Salicornia europaea* (Greenway, 1968; Chapman, 1974) and in other stem-succulents as *Arthrocnemum, Anabasis, Allenrolfea, Halocnemum* etc. The same is known from leaf-succulents, e.g. *Aster, Halimocnemis, Halogeton, Nitraria, Salsola, Suaeda*, etc.

Important structures in the salt economy of halophytic species are salt glands. They evolved convergently in many different families, such as Plumbaginaceae, Tamaricaceae, Primulaceae, etc. A similar function of salt recretion is ascribed to bladder trichomes of some Chenopodiaceae, e.g. *Atriplex* species. The accumulation of excessive salt in the bladder cell is a recretion process (according to terminology in Frey-Wissling, 1935), which is followed by desalting, that is the ultimate delivery of salts out of the plant body by abscission of mature bladders. The salt removal function of the bladders was discovered in 1959 (Berger-Landefeldt). Wood (1923) regarded bladder trichomes as water storage organs and Adriani (1958) called *Atriplex* a non-desalting succulent halophyte.

It is worth to look in detail on those bladders. They are found in several genera of the Chenopodiaceae (*Atriplex*, *Chenopodium*, *Halimione*, *Obione*, *Salsola*). Of about 400 species of *Atriplex* most are halophytes. Some are xerophytes and a few are ruderals. All *Atriplex* species develop bladders. These are thought to be the main structure for salt tolerance of halophilic *Atriplex* species, though their role in salt tolerance is not yet clearly understood. Reports on *Atriplex* species and their response towards salinity stress are numerous (Strogonov, 1975; Kelley et al., 1981), but bladders are often disregarded. Few data on halophytic species exist and none on ion concentrations in bladders of non-halophytic species.

Especially the behaviour of those species during high salt treatments could give a hint whether the recretion of ions in bladders is a special quality of *Atriplex* in saline environments or a general quality of the genus, independent of the salt tolerance. Another question is if there are similar relations in other genera. The discussion of these problems in the following sections are based on 1. a comparative literature survey of ion concentrations in bladders and mesophyll from different species, 2. on our own studies on *Atriplex hortensis* (Schirmer, 1980), a glycophyte, whose natural habitats are ruderal areas, and 3. on earlier studies on *Atriplex confertifolia* and *A. falcata*, two desert dwarf shrubs from Utah and Nevada (Moore et al., 1972; Breckle, 1974, 1976). We intend to gain some ideas of the role and importance of the bladder in the salt economy of the whole plant.

Besides the fact that bladders are outstanding structures in a few genera of chenopods, it will be important to know more on salt tolerance of *Atriplex* from the viewpoint of applied science (agriculture, environmental impacts etc., e.g. using the relatively high protein content). The usefulness of halophytes is important in many respects. Salinity problems have evolved in many, and not only arid, countries. *Atriplex* species seem to have an increasing importance (Breckle, 1978; Hoffman, 1980). An overall synopsis on *Atriplex* was given recently by Osmond et al. (1980).

216

2. Anatomy

Bladders occur in some genera of the Chenopodiaceae, essentially in all *Atriplex* species, but also in some species of *Salsola*, in most species of *Chenopodium* and in all species of *Obione* and *Halimione*. There is a great similarity in the appearance of the swollen epidermal hair or vesicles, examples are given in Figs. 1–4.

Generally the vesicle is stalked (1 to 4 stalk cells), bearing a swollen bladder cell, 80 to 200 μm in diameter (Fig. 7). Numerous vesicles are developed in young leaves of most species. In some species the vesicles break and the remnants form a mat over the surface of the mature leaf (West, 1970). Black (1954) described in detail the development of bladders for *Atriplex vesicaria* and *A. nummularia*. These hairs appear to originate as 'blowouts' from epidermal cells, the first cell division cutting off the initial cell for the vesicle. A stalk cell may be formed by a subsequent cell division of this initial cell, or, rarely, the initial cell itself may enlarge directly to the ovoidal vesicle which at maturity is usually 150 to 200 μm long. Generally the vesicles are either shortly stalked with one cell, or the stalk may be long and consist of up to four cells. This process of vesicle initiation commences from the dermatogen of the earliest stages in leaf primordia and stem apices. It apparently continues right through the ontogeny of the leaf up to the onset of senescence. It is not uncommon to see vesicle initial cells in mature leaves. Even more frequently, it may be observed, that below the older mature vesicle remnants on such leaves young vesicular trichomes develop which expel the remnants. This procedure (Fig. 5) as well as the starting time of vesicle formation can be confirmed for *Atriplex confertifolia* and *A. falcata* (Breckle, 1976) and for *A. hortensis* (Schirmer, 1980).

Most papers on structure and development of bladders dealing with *Atriplex* species show similar results. Divergence arises in numbers of bladders per leaf or leaf area and in time span for their formation. *Atriplex halimus* (Smaoui, 1971), *A. falcata* and *A. confertifolia* (Breckle, 1976) are shown to form new bladders during the entire life of the leaf, however, decreasing intensity with leaf age.

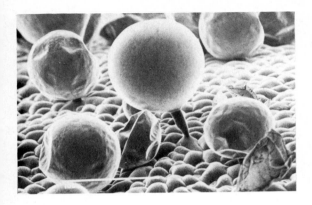

Fig. 1. Scanning electron micrograph (SEM) from the surface of a mature leaf of *Atriplex hastata*, showing various stages of development, from fully turgid bladders to remnants totally shrunk (Schirmer, 1980).

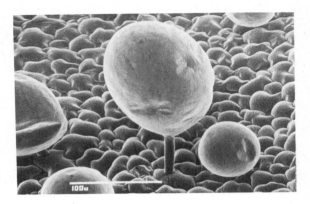

Fig. 2. SEM of a bladder trichome from an old leaf surface of *Atriplex hortensis*, the bladder cell showing some first signs of shrinking (Schirmer, 1980).

Fig. 3. SEM of bladder trichomes in *Atriplex hortensis* with a spherical bladder cell showing rough cuticular texture and a long slender stalk cell (Schirmer, 1980).

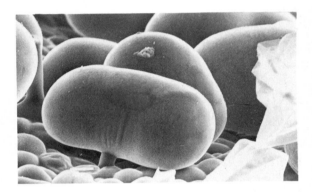

Fig. 4. SEM of bladder trichomes in *Atriplex hortensis* with irregular or partly cylindrical bladder cell and rather short stalk (Schirmer, 1980).

This results in several layers of bladders; an upper layer of squashed hair as Black (1954) reported from *A. vesicaria*. The leaves of *A. hymenelytra* in Nevada, in Arizona and in Death Valley Area appear almost white by a thick layer of bladder remnants. In this case the dead bladders serve as an additional protection against extreme radiation.

Other species of *Atriplex* exhibit only thin layers of bladders, e.g. *A. patula* (Hall and Clements, 1923), *A. buchananii* (Troughton and Card, 1974), *A. hortensis (*Schirmer, 1980) and *Halimione portulacoides* (Baumeister and Kloos, 1974). This latter species develops bladders only on young leaves where they can also form several layers, but with the expansion of the leaf the number of bladders per unit area declines. With increasing leaf age more and more bladders collapse and are probably dropped or washed away by rain. The formation of new bladders decreases with age in all species observed.

Osmond et al. (1980) tried to relate the thickness of bladder layers to the fact, whether a plant is annual or perennial. There also may be a correlation between salt tolerance of a species and the number of bladders, but an increase of bladders with increasing external salt concentration was insignificant (Breckle,1976; Schirmer, 1980). Variations in volume, size, shape, and bladder numbers seem to be rather independent from external salt conditions. The bladder anatomy of certain Chenopodiaceae is obviously genetically determined. The percentage

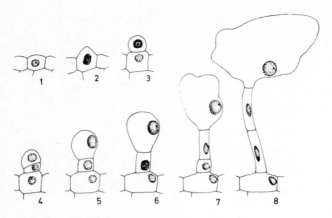

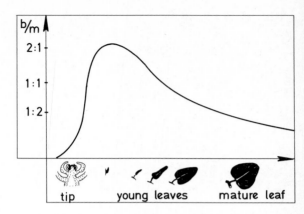

Fig. 5. Eight stages of the development of a bladder near the shoot tip in *Atriplex falcata* (*A. nutallii*), drawn after microscopic observations (Breckle, 1976).

Fig. 6. Ratio (b/m) of total bladder volume (b) to leaf lamina volume (m) (epidermis and mesophyll) during leaf ageing in *Atriplex confertifolia* (Breckle, 1976).

of bladder volume in comparison with the whole leaf is highest in the first stages of leaf differentiation. This is common for all species studied so far (Fig. 6).

3. Ecophysiology

In contrast to the numbers of theories about the importance of the bladders in the salt economy of *Atriplex* species, only few data about the ion concentration in bladders are available. Waisel (1972) indicates some doubt in bladder efficiency in the desalting process. Osmond et al. (1980) do not dare to give a final answer, though they give a very comprehensive description of the genus *Atriplex*. Difficulties in solving this problem arise from the specific response of various species. As far as we know all *Atriplex* species develop bladders, but on their function in different habitats (ruderal, halophytic, xerophytic stands) almost no data exist. Mozafar (1969) was one of the first to analyze the ion concentration of single bladders (10 bladders/analysis) of *Atriplex halimus* cultivated on different salt concentrations. He found the mesophyll concentration (Na^+, K^+, Cl^-) remaining almost constant over the whole range of applied external salt concentration range (osmotic potentials from -0.36 to -10 bars or 200 mM NaCl), while the ion concentration in the bladders increased with increasing salt treatment and reached

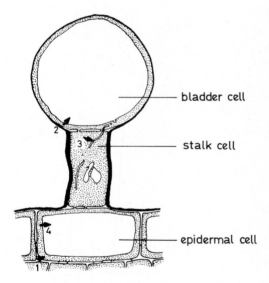

Fig. 7. Schematic diagram showing the bladder system structure and possible sites of active Cl^- transport (1–4) (Osmond et al., 1969; Pallaghy, 1970; Lüttge, 1971; West, 1970; Breckle, 1976). The bladder cell has a giant central vacuole. The stalk cell is characterized by a densely cutinized cell wall; this may be analogous to the Casparian strip in root endodermis (Lüttge, 1971). However, there are indications of some parts of the stalk cell wall being less cutinized (Smaoui, 1971; Thompson, 1975). 1. Active transport through plasmalemma, from apoplast to synplast; 2. Active transport through tonoplast, from cytoplasm to vacuole of bladder cell; 3. Active transport to the membrane system in cytoplasm of stalk cell, movement of small vesicles to bladder cytoplasm; 4. Active transport through tonoplast of epidermis cell.

concentrations of almost saturated NaCl solutions. In his opinion the vesiculated hairs play a significant role in removal of salts from the vascular tissues and parenchyma of leaves and thus prevent dangerous accumulation of toxic salt levels in the leaves.

A powerful salt pump must be operative in the stalk cell (Fig. 7) in order to maintain and produce such a high gradient of salt concentration between the bladder and the adjoining leaf cells. The light stimulated ion flux of $^{36}Cl^-$ to the bladder is of the order 1 meq $\cdot$ g^{-1} $\cdot$ h^{-1} at 5 mM KCl treatment for *Atriplex vesicaria* (Osmond et al., 1969). The same authors made similar studies with *A. spongiosa* and *A. inflata*, both annual plants, and with the perennials *A. vesicaria* and *A. nummularia*. Mozafar (1969, 1970) did similar studies with *A. halimus*. The results do not clearly reveal the function of the bladders. In Fig. 8 the Cl$^-$ concentration for *A. vesicaria*, *A. nummularia*, and *A. spongiosa* at salt levels in the nutrient solution up to 250 mM in bladders and lamina is given. A difference in the behaviour between annuals and perennials is noticeable. With increasing salinity treatments *A.*

nummularia shows reaction curves close to those of *A. halimus* with mesophyll concentrations increasing only slightly but bladder concentrations increasing rapidly. The concentration curves show the typical character of halophytes with already high ion values even at low salinity stress. The bladders exhibit something like a saturation behaviour which may indicate the functioning of the bladder is exhausted at higher salinity stress in *A. vesicaria* and *A. nummularia*. As can be seen for *A. spongiosa* (Fig. 8) the concentration of both the bladders and the lamina are increasing with increasing salinity stress. The concentration of the bladders always exceeds that of the lamina reasonably. It may be mentioned that the leaves of the annual *A. spongiosa* produce only one layer of bladders, while perennials such as *A. nummularia* produce several layers when growing under field conditions and thus may continue to recrete salt from the lamina. In any case the ion concentration of bladders and of the lamina increases more or less with age.

Atriplex confertifolia, an obligate halophyte, concentrates in the bladders ions against a gradient 10

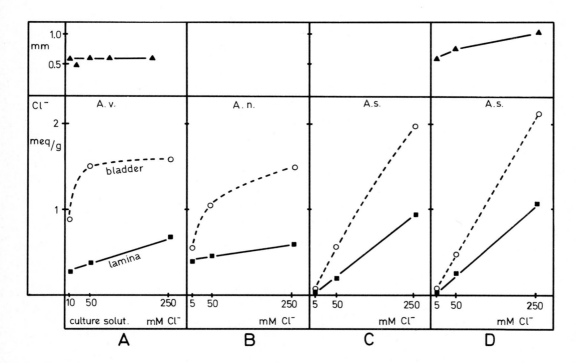

Fig. 8. Concentration of Cl$^-$ in lamina (■) and bladders (○) of *Atriplex vesicaria* (A); *A. nummularia* (B); and *A. spongiosa* (C,D); grown in culture solutions containing NaCl. In A and D thickness of leaf lamina (▲) is indicated (redrawn from Osmond et al., 1969, 1980).

times that of the mesophyll (at least for sodium, somewhat less for chloride). The ion concentration of the bladders reaches values of approximately saturated sodium chloride concentration. However, the lamina cells contain large amounts of salts too, despite the activity of the bladders. This means that the plasma of mesophyll cells must either possess a certain salt resistance or the salt is in the mesophyll cells and their organelles strictly compartmentalized. Bladders of *A. confertifolia* are not capable of keeping the salt concentration of the mesophyll constant, though their concentration is 3 to 12 times

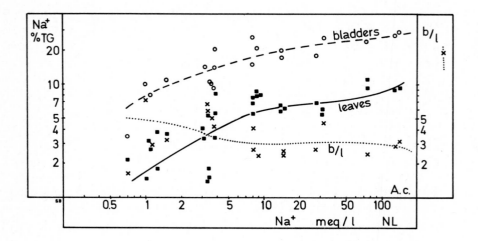

Fig. 9A

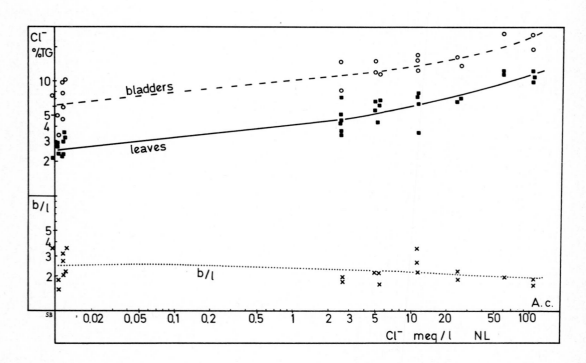

Fig. 9B

higher (Fig. 9) than the mesophyll concentration (Breckle, 1976). For *A. falcata* the same author finds rather low sodium and chloride ion concentrations, but an excess of potassium. Some ecotypes from *A. falcata* (aggr. inclusive *A. nutallii*) are potassiophilic (Table 1) in contrast to most other sodiophilic species of the genus. But there are apparently ecotypes of the *A. falcata* complex, which are sodiophilic (Albert, 1980/1981). The wide range of the *A. falcata* (*nutallii*) complex along an entire gradient from most saline communities with *Allenrolfea*, then *Sarcobatus* to less saline with

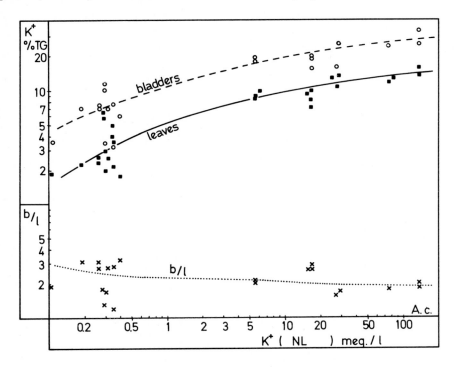

Fig. 9C

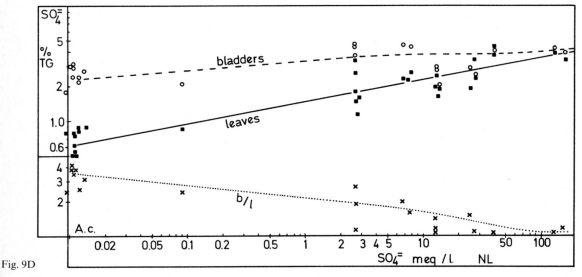

Fig. 9D

Fig. 9. Content (% of the dry matter) of Na^+ (A); Cl^- (B); K^+ (C); and SO_4^{2-} (D); in leaf lamina and bladders of *Atriplex confertifolia*, grown in culture solutions with increasing amounts of the respective ion (Breckle, 1976). and accumulation ratio bladder/lamina (b/l).

221

Ceratoides and lowest with *Artemisia tridentata* led to the conclusion, that several ecotypes are involved.

Goodman and Caldwell (1971) and Goodman (1973) could distinguish three genetically distinct populations by transplanting experiments. Ecotypic variation exists also in *A. lentiformis* (Hall and Clements, 1923; Pearcy, 1976; Pearcy et al., 1977). Chatterton et al. (1970) had investigated three populations from Californian *Atriplex polycarpa*, showing very high salinity tolerance with marked differences in growth response. In all the examples

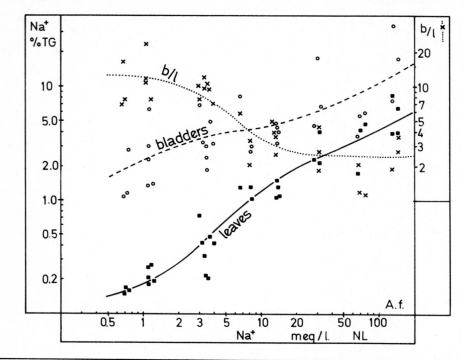

Fig. 10A

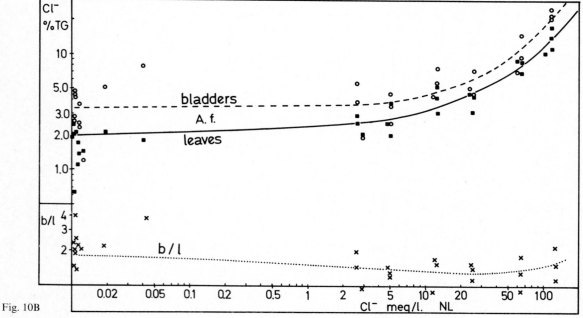

Fig. 10B

the bladder system was not examined. Formation, density, anatomy and activity of the bladder system would be worth checking. Some studies in Utah (Breckle, 1976) indicate, that the function of the bladder in potassiophilic populations of *A. falcata* remains the same; K^+ is recreted instead of Na^+ to the bladders and concentrated. The ratio of the concentration between bladders and lamina is

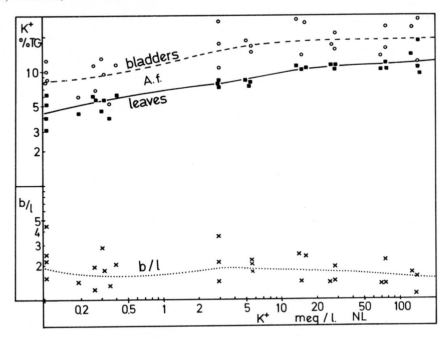

Fig. 10C

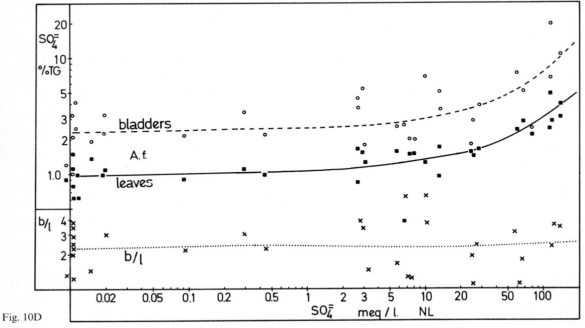

Fig. 10D

Fig. 10. Content (% of the dry matter) of Na^+ (A); Cl^- (B); K^+ (C); and SO_4^{2-} (D); in leaf lamina and bladders of *Atriplex falcata* (*A. nutallii*), grown in culture solutions (NL) with increasing amounts of the respective ion (Breckle, 1976), and accumulation ratio bladder/lamina (b/l).

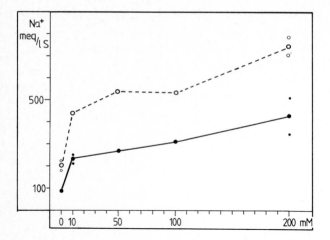

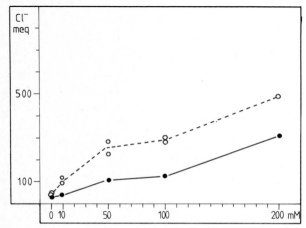

Fig. 11. Content (meq l⁻¹ cell sap) of Na⁺ (A); and Cl⁻ (B); in leaf lamina (●) and bladders (○) of *Atriplex hortensis*, grown in culture solutions with increasing salinity (Schirmer, 1980).

Table 1. Ion content (I.C.) in leaves (% of the dry matter) and percentage of loss of ions from leaves by thoroughly washing them twice with distilled water in two *Atriplex* species in Utah (Breckle, 1976).

	Na^+		K^+		Cl^-		SO_4^{2-}	
	I.C.	loss	I.C.	loss	I.C.	loss	I.C.	loss
Atriplex confertifolia	6.8	5	3.2	9	6.5	8	1.1	9
Atriplex falcata	0.29	15	5.4	7	2.1	11	0.49	9

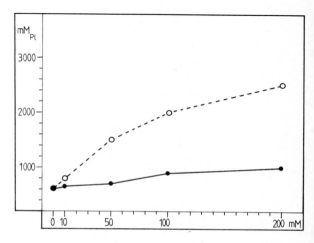

Fig. 12 Threshold of plasmolysis in bladder cells (○) and mesophyll cells (●) with increasing salinity in culture medium. Concentration of the plasmolytic medium is given in mM (Schirmer, 1980).

shown in Fig. 9 for *A. confertifolia* and in Fig. 10 for *A. falcata*. *Atriplex confertifolia* represents a species with several layers of bladders still producing new layers in mature leaves. Breckle (1976) was the first who took the age of the leaves into consideration when discussing the importance of bladders. As the ratio of bladders per lamina area declines with leaf age in a characteristic way (Fig. 6), he assumed that the function of the bladders is of primary importance in young developing leaves and buds. A similar conclusion can be drawn from data by Osmond et al. (1980), see Fig. 14.

Studies with *Atriplex hortensis*, an example of a glycophytic *Atriplex* species show results given in

Figs. 11 and 12. The Na⁺ and Cl⁻ concentration of both the bladders and the lamina are shown. The bladders again exhibit higher ion concentrations than the mesophyll cells, at least in plants under salinity stress. This means that a certain accumulation of ions in the bladders occurs, even in this species. Besides this, *A. hortensis* is capable to withstand 200 mM NaCl in culture solution, without any visible signs of damage. Again leaf age, e.g. from a series of leaf nodes and internodes, influences Na⁺ content distinctly, see Fig. 15 (Schirmer, 1980).

From other genera of the Chenopodiaceae there are almost no data on bladders available. The bladders of all *Chenopodium* species are assumed not to accumulate ions. Accordingly *C. bonus-henricus* is rather salt sensitive. We found a growth reduction at 50 mM NaCl concentration in the culture solution.

Baumeister and Kloos (1974) did detailed studies with *Halimione portulacoides*. The genus is closely related to *Atriplex* and the anatomy of the bladder system in both genera is equivalent. The physiological response of *Halimione* towards high salt levels is shown in Fig. 13. It indicates that the recreted sap (bladder cell sap) at high salinity levels always contains a higher concentration of ions than the sap pressed out of the mesophyll cells. The content of ions both of bladders and mesophyll sap are dependent on the concentration of the culture solution. The capacity of the bladders to recrete enough sodium seems to be exhausted above 500 mM NaCl in the culture solution. This is due to chloride as well, but to a lesser extent. These values are only representative for young leaves, as the bladders are here only active in young leaves. Mature leaves have a thin but tight film of collapsed bladders. Despite this fact, *Halimione* tolerates external salinity levels up to 1 M NaCl. This demonstrates that additional factors must be considered when discussing salt tolerance of species possessing bladder hairs. An important additional factor is succulence. *Atriplex* species show increasing succulence when growing under saline conditions, and *Halimione* does even more, though both genera are not regarded as typical succulent taxa. It is hardly possible to decide whether the salt tolerance is due to the function of the bladders or due to increasing succulence. Osmond et al. (1980) tried to distinguish between the two processes: 'There is no change in the thickness of the leaf lamina of *Atriplex vesicaria* in which Cl^- concentration increases only about twofold, but a 40% increase in *A. spongiosa* where Cl^- concentration increases more than tenfold.' Figure 8 shows that even in *Atriplex* species with relatively few epidermal trichomes, 'the thickness of the leaf lamina increases as the proportion of total leaf Cl^- in the trichomes declines'. Osmond et al (1980): 'It is clear from these experiments that ion recretion to the bladder cell of epidermal trichomes functions to

control salt concentration in the leaf lamina. When this control fails or if the species is poorly endowed with trichomes, a measure of regulation of salt concentration can be achieved by increased succulence. The latter response presumably involves a complex interaction between ion uptake, cell water relations and cell wall metabolism.' The increase of cell volume overcompensates the decrease in cell number (Black, 1958). So in *A. triangularis* (*hastata*) Black noted that the number of cells per leaf declined about six to eightfold when NaCl concentration in solution culture was increased from 0 to 600 meq . l^{-1}, and that the leaf thickness doubled over this range. These two processes, salt recretion and the development of succulence probably function to regulate the ion uptake by leaf cells in two different ways. Salt recretion presumably functions to regulate the extracellular accumulation of salt in the apoplast which occurs when transport into the leaf exceeds the capacity of salt uptake to the vacuoles of leaf cells. On the other hand, a halophyte which lacks effective salt recreting glands or trichomes, but has the capacity to respond to salinity with succulence may be equally capable of regulating apoplastic salt concentration.

The ion distribution within the plant body and specifically within the various organs is not at all the same in the species investigated. Many organs show specific ion distribution patterns. As already mentioned one main difference in the strategies of halophytes is whether the plant is a salt excluder or a salt accumulator. A species which behaves as a salt excluder should be less dependent on bladders in respect to salt tolerance. *Atriplex hastata* behaves somewhat like a salt excluder but nevertheless possesses bladders (Greenway and Osmond, 1970). Unfortunately there are no data on the ion concentration of bladders of this species. Within the large group of salt accumulators including probably most *Atriplex* species specific differences in the ion distribution in the plant organs can be found. The leaves are the place of ion accumulation especially for sodium. Chloride can also be present at high levels in leaves and stems. Distinct differences in ion patterns exist. Most *Atriplex* species prefer Na^+ over K^+. Few species prefer K^+ and accumulate this ion in concentrations like other halophytes accu-

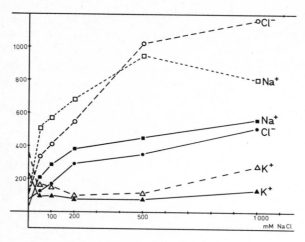

Fig. 13 The influence of increasing salinity in culture solution to Na^+, K^+ and Cl^- concentration in leaf cell sap (closed symbols) and bladder recreted fluid (open symbols) of *Halimione portulacoides* (redrawn from Baumeister and Kloos, 1974).

mulate Na^+. This phenomenon is parallel to many other ion patterns in Chenopodiacean genera, where a broad variability in specific ion pattern occurs (Breckle, 1975; Mirazai and Breckle, 1978). This is not yet clearly understood and should be studied in the future. The same is known from the genus *Salsola*, where only the section *kali* accumulates K^+, whereas all other sections are more or less sodiophilic.

Another point is that ion concentrations of the

organs normally increase with age. Future studies should give therefore always phenological or annual (or even diurnal) fluctuations in ion content. To judge the salt tolerance of a species, the germination under saline conditions should furthermore be considered. Particular stages in the life cycle of a plant show different sensitivity towards salinity stress (Hayward, 1956). The highest sensitivity is without doubt during germination (Waisel, 1972). Thus salt tolerance of a seed or seedling is the most important property of any species for its settlement in saline habitats (Ungar, 1981). It is well known that most halophytes have highest germination rates in destilled water (Lötschert, 1970; Rajpurohit, 1980) and tolerate higher salt concentrations only to some degree. With age, however, the euhalophytes develop mechanisms which enhance growth on saline over non-saline habitats.

The cotyledos of *Atriplex hortensis* do not de-

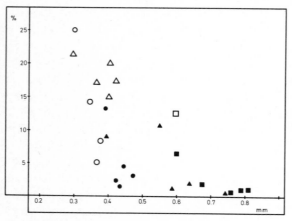

Fig. 14. Correlation between thickness of the leaf lamina and the percentage of total Cl^- found in the epidermal bladders of *Atriplex* species, triangles △: *A. triangularis;* circles ○: *A. rosea*; squares □: *A. glabriuscula*; in young (open symbols) and in old leaves (closed symbols) (redrawn from Osmond et al. 1980).

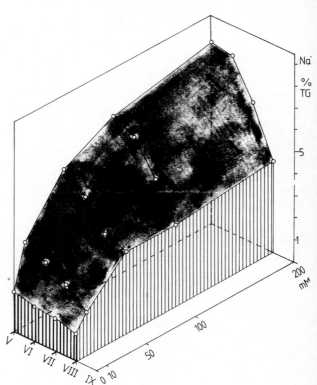

Fig. 15. Na^+ content (% of the dry matter) of leaves with increasing salinities in culture medium (mM NaCl) for different nodes of leaf pairs (the leaves are opposite; V = oldest leaf node; VIII = youngest leaf node) of *Atriplex hortensis* (Schirmer, 1980).

velop a single bladder. It would be interesting to investigate this for other halophytic species. Since seedling establishment is the most vulnerable stage in the life cycle, it may be profitable to examine closely the influence of nutrition, salinity tolerance, and especially photosynthesis according to C_3 or C_4 metabolism on germination (Osmond et al., 1980).

Tolerance for toxic substances was rarely discovered in *Atriplex*. Breckle (1976) has shown that *Atriplex confertifolia* and *A. falcata* have a high tolerance for boric acid in the culture solution. Concentration of boric acid in plants increases rapidly with increasing levels in the culture solution. Toxicity symptoms were observed above 500 ppm B in leaf dry matter (30 ppm in culture solution). These confirm the finding by Goodman (1973). It was not yet possible to study the role of the bladder system in connection with a possible B recretion and B tolerance.

Tolerance for metals was probably involved in studies in *Atriplex triangularis* growing on ash deposits near a coal firing power station. This ash, rich in manganese, aluminium, etc. excludes all common weeds, as was shown by Rees and Sidrak (1956). Osmond et al. (1980) assumed ecotypic adaptation or a general tolerance of this species towards such extreme conditions. Again the role of the bladder system as a secreting system would be worth studying;

4. Conclusions

Anatomical features of the bladders as shown for *Atriplex hortensis* (Fig. 1 to 4) can be found in the literature for xerophytic as well as for halophytic or glycophytic species. The development of the bladders in general seems to be a genetic character which does not leave much space for environmentally controlled adaptations. Ecological factors as high salt concentration (halophytes) or shortage of water (xerophytes) probably do not cause major changes in the development of structure or number of bladders. On the other hand significant and ecological relevant differences were demonstrated for the number of bladders per leaf area of some species, or for the continuing formation of new

bladders in mature leaves. As the formation of bladders starts in the very early stages of leaf development and as the percentage of bladder mass per the other leaf tissue mass is highest in this stage, it seems evident that the basic role of bladders is the protection of young developing shoots and leaves from toxic salt levels first in the apoplast and subsequently in the symplast.

Mozafar and Goodin (1970) show a 60–fold accumulation of the salt concentration of bladders over the mesophyll. Their analytic data (Mozafar, 1969) show sodium concentrations up to 14 M, which corresponds to somewhat less than − 700 bars osmotic potential. This is only realized in over-saturated NaCl solutions, respectively in solutions with crystallized NaCl. This needs investigation.

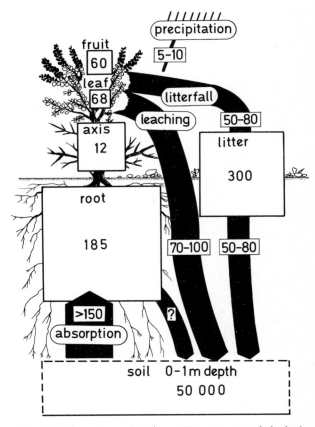

Fig. 16 Na^+ cycling and Na^+ standing mass on a halophytic stand (*Atriplex confertifolia* community/Northern Utah). The fluxes of Na^+ (figures in arrows) are given in kg $Na^+ \cdot ha^{-1} \cdot a^{-1}$; the standing mass (figures in squares) in kg $Na^+ \cdot ha^{-1}$. The compartment squares size reflects the biomass pool size (redrawn from Larcher, 1980; Breckle, 1976).

227

The fact that more than 50% of the total amount of ions of the leaf are present in the bladders, which constitute only 21% of the weight of the whole leaf leads to the conclusion that the bladders play an important role in salt removal. Results of investigations and precise data are few.

The fact, that below *Atriplex* shrubs the salt content in soil profiles is higher than between shrubs was demonstrated by Roberts (1950), and Sharma and Tongway (1973). We have shown, however (Breckle, 1976), that in Utah the Cl^- content in soils is of high significance and that Na^+ content is significantly lower below *Atriplex confertifolia* than between the shrubs. This controversial finding needs further examination. The cycling of sodium within the ecosystem is shown in Fig. 16. The amount of sodium within biomass equals about 460 kg . ha^{-1}. Annually at least 150 kg . ha^{-1} are cycling. The role of the bladders within the system is not yet clear. But a relatively high proportion of salt is deposited in the bladders and is slowly leached, respectively falling down with litter. Because of the cuticle the direct leaching from bladders is either hindered or almost zero (Mozafar, 1969). This is also true for other *Atriplex* species (Breckle, 1976); see Table 1. It may be assumed that in other *Atriplex* species or even in some typical chenopod shrublands (*Atriplex hymenelytra* in Nevada, *A. canescens* in South West USA., *A. nummularia* and *A. vesicaria* in Australia) cycling of salt by bladders is still more important. Osmond et al. (1969) and Lüttge (1971) assume that emission of salt is apparently the result of the rupturing and collapse of the bladder cells. But this may vary from species to species.

Within the *Atriplex confertifolia* community in Utah at least 30 to 40% NaCl is cycling by recretion through bladders. This can be concluded, though even the accumulation factor (bladder/mesophyll) in this species is lower (10 to 5) than that reported for *Atriplex halimus* by Mozafar (1969).

The effectiveness of the various strategies in coping with salinity stress was discussed as early as 1928 by Stocker. He compared occurrence and frequency of desalting halophytes in various halophytic vegetation types and showed that the desalting halophytes occur almost regularly, but play only a minor role in number of species and individuals. Xerohalophytic types in his comparison were much underrepresented. Genera with bladders at that time were not regarded to be desalting. In North America and Australia semideserts and deserts (Beadle, 1981), chenopod shrublands dominated by *Atriplex* play a major role. There might be some parallel to the desalting mangrove species which are common in the mangrove types especially in subtropical coast regions (Walsh, 1974).

Waisel (1972) concluded that desalinization in *Atriplex* is ineffective. The number of bladders seems too small, in his opinion. According to our data and to some Australian studies the activity of the bladders in salt removal within *Atriplex* communities is apparent, however. An overall study of bladder species including selectivity, anatomy (formation, replacement, density), physiological activity in salt recretion mechanisms, and efficiency is not at hand; thus ecological significance only can be estimated.

Annuals possess a lower effective protection of the mesophyll by bladders than perennials. But even in glycophytes (*A. hortensis*) the bladders can accumulate ions. The accumulation factor is lower than in perennials, around 2 to 5 (Schirmer, 1980). Anyhow, this might support the opinion of Collander (1941), that *A. hortensis* is originally a halophyte, although this small increase of ions in bladders can not play a significant role in the salt economy of this species. Besides, an appreciable number of bladders is formed in developing leaves only.

Bladders of the halophilic *Atriplex* species help these species to cope with high salt concentration. The importance varies from species to species. But to which extent they are responsible for the overall salt tolerance can not be answered yet. Chenopod species, lacking bladders, as e.g. *Eurotia lanata* = *Ceratoides lanata* (Caldwell, 1974) can compete similarly effective, with *Atriplex confertifolia* on moderately saline places using rather different physiological strategies. Bladders in genera other than *Atriplex* seem to be equally important only in *Halimione* (Kappen and Maier, 1973; Baumeister and Kloos, 1974). But almost any data on salt fluxes via bladders are lacking. *Halimione portulacoides* is known as a very salt resistant species along European coasts. Leaves can become very succulent. Bladder layers on older leaves appear shrunken and crumbled and thus ineffective.

In *Salsola* bladders are known only from a few species of this large genus. They play a minor role; they might be only an evolutionary relict there. Their density is low and their volume there is small (Bhadresa, pers. commun.).

In *Chenopodium* most species exhibit bladders. They occur mostly monolayered. Their physiological properties are not tested. But because *Chenopodium* species rarely occupy saline habitats, the conclusion may be that bladders in this genus are of minor ecological significance, at least in salt removal. This is stressed by the fact that e.g. *Chenopodium botryodes* in the Neusiedlersee area (Austria) on alkali soils has increased succulence only and not bladder density. But here too bladders are formed very early during leaf and shoot development close to the shoot tip. This very early formation of bladders is in contrast with similar structures in grasses, as in *Aeluropus*, *Spartina* and *Chloris* (Liphschitz and Waisel, 1981), where recretion of salt probably starts much later. Additionally the process of salt removal is not an accumulation in a giant vacuole, but a movement across the plasmalemma and cell wall to a distinct subcuticular space. As in all other salt glands (*Acanthus, Aegialitis, Aegiceras, Avicennia, Cressa, Frankenia, Glaux, Ipomoea, Limonium, Tamarix*), which are metabolically very active, the ultrastructural organelle pattern is characterized by numerous mitochondria and dense, partitioning membranes in a dense cytoplasm (Thompson, 1975) as well. Some recretion rates are known from salt glands which recrete fluid out of the gland via pores in the cuticular layer. A rate of $0.4 \, u \, eq . cm^{-2} . h^{-1}$ is reported for *Aegialitis* by Atkinson et al. (1967); this is one of the highest known to occur in biological systems (Jennings, 1968). Several studies on selectivity and ionic recretion activity have been done with *Tamarix* (Waisel, 1961; Berry and Thompson, 1967; Berry, 1970) and with *Limonium* (Hill, 1967; Ziegler and Lüttge, 1966). The only quantitative data for bladder activity in *Atriplex* is given by Osmond et al. (1969); it is several times lower than the above figure for gland activity.

Studies with *Atriplex* have shown that light induces membrane potentials in the mesophyll cells and similar changes in the gland cell below the bladder, but these transient changes were not observed in *Chenopodium* (Lüttge, 1971). Thus bladders from different species with apparent similar anatomy may function rather differently. That bladders play a certain role in salt metabolism can be deducted from the density of bladders and the salt recreting activity which is higher in salt tolerant species than in glycophytes.

But many problems are still open. One of many e.g. is if the bladders in *Atriplex* might function as a detoxifying system for high levels of toxic substances (boric acid, metals).

The bladder-bearing species use one of several successful strategies (see Fig. 17) to compete in saline environments. These strategies cannot be separated from each other clearly as some plants use more than one strategy. The different possibilities of salt fluxes through the plant body and its controlled reduction allow us to distinguish halophytic types (Fig. 17) by absorption, transport, accumulation, and recretion processes. Bladder-bearing species are mostly one type within the crino-halophytes, namely recretors.

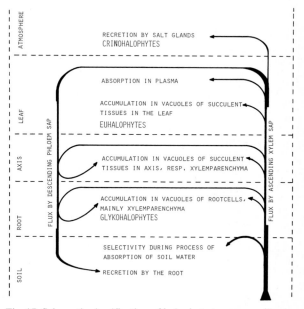

Fig. 17. Schematic classification of halophyte types according to strategies coping with internal salt content (redrawn from Henkel and Shakhov, 1945; Breckle, 1976).

Literature cited

Adriani, M.J. 1958. Halophyten. In *Hdb. der Pflanzenphysiologie*, Bd. 4, ed. W. Ruhland, pp 709–736, Berlin.

Albert, R. 1980/1981. *Halophyten*. Habilitationsarbeit, Wien.

Atkinson, M.R., Findlay, G.P., Hope, A.B., Pitman, M.G., Saddler, H.D.M. and West, K.R. 1967. Salt regulation in the mangroves *Rhizophora mucronata* Lam. and *Aegialitis annulata* R.Br. *Aust. J. Biol. Sci.* 20: 589–599.

Baumeister, W. and Kloos, G. 1974. Über die Salzsekretion bei *Halimione portulacoides. Flora* 163: 24–56.

Baxter, R.F. and Gibson, N.E. 1954. The glycerol dehydrogenases of *Pseudomonas salinasia*, *Vibrio costicolus*, and *E. coli* in relation to bacterial halophilism. *Can. J. Biochem. Physiol.* 32: 206–217.

Beadle, N.C.W. 1981. The Vegetation of Australia. In *Vegetationsmonographien der einzelnen Großräume*, Bd. 4, eds. H. Walter and S.-W. Breckle. Fisher, Stuttgart.

Berger-Landefeldt, U. 1959. Beiträge zur Ökologie der Pflanzen nordafrikanischer Salzpfannen. *Vegetatio* 2: 1–48.

Berry, W.L. 1970. Characteristics of salts secreted by *Tamarix aphylla. Amer. J. Bot.* 57: 1226–1230.

Berry, W.L. and Thompson, W.W. 1967. Composition of salt secreted by salt glands of *Tamarix aphylla. Can. J. Bot.* 45: 1774–1775.

Black, R.F. 1954. The leaf anatomy of Australian members of the genus *Atriplex*. I. *Atriplex vesicaria* Heward and *A. nummularia* Lindl. *Aust. J. Bot.* 2: 269–286.

Black, R.F. 1958. Effect of sodium chloride on leaf succulence and area of *Atriplex hastata* L. *Aust. J. Bot.* 6: 306–321.

Breckle, S.-W. 1974. Wasser- und Salzverhältnisse bei Halophyten der Salzsteppe in Utah, USA. *Ber. Dtsch. Bot. Ges.* 87: 589–600.

Breckle, S.-W. 1975. Ionengehalte halophiler Pflanzen Spaniens. *Decheniana* (Bonn) 127: 221–228.

Breckle, S.-W. 1976. Zur Ökologie und zu den Mineralstoffverhältnissen absalzender und nicht absalzender Xero-Halophyten. *Dissertationes Botanicae* Bd. 35.

Breckle, S.-W. 1978. Salinity as an ecological factor promoting desertification. Manuscript of paper presented at the postplenary session of the Xth Intern. Congress on Anthropological and Ethnological Sciences, A Symposium on 'Anthropology and Desertification' in Jodhpur, India, Dec. 1978.

Breckle, S.-W. 1981. Cool deserts and shrub semideserts in Afghanistan and Iran. In *Ecosystems of the world*, Vol. 5: *Shrub steppe and cold desert*, ed. N. West, in press. Elsevier, Amsterdam.

Caldwell, M.M. 1974. Physiology of desert halophytes. In *Ecology of Halophytes*, eds. R.J. Reimold, and W.H. Queen, pp. 355–377. Acad. Press, New York.

Chapman, V.J. 1974. Salt marshes and salt deserts of the world. In *Ecology of Halophytes*, eds. R.J. Reimold and W.H. Queen, pp. 3–32. Academic Press, London.

Chatterton, N.J., McKell, C.M., Bingham, F.F. and Clawson, W.J. 1970. Absorption of Na, Cl, and B by desert saltbush in relation to composition of nutrient solution culture. *Agron. J.*
62: 351–352.

Collander, R. 1941. Selective absorption of cations by plants. *Plant Physiol.* 16: 691–720.

Frey-Wissling, A. 1935. *Die Stoffausscheidung der Höheren Pflanzen*. Springer, Berlin.

Goodman, P.J. 1973. Physiological and ecotypic adaptations of plants to salt desert conditions in Utah. *J. Ecol.* 61: 473–494.

Goodman, P.J. and Caldwell, M.M. 1971. Shrub ecotypes in a salt desert. *Nature* 232: 571–572.

Greenway, H. 1968. Growth stimulation by high chloride concentration in halophytes. *Israel J. Bot.* 17: 169–177.

Greenway, H. and Osmond, C.D. 1970. Ion relations, growth and metabolism of *Atriplex* at high external electrolyte concentrations. In *The Biology of Atriplex*, ed. R. Jones, pp. 49–56. CSIRO, Deniliquin.

Hall, H.M. and Clements, F.E. 1923. The phylogenetic method of Taxonomy: The North-American species of *Artemisia*, *Chrysothamnus* and *Atriplex*. Publ. Carnegie Inst. No. 326.

Hayward, H.E. 1956. Plant growth under saline conditions. UNESCO, *Arid Zone Research, Utilization of Saline Water* 4: 37–71.

Hedenström, H.V. and Breckle, S.-W. 1974. Obligate halophytes? A test with tissue culture methods. *Z. Pflanzenphysiol.* 74: 183–185.

Henkel, P.A. and Shakhov, A.A. 1945. The ecological significance of the water regime of certain halophytes (Russian). *Bot. Jh.* 30: 154–166.

Hill, A.E. 1967. Ion and water transport in *Limonium*. II. Shoot circuit analysis. *Biochem. Biophys. Acta* 135: 461–465.

Hoffmann, A. 1980. Der Einfluß winterlichen Streusalzes auf die Vegetation am Autobahnrand. Staatsexamensarbeit, Univ. Bonn.

Jennings, D.H. 1968. Halophytes, succulence and sodium in plants – a unified theory. *New Phytol.* 67: 899–911.

Kappen, L. and Maier, M. 1973. Bedeutung einiger nichtflüchtiger Carbonsäuren für die Frostresistenz des Halophyten *Halimione portulacoides* unter dem Einfluß verschieden hoher Kochsalzbelastung. *Oecologia* 12: 241–250.

Kelley, D.B., Goodin, J.R. and Miller, D.R. 1982. Biology of *Atriplex*. In *Contributions to the Ecology of Halophytes*, eds. D.N. Sen and K.S. Rajpurohit, pp. 79–107. Dr W. Junk Publ., The Hague.

Larcher, W. 1980. *Ökologie der Pflanzen auf physiologischer Grundlage*, 3. Aufl. Ulmer, Stuttgart.

Levitt, J. 1956. *The hardiness of plants*. Acad. Press, New York.

Liphschitz, N. and Waisel, Y. 1982. Adaptation of plants to saline environments: salt excretion and glandular structure. In *Contributions to the Ecology of Halophytes*, eds. D.N. Sen and K.S. Rajpurohit, pp. 197–214. Dr W. Junk Publ., The Hague.

Lötschert, W. 1970. Keimung, Transpiration, Wasser- und Ionenaufnahme bei Glykophyten und Halophyten. *Oecol. Plant.* 5: 287–300.

Lüttge, U. 1971. Structure and function of salt glands. *Ann. Rev. Plant Physiol.* 22: 23–44.

Mirazai, N.A. and Breckle, S.-W. 1978. Untersuchungen an

afghanischen Halophyten. I. Salzverhältnisse in Chenopodiaceen Nord-Afghanistans. *Bot. Jahrb. System.* 99: 565–578.

Moore, R.T., Breckle, S.-W. and Caldwell, M.M. 1972. Mineral ion composition and osmotic relations of *Atriplex confertifolia* and *Eurotia lanata. Oecologia* 11: 67–78.

Mozafar, A. 1969. Physiology of salt tolerance in *Atriplex halimus* L.: Ion uptake and distribution, oxalic acid content, and catalase activity. Doct. Diss., Univ. Calif. Riverside.

Mozafar, A. 1970. Vesiculated hairs: a mechanism for salt tolerance in *Atriplex halimus* L. *Plant Physiol.* 45: 62–65.

Osmond, C.B., Troughton, J.H. and Goodchild, D.J. 1969. Physiological, biochemical and structural studies of photosynthesis and photorespiration in two species of *Atriplex. Z. Pflanzenphysiol.* 61: 218–237.

Osmond, C.B., Björkman, O. and Anderson. D.J. 1980. *Physiological Processes in Plant Ecology; towards a synthesis with Atriplex. Ecol. Studies* No. 36. Springer, Berlin.

Pallaghy, C.K. 1970. Salt relations in *Atriplex* leaves. In *The Biology of Atriplex*, ed. R. Jones, pp. 57–62. CSIRO, Deniliquin.

Pearcy, R.W. 1976. Temperature effects on growth and CO_2 exchange in coastal and desert races of *Atriplex lentiformis. Oecologia* 26: 245–255.

Pearcy, R.W., Berry, J.A. and Fork, D.C. 1977. Effects of growth temperature on the thermal stability of the photosynthetic apparatus of *Atriplex lentiformis* (Torr.) Wats. *Plant Physiol.* 59: 873–878.

Rajpurohit, K.S. 1980. Soil salinity and its role on phytogeography of western Rajasthan. Ph.D. Thesis Univ. Jodhpur.

Rees, W.J. and Sidrak, C.H. 1956. Plant nutrition on fly ash. *Plant and Soil* 8: 141–159.

Repp, G. 1961. The salt tolerance of plants; basic research and tests. *UNESCO, Arid Zone Res.* 14: 153–161.

Roberts, E.C. 1950. Chemical effects of salt-tolerant shrubs on soils. 4th Int. Congr. Soil Sci. Amsterdam, I: 404–406.

Schimper, A.F.W. 1935. *Pflanzengeographie auf physiologischer Grundlage*, 3.Aufl. Fischer Verlag, Jena.

Schirmer, U. 1980. Blasenhaare und halophiles Verhalten bei *Atriplex hortensis* L. Diplomarbeit, Universität Bonn.

Sharma, M.L. and Tongway, D.J. 1973. Plant induced soil salinity patterns in two saltbush (*Atriplex* spp.) communities. *J. Range Managem.* 26: 121–125.

Smaoui, A. 1971. Différenciation des trichomes chez *Atriplex halimus*. L. *C.R. Acad. Sci.*, sér. D 273: 1268–1271.

Stocker, O. 1928. Das Halophytenproblem. *Ergeb. Biol.* 3: 265–353.

Strogonov, B.P. 1975. Salt tolerance and salt metabolism of plants. Cumulative index of literature, published in Russian in 1875–1975. USSR Academy of Science, Moscow.

Thompson, W.W. 1975. The structure and function of salt glands. In *Plants in saline environments*, eds. A. Poljakoff-Mayber and J. Gale, pp. 118–146, Springer, Berlin.

Troughton, J.H. and Card, K.A. 1974. Leaf anatomy of *Atriplex buchananii. New Zealand J. of Bot.* 12: 167–177.

Ungar, I.A. 1982. Germination ecology of halophytes. In *Contributions to the Ecology of Halophytes*, eds. D.N. Sen and K.S. Rajpurohit, pp. 143–154. Dr W. Junk Publ., The Hague.

Waisel, Y. 1961. Ecological studies on *Tamarix aphylla* (L.) Karst. III. The salt economy – *Plant and soil* 13: 356–364.

Waisel, Y. 1972. *Biology of Halophytes*. Acad. Press, New York.

Walsh, G.W. 1974. Mangroves: a review. In *Ecology of Halophytes*, eds. R.J. Reimold and W.H. Queen, pp. 51–174. Academic Press, New York.

West, K.R. 1970. The anatomy of *Atriplex* leaves. In *The Biology of Atriplex*, ed. R. Jones, pp. 11–15. CSIRO, Deniliquin.

Wood, J.G. 1923. On transpiration in the field of some plants from the arid portions of South Australia, with notes on their physiological anatomy. *Trans. R. Soc. S. Aust.* (Adelaide) 47: 259–278.

Ziegler, H. and Lüttge, U. 1966. Die Salzdrüsen von *Limonium vulgare* L. I. Die Feinstruktur. *Planta* 70: 193–206.

PART THREE

Potentialities and uses of halophytic species and ecosystems

DAVID N. SEN and KISHAN S. RAJPUROHIT

Introduction

Poljakoff-Mayber and Gale (1975) opined that the applied aspect of basic research on salinity at times seems a little academic since many of the agrotechnical solutions devised to cope with salinity were and are based on trial and error. Nevertheless, an intelligent approach to problem solving and a thorough understanding of the effect of man on the environment requires this basic information. An enormous area of the world is too saline to produce economic crop yields, and newer areas are becoming unproductive by human interference due to salt accumulation. Salinity problems in agriculture are pronounced in arid and semi-arid regions where rainfall is too low to leach salt from the plant root zone. Drainage water that has passed through the soil has a higher salt concentration than the irrigation water (Wilcox and Resch, 1963; Carter et al., 1971). When such a drainage water returns to the natural stream or channel, problems of salinity are created for agriculture.

Generally, plants are more sensitive to salinity during germination or early seedling growth. Well established plants are more tolerant than new transplants (Bernstein, 1964). Therefore, crop selection is an important management decision in salinity-effected areas. It is both worthwhile and of imminent importance that ways be considered for plants occurring naturally in saline areas to be put to economic use. With present research concentrating on reclaiming saline areas, it is desirable to choose those species which are well established in these areas. New knowledge which may lead to more effective solutions of the salinity problems in agriculture and potential uses of halophytic species in saline ecosystems is extremely essential.

Literature cited

Bernstein, L. 1964. *Salt Tolerance of Plants.* USDA Agr. Inf. Bul. 283.

Carter, D.L., Bondurant, J.A. and Robbinson, C.W. 1971. Watersoluble NO_3-nitrogen, PO_4-phosphorus, and total salt balances on a large irrigation tract. *Soil. Sci. Soc. Am. Proc.* 35: 331–335.

Poljakoff-Mayber, A. and Gale, J. (eds.) 1975. *Plants in Saline Environments.* Springer-Verlag, Berlin.

Wilcox, L.V. and Resch, W.F. 1963. *Salt Balance and Leaching Requirements in Irrigated Lands.* USDA Tech. Bul. 1290.

CHAPTER 1

Halophytes and human welfare

M.A. ZAHRAN* and AMAL A. ABDEL WAHID

Botany Department, Faculty of Science,
Mansoura University, Mansoura, Egypt

1. Introduction

The worldwide salt affected soils hinder the development of agriculture production. Their distribution is closely related to environmental factors, such as arid and semi-arid climates, accumulation of product of weathering in ground water near the surface, etc. Szabolis (1976) states: 'These types of soil cover such an area in many countries in arid and semi-arid regions of Asia, Africa end South America that they cause considerable problems regarding not only the natural environment of these areas, but also the national economy.' Information about their exact areas is not available for all the countries. According to Massoud (1977) about 7% of the total surface area of the world is covered by salt affected land. Although the arable portion of this sizeable area is not exactly known, it is estimated that more than 50% of all irrigated land of the world has been damaged by secondary salinization and/or sodification and water logging.

In many countries the frequency of salt affected soils make their utilization necessary, mainly by irrigation, application of chemical amendments and also by planting salt tolerant plants (biological desalination). Survival problems in arid, semi-arid and sub-humid regions are a century old. The limited resources of these regions have been threatened by climatic degradation.

* Present address: Institute of Meteorology and Arid Lands Studies, King Abdulaziz University, Post Box 1540, Jeddah, Saudi Arabia.

With the increasing world population and the need for increased crop production, the non-productive lands, many of them salt-affected, may be used to produce non-conventional crops of economic value. The basic idea of our studies is to take the advantage of the halophytic nature of certain plants that proved to have agricultural and/or industrial economic potentialities and to grow them on saline non-productive soils. The cultivation of the saline lands with halophytic plants, which will be irrigated regularly, may be considered, also, as a biological way for soil desalination and reclamation.

The details in the following pages will show the economic potentialities of the halophytes as a source of raw materials in paper (*Juncus* spp.), drug (*Salsola tetrandra*), and as a range plant (*Kochia* spp.) for livestock.

2. *Juncus* spp. in paper industry

Ecologists claim that the green areas are getting smaller as a result of the uncontrolled felling of forest trees and ill-advised land and water management. This may indicate that the pulp importing countries may face some future difficulties. The pulp exporting countries may stop cutting their woods to maintain environmental equilibrium. Accordingly, pulp importing countries, mostly in arid and semi-arid regions, must search for sufficient local available raw materials for their paper industry.

Paper can be produced from materials other than wood, mainly various kinds of grasses, grown either as a crop for pulping, e.g. esparto or bamboo, or left as a residual from agricultural crops as sugarcane bagasse and cereal straw. About 6% of the total pulp produced in the world originates from these sources (Rydholm, 1965).

In Egypt, rice straw, bagasse and waste paper are the main local raw materials for the production of paper. The disadvantage of using grasses for paper production is to some extent technical but mainly economical. One of the technical disadvantages of grasses as a raw material for pulping is their heterogenous composition and structure as compared with wood used for pulping. From the economic point of view it is costly to handle, collect and transport them. Also, they are produced in scattered and fairly small areas, in typical seasonal variations, which tend to deteriorate rapidly on storing without special protection.

Paper mills of Egypt import wood pulp to produce paper of higher strength as well as to improve the strength of paper produced from locally available raw materials. The amount of wood pulp imported is increasing considerably e.g. from 22 tonnes in 1954 to 39, 551 tonnes in 1965 (Zahran, et al., 1972). Consequently, the search for local material for paper industry is necessary.

Those plants which are salt tolerant and fiber producing, if successfully managed to produce good quality paper, would provide a non-conventional crop to be cultivated on saline soils of Egypt. Two advantages can be obtained from this: a) cultivation of the saline soils which are usually left barren or covered with unwanted wild halophytic vegetation; and b) production of local raw material for paper industry.

Juncus rigidus and *J. acutus* (salt tolerant and fiber plants; Boyko 1966; Zahran et al., 1972; El-Bagouri et al., 1976), were selected to be the first group of fiber plants for transplantation on saline soils. Such a study might help in establishing and developing agricultural environment. It would make possible the extension of reclaimed areas and

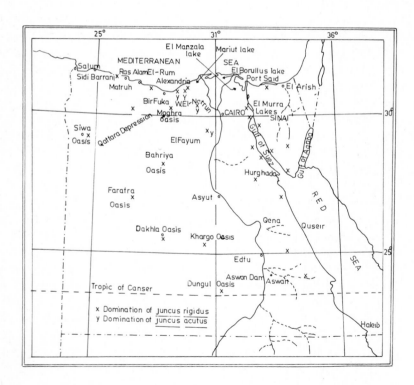

Fig. 1. Map showing the general distribution of *Juncus rigidus* C.A. Mey and *J. acutus* L. in Egypt.

236

help provide for use of land areas that are too saline to be cultivated with ordinary crops and require new crops to the country.

I. Ecological characteristics

Nine species of *Juncus* are recorded in Egypt (Täckholm and Drar, 1950; Täckholm, 1974), namely: *J. littoralis*, *J. bufonius*, *J. punctorius*, *J. fontenesii*, *J. acutus*, *J. rigidus*, *J. subulatus*, *J. inflexus* and *J. effusus*. *J. rigidus* and *J. acutus* are the subject matter of the present study.

J. rigidus C.A. Mey and *J. acutus* L. are closely related euhalophytes of the cumulative type (Walter, 1961). They accumulate excess salts in the upper parts of their green culms, an advantage that characterizes *Juncus* spp., and may lead to desalination of saline soils. Referring to salt accumulating crop plants, Boyko (1966) said, 'each harvest is diminishing the salt content in soil and/or ground water.'

J. rigidus is a densely tufted rush with slender pungent nodeless culms more than one meter high. It has sympodial creeping rhizomes developing leafy shoots (culms) every year. These rhizomes extend horizontally exploiting wide areas and may produce a dense plant growth within a few years. In Egypt, *J. rigidus* is a common rush that flourishes in the saline and alkaline soils with total soluble salts up to 3–3.8% in rhizosphere layer. Salts are mostly chlorides (>80%) and partly (<20%) sulphates and bicarbonates (Zahran et al., 1972). The biogeography of *J. rigidus* in Egypt (Fig. 1) shows that the plant is not only salt tolerant, but it also tolerates a wide range of climatic conditions. It is recorded in the semi-arid region of Egypt: salt marshes of the Mediterranean coast and those of the northern lakes (Tadros, 1956; Montasir, 1937), where the climate is mild, and in the arid and extremely arid regions represented by the littoral salt marshes of the Red Sea (Kassas and Zahran, 1967) and the inland salt marshes of Egypt (Girgis et al., 1971; Zahran and Girgis, 1970; Zahran, 1972, 1979), where the climate is hot and dry.

The world distribution of *J. rigidus* is as follows (Täckholm and Drar, 1950): It is recorded in Egypt, Palestine, Jordan, S. Turkistan, Arabia, Cyrenaica, Tripoli, Fezzan, Tunisia, Algeria, Morocco, Central Sahara, South Africa, Cape Province, S.W. Africa, Angola and North Rhodesia.

J. acutus has the same morphological aspects as *J. rigidus*, but its rhizomes grow in a special way giving rise to a great number of green culms that form circular patches of variable size. The community type dominated by *J. acutus* is widespread in the salt marshes of the semi-arid and arid (Mediterranean coast, Wadi El-Natrun and El-Fayum salt marshes), but it is either rarely recorded or absent from the salt marshes of the extremely arid parts of Egypt (Zahran et al., 1977). The habitat of *J. acutus* is usually highly moistened soil in the downstream zone of the sedge meadow close to the reedswamps. The total soluble salts of the rhizosphere layers range between 1.2 to 1.7%, mostly (>75%) chlorides.

J. acutus is a warm temperate species with its largest distribution area within the Mediterranean region, West Europe and S.W. Asia (Täckholm and Drar, 1950). Its presence extends tó 40° N and S (in Europe further north). It is recorded in Egypt, Cyrenaica, Tripoli, Fezzan, Tunisia, Algeria, Morocco, Canaris, Azores, Ireland, England, Sardinia, the coast of the Adriatic Sea, Balkan and islands of Crete, Cyprus, Syria, Palestine, Iran, Afghanistan, Caucasia, Cape Province, California, South America (Argentina, Uruguay, Chile).

II. Economic importance

Juncus plants have many important uses both in the old and recent times. Täckholm and Drar (1954) described the different uses of *Juncus* in the older days as follows: *J. acutus* and *J. arabicus* (syn. of *J. rigidus*), especially the last species, are frequently used for making mats. The mat industry of *Juncus* is described already by Abu Hanifa (895 A.D), and Ibn El-Beitar (1248 A.D). The former also adds the name Kawlan for the plant, which is usually called Samar Murr (arabic translation of bitter sedge) to be distinguished from Samar Helw (Arabic translation of sweet sedge) or *Cyperus alopecuroides*, which is also made into mats. In addition, the stems of Kawlan are beaten and made into ropes.

Täckholm and Drar (1954) have quoted that rush

mats were even exported, Cairo was the center of rush mat industry and the economic importance of *Juncus* culms as a raw material for mat making were all recognized in early and middle nineteenth century. Also, the seeds of *J. rigidus* as well as other *Juncus* spp. were employed in oriental medicine as diuretic and as a remedy for diarrhoea, etc. It may be added that during several thousands of years from the very beginning of the art of writing until about the third century B.C., the Ancient Egyptian writing implement was rushes (mostly *J. rigidus*). Baskets and sandals were also made from *Juncus* culms.

Recently, it was found by Osman et al. (1975), that the seeds of *J. rigidus* and *J. acutus* are rich in fatty acids, especially palmitic, oleic and linoleic, lauric, myrestic, stearic and linolenic acids were also detected in their seeds (Table 1). Computation of the glycerides of the oils extracted from the seeds of *J. rigidus* and *J. acutus* revealed the presence of 48 and 52 glycerides, respectively. These findings might suggest the possibility of using these seeds as potential oil sources of diggerent purposes.

Preliminary phytochemical investigation on *Juncus* seeds and culms (Zahran and El-Habibi, 1979) indicate the presence of 13 amino acids (histidine, arginine, serine, glycine, aspartic acid, glutamic acid, threonine, proline, valine, lysine, phenyl alanine, isoleucine and leucine) (Table 2). Flavonoides, glycosides, saponin, tannin and unsaturated steroles were also detected in the seeds of *J. rigidus* and *J. acutus*.

The most important industrial use of *Juncus* plants is that from their culms raw materials for paper industry can be obtained. 'This rush (*Juncus*) could be developed as raw material for printing paper of high quality' (Boyko, 1966). This was confirmed by studies on fiber length and on chemical characteristics of *J. rigidus* culms carried by Zahran et al. (1972). The results of the latter studies showed that the fiber length of *J. rigidus* culms range between 407 to 2421 micra with an average length of 1484 and width of 16 micra (ratio of length to width = 92.7: 1).

The thickness of fiber cell wall is 6.5 micra and the ratio of width to thickness of cell walls is 2.45:1. The percentage of fiber cells is longer than 1000 micra (1 mm), which is preferred in paper industry.

The chemical analyses and pilot plant experiment (carried in the National Company for Paper Industry, Alexandria, Egypt) proved that *Juncus* culms contain low ash content (6.5%), low percentage of lignin (13.3%), high percentage of alphacellulose (39.8%) and high yield of unbleached pulp (36.8%). The strength properties of the depithed unbleached *Juncus* pulp are much higher than those of rice straw (grade index = 24%) and bagasse (grade index = 42%) and gives a grade index of 73% compared to imported softwood long fiber unbleached kraft pulp (grade index = 100%). Apart from that, the bleaching of *Juncus* pulp gave a bleached pulp with a degree of brightness of 76 photo volts and good strength properties. *Juncus* pulp may be used alone to produce paper, while the other local raw materials for Egypt (rice straw and bagasse) should be mixed with wood pulp before

Table 1. Fatty acid composition of *J. rigidus* and *J. acutus* seeds (mole %).

Fatty acids	J. rigidus			J. acutus		
	mg	tg	pl	mg	tg	pl
Lauric	2.4	1.4	—	5.4	1.8	—
Myristic	1.5	0.5	—	14.7	4.9	—
Palmitic	16.2	23.4	56.9	8.5	20.7	70.9
Stearic	3.3	1.1	6.7	—	1.2	2.0
Oleic	—	38.6	29.3	5.9	37.5	18.2
Linoleic	71.9	32.1	7.1	51.4	29.2	8.9
Linolenic	2.9	2.9	—	14.1	4.7	—

mg = 2-monoglycerides, tg = triglycerides, and pl = phospholipids.

Table 2. Free amino acids and protein in *J. rigidus* and *J. acutus* seeds (mg %).

Plant species	Protein	Amino acids												
		Histidine	Arginine	Serine	Glycine	Aspartic acid	Glutamic acid	Threonine	Proline	Valine	Lysine	Phenyl alanine	Isoleucine	Leucine
J. rigidus	11.12	2.36	4.72	2.75	2.10	5.24	1.96	2.19	5.37	1.24	2.75	4.42	1.74	2.80
J. acutus	10.54	3.51	2.14	1.86	4.68	1.87	1.03	0.00	4.94	3.09	0.00	4.12	2.05	1.64

producing paper.

Consequently, in Egypt, if sufficient amounts of *Juncus* culms are made available to paper mills, the quantity of paper pulp imported may be changed. However, large scale economic production of paper entails large production of *Juncus* plants from homogeneous natural population of *Juncus* in Egypt. Hence, it will be of importance to study the main factors affecting the productivity of such plants under agricultural practices on saline soils. This will provide raw material for paper industry, through management and use of areas that are otherwise not economically productive. The effect of ecological factors on fiber qualities, cellulose and lignin contents, etc. of *Juncus* culms is of equal importance.

III. Field experiments

Juncus rigidus and *J. acutus* can be planted either from seeds (Zahran, 1975) or rhizomes (El-Bagouri et al., 1976). The latter way is preferable as rhizomes produce new culms in shorter periods than seeds.

The site of experiment was selected in a part of badly drained soil under the influence of El-Manzala lake (Fig. 1). The depth of the underground water table (brackish) was shallow (at about 10 to 15 cm in summer and at about 50 to 70 cm in winter below soil surface). The soil of the experiment was usually saturated with water, a factor that devaluated its productivity. Analyses of representative soil samples collected, at random, from the experimental site before transplanting *Juncus* showed the following characteristics: clayey in texture, black in colour, alkaline in reaction (pH = 7.8 to 8.4),

organic carbon content = 0.9 to 1.21%, calcium carbonate content = 3.85 to 4.4%, total soluble salts, = 0.88 to 1.85% in the surface layer and 0.28 to 0.65% in the subsurface layer.

Two experiments have been carried out on the above described site:

(i) Growth of Juncus on saline soil

The objective was to see how far the rhizomes of *Juncus* collected from the marsh would produce vegetative yield on another saline land.

a. Material and method. An area of 4,032 m² of the above mentioned land was prepared and divided into 96 plots (7 × 6 m each) for *Juncus* growth experiment (48 plots each for *J. rigidus* and *J. acutus*). On September 30, 1974, rhizomes of the two *Juncus* spp. were collected from their natural domination in Mariut salt marshes (Mediterranean coast, near Alexandria, Fig. 1). Except for rainfall which is about three times higher in Mariut (mean annual rainfall = 192 mm) than in Manzala (mean annual rainfall = 66.3 mm), climatic conditons of Mariut and Manzala areas are almost comparable (mean maximum temperature of 24.9° C and 24.6° C; mean minimum temperature of 15.9° C and 16.5° C; mean relative humidity of 70% and 72%; mean evaporation of 5.2 mm/day and 6.7 mm/day, respectively). In both areas rainfall is usually confined to winter months (December to February).

On the first of October 1974, the experiment started by transplanting 12, more or less equal sized pieces of *J. rigidus* and *J. acutus* rhizomes at equal distances and at about 5 to 10 cm depth in each plot. Irrigation was carried out every 5 days in summer, but during winter (wet season) irrigation intervals were extended to 7 days. The irrigation water of Manzala area, being near the saline lake Manzala and being taken from the most downstream part of the River Nile (Damietta Branch of the Nile Delta), contains a relatively high proportion of soluble matter (E.C. of 529 micromohs/cm, Cl of 71 ppm and SO_4 of 44 ppm). Harvest (cutting of aerial green culms) was carried out on October 1, 1975. The following parameters were taken:

(1) Number of *Juncus* rhizomes succeeded to produce new tussocks; (2) Mean fresh and mean oven dry weights of *Juncus* culms per plot; (3) Mean height of *Juncus* culms.

b. Results. The results of growth experiment (Table 3) show that both *Juncus* spp. may be cultivated on badly drained saline soil. The growth of *J. rigidus* seems to be better than that of *J. acutus*. This can be deduced from the following:

Table 3. Growth parameters in two *Juncus* spp. at Manzala.

Parameters	J. rigidus	J. acutus
Number of new culms		
(out of 12 rhizomes/plot)	10.46	8.95
Height of culms (cm)	162.00	85.00
Fresh weight of culms/plot (kg)	4.96	2.81
Oven dry weight of culms/plot (kg)	1.96	1.11
Water content of culms (%)	63.50	60.23

1. Mean number of rhizomes that produced new tussocks was higher for *J. rigidus* (80%) than for *J. acutus* (74%).

2. Mean height of *J. rigidus* culms was more (162 cm) than that of *J. acutus* (85 cm).

3. Mean fresh weight of *J. rigidus* culm (4.96 kg/plot) was higher than that of *J. acutus* (2.81 kg/plot).

4. Though the mean water content of *J. rigidus* culms was slightly higher (63.50%) than that of *J. acutus* (60.23%), yet the mean oven dry weight of the former (1.96 kg/plot) was higher than that of the latter (1.11 kg/plot).

(ii) Fertilization of Juncus plants growing on saline soil
The objective was to find out the effects of the different macro and micro elements on the vegetative yield, fiber length and amount of cellulose and lignin in the culms of *J. rigidus* and *J. acutus* under experiments.

Nitrogen and phosphorus are present in soils in both inorganic and organic forms. Nitrogen occurs as nitrates, nitrites, ammonia salts and free ammonia dissolved in the soil solution. Humus contains phosphorus as constituent of various organic molecules through the activity of microorganisms (Sutcliffe and Baker, 1974). Fink (1977) stated that as plants on saline soils suffer from nutritional stress, the study of the effect of nitrogen/ phosphorus manurings on the productivity of saline soil cultivated with *Juncus* plants is of importance.

a. Material and method. The growth experiment of *Juncus* was extended to find out the effect of different fertilizers (macro and micro elements) on the vegetative yield, fiber length and cellulose and lignin contents of both *J. rigidus* and *J. acutus* culms. Accordingly, after the harvest of growth experiment, the following fertilization scheme was applied at random to the plots of the experiments. Six replicates were taken for each treatment, as given below. This represents the first stage of the fertilization experiment (El-Demerdash, 1978).

A	Control	:	No addition of fertilizer	
B	N_1P_1	=	20 kg N[+] + 25 kg P[++]	per Feddan[+++]
C	N_2P_1	=	40 kg N + 25 kg P	per Feddan
D	N_3P_1	=	80 kg N + 25 kg P	per Feddan
E	N_1P_2	=	20 kg N + 50 kg P	per Feddan
F	N_2P_2	=	40 kg N + 50 kg P	per Feddan
G	N_3P_2	=	80 kg N + 50 kg P	per Feddan
H	N_2P_2		+ Micro nutrients	

N[+] = $(NH_4) SO_4$; P[++] = P_2O_5; Feddan[+++] = 4,200 m^2.

For treatment 'F', in which N_2P_2 was applied, 12 plots were treated (6 for the microelements treatment + N_2P_2 in the second stage of this experiment). Plants of the experiment were irrigated directly after the application of fertilizers (October 3, 1975), thence irrigation was at weekly intervals. The experiment continued for about 6 months, then culms were harvested (March 29, 1976). This is the harvest of the first stage of this experiment. On the day of harvest, the same N–P fertilizer treatments were applied to the same plots. Exactly after 2 months growth and on 29th May, 1976, H treatment (N_2P_2) + microelements[1] was applied to the new culms of *Juncus* plants. Irrigation was done regularly every week, 10 days and 12 days during summer (and spring), autumn and winter seasons, respectively. On 29th June, 1977 (after 15 months growth) the area covered with *Juncus* tussocks, was measured. The culms were harvested and their fresh weights were taken. Samples of *Juncus* culms and seeds were collected for various laboratory investigations.

b. Results. The individual estimation of area (m^2) and fresh and dry weights (kg) of *Juncus* tussocks and culms, respectively (Table 4) and the statistical analyses of these results were made according to Snedecor (1956), (Table 5) to show the following:

1. The areas covered with *J. rigidus* in plots treated with B and C fertilizers increased significantly while those of D, E, F, G and H increased with high significance.

2. The areas covered with *J. acutus* treated with B, E and G fertilizers, increased insignificantly; while those of D and F treatments showed significant increase. High significant increase was observed in plots treated with H fertilizer (having micro elements).

3. Fresh weight of *J. rigidus* insignificantly increased in plots treated with B and C fertilizers. High significant increase was recorded in plots of D, E, F, G and H treatments.

4. The fresh weights of *J. acutus* culms under B, C, D, and E treatments increased significantly. High significant increases were detected in plots of F, G and H treatments.

5. The dry weights of *J. rigidus* culms treated with B and E treatments increase significantly while those of D, increased insignificantly. High significant increase was detected in the dry weights of *J.*

Table 4. Vegetative yields of *Juncus* spp. at Manzala.

Plant species	Parameter	A	B	C	D	E	F	G	H
J. rigidus	at	0.93	1.10	1.15	1.45	1.50	1.50	1.50	1.60
	fw	22.70	28.90	28.40	34.90	32.5	32.40	34.80	37.50
	dw	8.90	10.50	10.30	12.70	11.80	11.70	12.5	14.90
J. acutus	at	0.77	0.94	1.30	1.22	0.88	1.02	1.12	1.60
	fw	15.00	21.10	24.90	20.50	21.00	26.60	23.80	30.90
	dw	5.96	8.40	9.50	9.70	8.40	10.60	9.40	12.30

A to H = fertilizer treatments, at = area of tussocks (m^2), fw = fresh weight and, dw = dry weight (kg).

Table 5. Statistical differences (significant values) between the mean values of the area covered by tussocks, and the fresh and dry weights of *Juncus* culms under different fertilizer treatments.

Plant species	Parameter	A–B	A–C	A–D	A–E	A–F	A–G	A–H
J. rigidus	at	+ +	+ +	+ + +	+ + +	+ + +	+ + +	+ + +
	fw	+ + +	+ + +	+ + +	+ + +	+ + +	+ +	+ + +
	dw	+	+ +	N.S.	+	+ + +	N.S.	+ +
J. acutus	at	N.S.	N.S.	+	N.S.	+	+	+ + +
	fw	+	+	+	+	+	+	+ + +
	dw	N.S.	N.S.	+ +	+ + +	+ +	+ + +	+ + +

N.S. = not significant (0.1 p), + = almost significant (0.5–0.01 p).
+ + = significant (0.01–0.005 p), + + + = highly significant (0.005–0.001 p).
at = area of tussocks, fw = fresh weight, and dw = dry weight (kg).

1. The microelement solution (Arnon nutrient solution diluted to 1/5) was sprayed to the green culms of *Juncus*.

rigidus culms under C, F, G and H treatments.

6. The dry weights of *J. acutus* culms in plots treated with B and C fertilizers increased significantly, while those of D, E, F, G and H showed highly significant increase.

7. Generally, the total vegetative yields of *Juncus* culms (fresh and dry weights) were, relatively, higher in plots treated with excess amounts of nitrogen in fertilizer added, i.e. D and G treatments. Nitrogen is concerned mainly with the building up of plant body.

8. It was observed that the culms of *J. acutus* were susceptible to infection by *Puccinea rhimosa* (fungal disease), a factor that prohibits their normal growth. This may explain the great drop in the area, height, fresh and dry weight of *J. acutus* in plots treated wth E fertilizer (N_1P_2) since greatest number of infected plots were recorded in those treatments. *J. rigidus* culms were not infected.

(iii) Juncus spp. and soil biological desalination

The success of *Juncus* plantation on saline soil may declare its importance as a non-conventional crop. This was associated with another agricultural advantage that increases its economic potentialities. *Juncus* are cumulative halophytes (Walter, 1961), i.e. accumulating excess salts which they absorb from soil in the upper parts of their culms. This was

detected in the saline soil of the *Juncus* experiment at El Manzala.

a. Material and methods. Twelve soil samples (1–12, profiles A–F, Table 6) were collected at random from 6 plots of the experiment before transplanting *Juncus* rhizomes (on October 1, 1974). A profile (0 to 30 cm) was dug in each of the 6 plots. Each profile was represented by two soil samples, surface (0 to 5 cm depth) and subsurface (5 to 30 cm depth). The soil samples were analysed chemically (Jackson, 1962) and their total soluble salts, calcium carbonate contents as well as pH values were estimated. On October 1, 1976, another set of soil samples (13–24, profiles G–L, Table 6) were collected from the same plots from the soil under *Juncus* tussocks and were also analysed.

b. Results. The results of soil chemical analysis before and after two years of *Juncus* growth (Table 6), indicate that *Juncus* plants may play a considerable role in diminishing the salt content of soil. Recognizable differences in the total soluble salts of the soil samples collected before and after *Juncus* growth were observed. Reductions were relatively higher in the plots of *J. rigidus* than in those of *J. acutus*. This was associated with slight increase in the organic carbon content and decrease in pH

Table 6. Soil chemical analyses from the *Juncus* experiment site at Manzala.

Plant species	Profile No.	Depth (cm)	Before transplantation					After transplantation			
			TSS (%)	pH	CaCO$_3$ (%)	Organic carbon (%)	Profile No.	TSS (%)	pH	CaCO$_3$ (%)	Organic carbon (%)
J. rigidus	A	0–5	1.85	8.3	4.26	1.2	G	0.77	8.18	4.24	1.22
		5–30	1.32	8.4	4.32	1.16		0.92	8.02	4.31	1.20
	B	0–5	0.95	8.3	3.85	1.002	H	0.32	8.01	3.97	1.10
		5–30	0.67	8.3	4.10	0.998		0.42	8.3	3.98	1.00
	C	0–5	1.02	7.9	3.92	1.21	I	0.54	7.8	3.96	1.25
		5–30	0.87	7.8	3.88	1.06		0.52	7.8	3.75	1.10
	D	0–5	0.99	8.2	4.38	1.18	J	0.88	8.1	4.3	1.20
		5–30	0.86	8.3	4.31	1.02		0.82	8.0	4.26	1.10
J. acutus	E	0–5	0.98	8.1	4.3	0.90	K	0.81	7.9	4.27	1.00
		5–30	0.82	8.1	4.4	1.10		0.72	8.0	4.29	1.12
	F	0–5	1.35	8.4	3.99	1.21	L	0.94	8.3	4.87	1.25
		5–30	0.97	8.3	4.21	1.06		0.72	8.5	4.15	1.08

TSS = total soluble salts.

242

value of soil. Calcium carbonate contents did not show appreciable changes.

(iv) Fiber length measurements

How far the different fertilizers would affect the length of the fiber of *Juncus* culms? This is a question that is most important in paper industry.

a. Material and methods. Samples of *Juncus* culms treated with various fertilizers were collected, air dried and cut into 3–4 cm pieces. These pieces were treated with hot concentrated NaOH solution till complete maceration was achieved. Microscopic measurements on 200 fibers for each fertilization treatment were carried out for each *Juncus* species.

b. Results. The results of fiber length measurement (Table 7), show that the addition of N-P fertilizers had positive effects on this parameter as follows (Zahran et al., 1979):

1. Reduction of the number of very short (< 500 micra) and short (500 to 1000 micra) fibers of both *Juncus* spp. from 11% to 0.5% and from 12% to zero% in *J. rigidus* and *J. acutus* culms, respectively, was determined.

2. The percentages of the fibers having lengths ranging between 1500 to 2000 micra increased by the addition of fertilizers from 18.5% to 36.5% and from 15% to 36.5% for *J. rigidus* and *J. acutus*, respectively.

3. Though the numbers of fibers having length

between 1500 to 2000 and >2000 micra were relatively low yet their amount increased by the addition of fertilizers.

4. Plants under treatment H, i.e. N_2P_2 + microelements showed the highest percentage of fiber lengths.

5. Under all treatments the fiber length of *J. rigidus* and *J. acutus* culms were almost comparable.

(v) Chemical investigations

These investigations were carried out to determine the most suitable mixture of N-P fertilizer that helps to produce *Juncus* culms having the best chemical qualities related to paper industry.

a. Materials and method. Samples of *J. rigidus* and *J. acutus* culms collected from the field experiment at El Manzala were chemically analysed (TAPPI Standard Method, 1960). Their ash, lignin, α-cellulose and pentosan contents as well as their hot water and alcohol benzene solubilities were determined (Zahran, et al., 1979).

b. Results. The results of the chemical analyses of *Juncus* culms (Table 8) elucidate the following:

1. The ash contents of the control samples of *J. rigidus* (6.42–6.67%) were relatively higher than those of *J. acutus* (4.92–5.84%) culms. Addition of fertilizers showed no appreciable effect.

2. Hot water and alcohol benzene solubilities of

Table 7. Mean and percentage of fiber lengths of two *Juncus* spp., culms at different fertilizer treatments.

Fertilizer treatments	Mean μ	*J. rigidus* Percentage of fiber length					Mean μ	*J. acutus* Percentage of fiber length				
		I	II	III	IV	V		I	II	III	IV	V
A	849.1	11.0	62.5	23.5	2.0	1.0	791.0	12.0	71.0	15.0	1.5	0.5
B	887.8	5.0	72.0	22.5	5.0	0.0	840.6	9.5	67.5	20.5	2.0	0.5
C	905.8	4.0	69.0	23.0	4.0	0.0	844.8	6.5	65.5	25.5	1.5	1.5
D	925.7	3.5	67.0	24.5	4.5	0.5	906.0	4.0	63.5	28.5	3.0	1.0
E	937.6	1.5	59.0	32.0	5.0	0.5	914.0	2.0	58.5	32.5	3.5	1.5
F	993.2	1.5	58.0	35.0	5.0	0.5	977.0	0.5	58.5	35.5	4.0	1.5
G	995.2	0.5	56.5	36.0	5.0	1.5	985.2	0.0	58.0	36.0	4.5	1.5
H	1451.7	0.5	54.5	36.5	6.5	2.0	1592.6	0.0	56.0	36.5	5.5	2.0

I = 500 μ, II = 500–1000 μ, III = 1000–1500 μ, IV = 1500–2000 μ, and V = >2000 μ.

Table 8. Chemical analyses of the culms of *Juncus* spp. after fertilizer treatments.

Fertilizer treatments	Plant species	Solubilities		Lignin	a-Cellulose	Pentosan	Ash
		Hot water	Alc. Benz.				
A	*J. rigidus*	17.55	3.83	13.80	42.85	25.28	6.54
	J. acutus	17.75	3.41	19.53	46.50	23.42	5.38
B	*J. rigidus*	16.32	2.75	12.76	46.40	26.61	6.20
	J. acutus	17.27	3.17	19.51	41.55	23.02	6.19
C	*J. rigidus*	16.77	3.55	13.24	47.15	27.95	8.51
	J. acutus	16.80	3.88	17.13	46.70	22.91	5.17
D	*J. rigidus*	18.05	3.06	13.88	54.80	27.07	7.67
	J. acutus	16.21	3.23	20.23	55.70	27.59	6.72
E	*J. rigidus*	17.99	2.72	13.41	53.60	28.29	7.11
	J. acutus	16.75	3.29	16.84	54.30	28.17	5.71
F	*J. rigidus*	22.82	2.90	14.32	54.65	32.50	6.61
	J. acutus	17.45	2.86	19.02	67.70	24.37	5.22
G	*J. rigidus*	21.42	3.39	13.25	55.25	27.37	6.77
	J. acutus	19.73	3.57	16.45	55.45	24.83	5.76
H	*J. rigidus*	20.71	3.52	13.60	57.25	27.27	5.36
	J. acutus	16.50	3.19	16.50	53.10	28.85	6.67

the control and treated samples were almost comparable. *J. rigidus* culms showed relatively higher proportions of material solubility in hot water than *J. acutus*.

3. The lignin contents of the control culms of *J. rigidus* (13.66–13.94%) were relatively lower than those of *J. acutus* (18.6–20.47%). Fertilization has no appreciable effects on this parameter.

4. The a-cellulose contents of the control *J. rigidus* culms (42.2–43.5%) were, relatively, lower than those of *J. acutus* (45.1–47.9%). The reverse was true for the pentosan contents. Addition of fertilizers showed appreciable effects on these two parameters. Increased amounts of phosphorus in the mixtures of fertilizers were associated with an increase in the percentage of pentosan and a-cellulose in the culms of both *Juncus* species. Phosphorus plays the main role in the formation of carbohydrates of plants through phosphorylation processes. Accordingly, pentosan and a-cellulose, the major constituents of carbohydrates (polysaccharides) of plants, were increased when the amount of phosphorus was increased in the fertilizers added.

IV. Conclusions

The preceding pages reveal the various agro-industrial potentialities of two halophytic rushes, namely: *J. rigidus* and *J. acutus*. These rushes can be cultivated successfully on saline non-productive soils. Their growth rates were so fast that enlarged vegetational cover of cultivated saline lands with non-conventional and economic crop may be expected. Addition of nitrogen-phosphorus fertilizers to *Juncus* plants was of prime importance from both agricultural and industrial point of view. The best mixture of N-P fertilizers was that formed from : 8 kg nitrogen fertilizer in the form of ammonium sulphate + 50 kg phosphorus fertilizer (in the form of P_2O_5 super phosphate) + micro nutrient. This fertilizer not only acted to increase the total vegetation yields, but it also helped to increase the amounts of long fibers (1500– >2000 micra) and the amount of a-cellulose and pentosan in *Juncus* culms. This means that this fertilizer develops the vegetative yield, and the properties (physical and chemical) of *Juncus* culms that increase the economic potentialities of these plants in paper industry.

Achievement of soil desalination was also proved

by *Juncus* plantation. Within few years, the saline soils cultivated with *Juncus* might become less saline and more productive.

Juncus seeds were found to be rich in oils, proteins, amino acids, carbohydrates, etc. These findings might suggest the possibility of their use as raw material in various chemical industries e.g. oil, drug, etc.

Accordingly, one may conclude that both *J. rigidus* and *J. acutus* have agro-industrial economic potentialities. However, *J. rigidus* is preferable to *J. acutus* for the following reasons:

1. *J. rigidus* seemed to be resistant against fungal infection, an advantage that increases its vegetative yields.

2. The salt tolerance and desalination effect of *J. rigidus* were higher than those of *J. acutus*.

3. The seeds of *J. rigidus* were, relatively, richer in their chemical constituents than those of *J. acutus*.

Consequently, it is possible that *Juncus* halophytes may play an effective role for human welfare, especially in the countries of arid and semi-arid regions.

3. *Salsola tetrandra* in drug industry

During the chemo-taxonomic studies for the natural wealth of the flora of Egypt for finding out their potentialities in drug industry, *Salsola tetrandra* was selected to be the subject of this study. The plant was studied ecologically, and chemically; in view of its pharmacological characteristics.

I. Ecological characteristics

S. tetrandra is a plant of wide ecological amplitude. In Egypt, it grows both in the arid, e.g. Sinai (Boulos, 1960; Zahran, 1967), Siwa Oasis (Zahran, 1972), Isthmic Desert (Täckholm, 1974), and in semi-arid i.e. Mediterranean coastal land (Tadros, 1956; El-Batanouny and Zaki, 1969; El-Batanouny and Abul Soud, 1972) areas, Fig. 2. In the arid part of Egypt where the climate is hot and dry with a mean annual rainfall of less than 32 mm; *S. tetrandra* is recorded as an associate species with other dominants e.g. *Zygophyllum album*, *Nitraria*

retusa, etc. (Zahran, 1979) where it has a negligible plant cover. On the other hand, in the semi-arid part of Egypt, where the climate is mild and wet with a mean annual rainfall up to 300 mm, *S. tetrandra* flourishes and dominates as one of the main communities of the littoral salt marsh ecosystem of the Mediterranean Sea. Its community occupies a continuous strip of the saline land parallel to the sea shore. Its area extends farther southwards than the general limit of the downstreams of wadis (valleys) crossing the area from south to north. It was estimated by Migahid et al. (1971) that in one sector, 25 km east of Sidi Barani (about 540 km west of Alexandria) the pure *S. tetrandra* community covers a total area of 527 acres, while there is a mixture of *S. tetrandra* and *Anabasis articulata* occupying another area of 684.8 acres. Generally, the total plant cover of the stands where *S. tetrandra* dominates is relatively high (30 to 45%).

The floristic composition of the community dominated by *S. tetrandra* varies according to the habitats. In the stands nearer to the influence of seawater spray and seepage, soil salinity is relatively high and the associate species are mainly halophytes, e.g. *Suaeda fruticosa*, *S. pruinosa*, *Limonium pruinosum*, *L. tubiflora*, *Reaumuria mucronata*, *Frankenia revoluta*, etc. In the stands nearer to the margins of the inner desert far from the reach of the influence of seawater, soil salinity is lower and consequently the non-salt tolerant species are the common associates, e.g. *Noea mucronata*, *Thymelaea hirsuta*, *Gymnocarpos decandrum*, *Atractylis flava*, *Pituranthus tortuosus*, *Haplophyllum tuberculatum*, *Salvia lanigera*, *Astragalus spinosus*, etc.

S. tetrandra is a sand binder plant that builds sand mounds of moderate size (1 to 4 m^2 in area and 50 to 70 cm high) around its stems and branches. The supporting soil of *S. tetrandra* domination is compact (alluvial consolidated soil) formed mainly of fine sand and silt (> 75%) mixed with coarse material (<25%). The chemical properties of soil samples collected from three profiles representing the soil of *S. tetrandra* are given in Table 9.

The results of table 9 elucidate the following facts:

1. *S. tetrandra* tolerates salinity in soil up to 2.56% and the soluble salts are mainly chlorides and

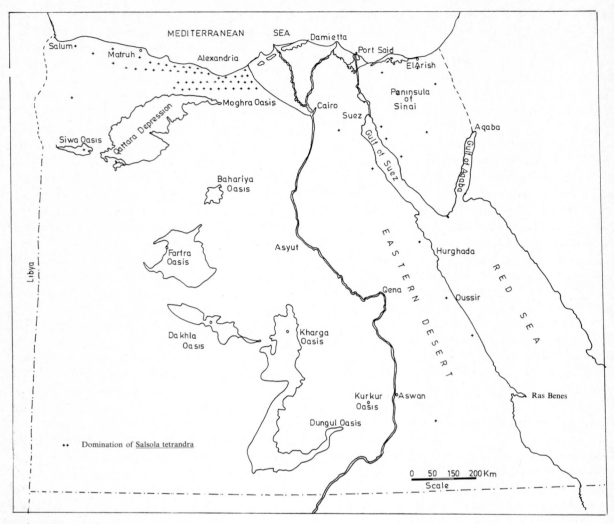

Fig. 2. Distribution of *Salsola tetrandra* in Egypt.

partly sulphates and bicarbonates.

2. Soil is alkaline in reaction.

3. *S. tetrandra* soil contains a considerable amount of calcium carbonates and is rich in organic matter.

II. Experimental studies

(i) Chemical and pharmacological investigations

The chemical screening of *S. tetrandra* indicates that it contains: alkaloids, glycosides, saponins, tannins, sterols as well as organic acids (Zahran and Negm, 1973).

The biological tests were performed to determine its action on the intestine, uterus, heart, blood pressure and respiration and also for testing its anthelmintic properties and toxicity (Zahran and Negm, 1973).

a. Material and methods. A 1% solution of the total alcoholic extract of the plant was prepared by dissolving the residue in the physiological solution used in the different tests. The effects were tested, in vitro, on the intestine of rat, ileum of rabbit and uterus of rats at various stages of sex cycle using an electrically glass-jar bath with an inner vessel of 50 ml capacity.

246

Table 9. The chemical properties of soil samples from three profiles representing *Salsola tetrandra*.

Profiles	Depth (cm)	pH	Organic carbon %	CaCO₃ %	Soluble salts (%)			
					Total	Cl	SO₄	HCO₃
A	0–20	8.25	0.09	27.5	0.36	0.139	tr.	0.07
	20–70	8.30	0.14	40.0	1.55	0.890	0.032	0.06
B	0–30	7.80	0.66	35.5	0.70	0.280	0.210	0.80
	30–75	7.90	0.21	35.8	2.56	1.690	0.450	0.17
C	0–15	8.00	0.41	13.3	0.14	0.026	tr.	0.03
	15–60	8.70	0.81	33.5	1.50	1.170	tr.	0.08

The action on the blood pressure and respiratory movements was examined in dogs anesthetised by rectal administration of pentobarbitone sodium (nembutal), using slowly moving kymographic drum paper for recording the blood pressure from the cartoid artery. Respiratory movements were recorded by a sphygomograph composed of a rubber belt to fit around the chest of the animal connected with a tambour by a rubber tubing and the movements were then recorded by a lever on a kymographic smoked paper. The effect of the heart was tested on toad's heart perfused with the drug solution.

Toxicity experiment was carried out on mice of the same age and weight to elucidate its possible toxicity. Five groups of 6 mice each were injected in different doses for determining the M.L.D. (mild lethal dose), L.D.$_{50}$ (lethal dose, 50%) and the P.M. (postmortem) findings. The L.D.$_{50}$ and its fiducial limit were statistically calculated.

b. Results. It was found that the alcoholic extract applied in different doses inhibited the uterine activity at all stages of sex cycle and relaxed the intestinal musculature of both rats and rabbits, an effect which indicates its possible antispasmodic and constipating value.

As to the anthelmintic activity of *S. tetrandra* alcoholic extract, it depressed the intestine and sheep taenia. This may indicate that it possesses some anthelmintic properties. But if it will be used in this respect purge should be given to expel the depressed parasite.

When *S. tetrandra* alcoholic extract was injected (in different doses) intravenously into the femoral vein of dogs, a sudden transit rise in the blood pressure was observed followed by gradual fall to the original level. No effect was noticed on the respiratory movements. On the other hand, the extract markedly decreased the force of the cardiac contraction when tested on toad's heart.

Toxicity experiment (Table 10) shows that the M.L.D. and L.D.$_{50}$ of 4% alcoholic extract in water are 0.4 g/kg B.W. (body weight) and 0.8 g/kg B.W. respectively when determined in mice.

The L.D.$_{50}$ as calculated statistically using the data of Table 10, was found to be 0.8 g, and its fiducial limit = 0.8 g = 0.26 at the probability 0.05.

Table 10. Effect of alcoholic extract (4%) of *S. tetrandra* on mice.

Group of mice	Dose injected in g/kg B.W.	Mortality	Probability	Remarks
1	0.2	0	0.03	—
2	0.4	17	4.05	M.L.D.
3	0.6	33	4.56	—
4	0.8	50	5.00	L.D.$_{50}$
5	1.0	100	5.97	L.D.$_{100}$

B.W. = body weight.

The P.M. findings were retardation and depth of breathing, convulsion and death. Congestion of the internal organs was observed.

III. Conclusion

The therapeutic possibilities of *S. tetrandra* in view of its pharmacological characteristics, may draw the attention to its use as intestinal antispasmodic and anthelmintic source. Accordingly, a sufficient amount of the plants are needed to make use of it in drug industry. The wide areas occupied by the plant along the Mediterranean coastal land of Egypt may be of economic importance. To be quite sure of a permanent supply of *S. tetrandra*, it is necessary to protect and develop its present wild vegetation and also to transplant it in new non-productive areas.

4. *Kochia* spp. as a prospective forage plant

The natural vegetation of arid and semi-arid countries comprises four principal vegetation forms, namely: accidental, ephemeral, suffrutescent perennial and frutescent perennial (Kassas, 1966). The floristic compositions of these vegetation forms are rich and characterized by considerable number of palatable species belonging to Gramineae, Leguminosae, Cruciferae, Compositae, Labiatae, Nitrariaceae, Chenodiaceae, etc. These palatable plants can be considered as reliable local natural range plants when their total vegetative yields are high enough to maintain continuous fodder supply for the local livestock. Unfortunately this is not ascertainable under the arid or semi-arid environmental conditions. As rainfall is generally low, the density and total vegetation cover of range plants is not enough to face the requirements of livestock all the year around. Also, it will not be possible to get homogenous natural vegetation with palatable plants, other unwanted and/or poisonous weeds will be present. The alternative promising solution, is to propagate certain palatable xerophytic and/or halophytic plants which may prove to have high nutritive values, under dry or saline conditions. Among such plants mention may be made of: *Panicum* spp., *Aristida* spp., *Stipa* spp., *Pennisetum*

spp., *Poa* spp., *Vicia* spp., *Trifolium* spp., *Trigonella* spp., *Vigna* spp., *Lathyrus* spp., *Astragalus* spp., *Artemisia* spp., *Noea* spp., etc. (xerophytes); *Kochia* spp., *Atriplex* spp., *Sporobolus* spp., *Nitraria* spp., etc. (halophytes).

Kochia has attracted the attention of many workers in U.S.A., U.S.S.R., Turkey, Morocco, India, Egypt, etc., (Borkowski and Drost, 1965; Coxworth and Salmon, 1972; Drar, 1952; Draz, 1954; El-Shishiny, 1953; El-Shishiny and Thoday, 1953; El-Shishiny et al., 1958; Golovchenko and Makhamadzhanov, 1972; Golyadkin et al., 1974; Kernan et al., 1973; Khamdamov, 1976; Rasulov, 1971; Sadek, 1974; Salem, 1960; Shiskina and Tegisbaev, 1970; Tandon and Agarwal, 1966; Thoday et al., 1956; Durham and Durham, 1979). Recently, in Egypt, the authors co-supervised studies on *Kochia indica*, for M.Sc. Degree carried by M.E. Ziada, Botany Department, Faculty of Science, Mansoura University (1978). The plant was studied ecologically, eco-physiologically and phytochemically to determine its favourable environmental conditions, probability of its growth under saline conditions and to evaluate its potentialities as fodder plant.

I. Ecological characteristics

Genus *Kochia* belongs to family Chenopodiaceae (Täckholm, 1974).

Kochia indica is the only species recorded in Egypt. It is an annual herb having principal tap root, which carries many secondary roots. The stem is long (under favourable conditions its height attains more than 2 meters), cylindrical, solid, erect and richly branched (Plate I). It carries an active apical bud which gives the vegetative sprouts. The leaves of the vegetative branches are sessile, simple, cauline, alternate, exstipulate, linear-lanceolate, entire, acute and 2 to 3 cm long. Fruit is a one-seeded nut. Seed is oval in shape, small in size and very buoyant. It starts to germinate during the second half of February and the plant attains its maximum vegetative growth during June/July (i.e. summer forage). On the first week of August, the plant blossoms (flowers small, bisexual, regular with 5-toothed perianth). Fruiting stage starts at the end of August and at the beginning of September. Dryness

Plate I. Dense growth of *Kochia indica* in one of its stands, Mariut salt marshes, Mediterranean coast, Egypt.

of plant (symptoms of the end of its life cycle) begins during October. The plant is reddish in colour during its dry condition.

Phytogeographically, *K. indica* belongs to the Irano-Turanian Saharo-Sindian regions (Täckholm, 1954). Its native home is India (Drar, 1952) and the plant was recorded also in Punjab (Bamber, 1916), Morocco (Dahadiez and Maire, 1934), Lahore (Kashyap and Joshi, 1936), Karachi and Sind (Hasanain and Rahman, 1957), Palestine, Israel and Jordan (Zohary, 1962), Algeria (Quezel and Santa, 1963), West Pakistan and Kashmir (Steward, 1972), Saudi Arabia (Migahid, 1978).

'Since the introduction of *K. indica* to Egypt, it has attracted attention owing to its local abundance in parts of the coastal region. Moreover, as it is a useful fodder plant in Australia and elsewhere in dry areas, it has been suggested that this species might be valuable in the Egyptian Deserts'. (Thoday et al., 1956). Nowadays, *K. indica* is a widely spread plant in Egypt especially in the Nile Delta, Nile valley, Mediterranean coast and northern oases (Abu Ziada, 1978) (Plate II).

Regarding the characteristics of soil supporting

K. indica growth in Egypt, soil samples were collected from two localities representing two different types of soils: calcareous and alluvial. The calcareous soils were collected from Alexandria district on the Mediterranean coast (Table 11) and the alluvial soils were collected from the Nile Delta area, Mansoura district (Table 12). The soil samples were analysed physically (particle size, porosity, water holding capacity and moisture content) and chemically (total soluble salts, chlorides, sulphates, calcium carbonates, carbonates, bicarbonates, organic carbon and pH value) following Jackson (1962).

Results shown in Tables 11 and 12 indicate that soil supporting the growth of *K. indica* in Alexandria district is sandy in texture (88% sand and fine sand) while in Mansoura district soil contains a relatively high percentage of silt (up to 67%). Porosity of soil samples does not vary greatly in case of calcareous soil as it ranges between 43.16 to 51.61%, while in the alluvial soil there is a wide range between 30% (thlowest) and 58.9% (the highest) of pore spaces. Water holding capacity is generally high (up to 81.7%) in case of the alluvial

Table 11. Analyses of soil samples supporting *Kochia indica* growing at Mediterranean coastal semi-arid region in Alexandria district, Egypt.

Localities	Depth (cm)	Physical characteristics							Mean porosity	Mean WHC
		% of soil fractions								
		>2.057 mm	2.057–1.003	1.003–0.5	0.5–0.211	0.211–0.104	0.104–0.053	<0.053 mm		
1 km south	0–5	0.000	1.213	8.840	41.200	39.030	6.900	2.811	45.40	32.45
of Alex.	5–25	0.000	2.390	8.350	45.990	35.000	5.590	2.665		
11 km south	0–5	0.000	1.780	13.430	45.020	26.130	8.353	5.280	43.16	56.50
of Alex.	5–25	0.000	1.910	14.080	48.550	23.530	9.530	2.370		
12 km south	0–5	0.000	0.000	6.960	37.960	49.270	4.120	1.690	51.60	52.90
of Alex.	5–25	1.600	8.600	26.390	32.100	17.710	10.420	3.160		

Localities	Depth (cm)	Chemical characteristics							
		$CaCO_3$ %	Organic carbon	TSS %	Analysis of 1:5 water extract				
					Cl %	SO_4 %	CO_3 %	HCO_3 %	pH
1 km south	0–5	89.5	0.7	1.115	0.059	0.234	0	0.060	8.00
of Alex.	5–25	89.5	0.8	1.209	0.056	0.408	0	0.075	8.11
11 km south	0–5	69.0	2.1	1.083	0.157	0.551	0	0.060	7.18
of Alex.	5–25	70.5	1.1	1.461	0.073	0.447	0	0.060	7.35
12 km south	0–5	17.0	1.0	0.807	0.095	0.703	0	0.045	7.62
of Alex.	5–25	21.0	1.6	2.375	0.175	1.326	0	0.060	7.73

TSS = total soluble salts, and WHC = water holding capacity.

soils when compared with that of calcareous ones (up to 56.5%). Moisture contents of soil(up to 43.9%) shows that *K. indica* seems to flourish only in soil with high water content (lowest = 18.1%).

Soil reaction is generally alkaline (pH = 7.08 to 8.49 in Mansoura district and 7.18 to 8.11 in Alexandria district). In both areas organic carbon contents are generally higher in the surface layers (0.4 to 3.4%) than in the subsurface ones (0.1 to 1.6%) with few exceptions. Obviously, the calcium carbonate contents are relatively higher (more than 4 times) in the soils of Alexandria area (up to 89.5%) than those of Mansoura area (up to 17.5%).

Generally, the soil supporting *K. indica* in Egypt is saline in nature. In most profiles of Mansoura are the surface layers contain relatively higher salt content (1.31 to 4.77%) than the subsurface ones (0.200 to 3.358%). The reverse is true in the soil samples of Alexandria profiles where salt contents are relatively higher in the subsurface layers (1.209, 1.461 and 2.375%) than in the surface ones (1.115, 1.083 and 0.807%). The soluble matters are mainly chlorides and partly sulphates and bicarbonates. Soluble carbonates are absent. These results may indicate how far *K. indica* is a salt tolerant plant.

II. Experimental studies

Depending upon the results of the ecological studies, *K. indica* can be considered as a salt tolernt plant. But, will it be possible to propagate this wild plant under saline conditions? Shall we get a non-conventional forage crop growing on non-productive soils? To answer these questions, two types of experiments had been carried out on *K. indica*: germination and growth under different salinity levels. Response of *Kochia* plants to mineral nutrition was also experimented.

(i) Germination experiment

Some halophytes may be quite sensitive to salinity during germination – which usually takes place during or after rainy season. Chapman (1966) found that a great majority of salt marsh plants

Table 12. Analyses of soil samples supporting *Kochia indica* inhabiting Mansoura district and Mansoura-Damietta road, Egypt.

Localities	Depth (cm)	Physical characteristics					Chemical characteristics							
		% of soil fractions			Mean porosity	Mean WHC	$CaCO_3$ %	Organic carbon %	Analysis of 1:5 water extract					
		Sand	Silt	Clay					TSS %	Cl %	SO_4 %	CO_3 %	HCO_3 %	pH
Bilqas	0–5	63.00	30.40	6.50	43.9	74.2	10.0	2.3	1.312	0.213	1.296	0	0.090	8.12
	5–25	40.20	33.75	26.00			9.5	1.1	0.583	0.050	0.673	0	0.090	8.22
Kafr-sa'd	0–5	41.75	53.50	4.60	33.8	62.9	10.5	1.0	2.027	0.877	0.314	0	0.081	7.95
	5–25	32.00	63.25	4.50	30.0		10.5	0.3	0.863	0.455	0.360	0	0.075	8.20
Basandila	0–5	49.20	46.00	4.75		60.3	7.5	0.4	0.476	0.250	0.376	0	0.061	7.54
10 km from Bilqas	5–25	47.80	41.60	10.20			7.5	0.5	0.342	0.120	0.158	0	0.075	8.49
El-Sawalim 23 km from Damietta	0–25	28.00	67.00	4.80	35.1	71.6	11.5	1.5	4.437	1.300	0.783	0	0.060	8.10
Near the Fac. of Pharm.	0–5	63.00	34.00	2.80	44.4	55.7	12.0	0.8	1.449	0.672	0.446	0	0.045	7.95
	5–25	42.00	55.20	2.75			12.5	0.7	0.683	0.357	0.303	0	0.060	7.91
Near the Fac. of Agric.	0–5	53.60	42.20	4.00	43.5	61.3	14.5	1.1	2.268	1.040	0.631	0	0.045	7.60
	5–25	44.36	51.60	4.00			15.5	1.0	1.028	0.437	0.353	0	0.060	7.75
Near the Fac. of Sci.	0–5	61.10	36.20	2.60	48.0	63.1	17.5	0.2	1.470	0.597	0.252	0	0.045	7.32
	5–25	59.10	38.00	2.75			15.0	0.1	0.343	0.547	0.296	0	0.060	7.57
	0–5	58.00	33.00	8.90	50.9	81.7	14.5	3.2	3.230	1.705	0.858	0	0.090	7.89
	5–25	46.00	51.30	2.50			15.5	1.5	1.686	0.500	0.421	0	0.090	7.87
Mit-Badr Khamis	0–5	54.00	28.25	17.70	63.0	59.0	16.0	3.4	3.479	1.300	1.335	0	0.090	7.68
	5–25	50.00	45.50	4.40			15.5	3.0	3.358	1.225	0.702	0	0.080	8.23
Sandub	0–5	90.50	4.80	4.50	42.4	62.0	12.0	3.5	1.665	3.600	0.362	0	0.054	7.79
	5–25	86.00	8.00	5.80			13.5	2.3	0.959	0.175	0.161	0	0.060	7.66
Qulungil	0–5	63.20	26.50	10.00	47.6	54.0	13.0	1.9	4.770	1.670	0.943	0	0.060	7.08
	5–25	65.00	29.50	5.40			13.0	1.0	2.070	0.625	0.667	0	0.060	7.35

Mansoura-Damietta road

Mansoura district

Plate II. *Kochia indica* domination, Mariut salt marshes, Mediterranean coast, Egypt.

showed maximum rate of germination under fresh water conditions. He added 'it is evident that very few species will tolerate more than 2% of NaCl solution'. This was confirmed by the successful germination of *Juncus rigidus* seeds watered with 3% NaCl solution (Zahran, 1975). Accordingly, the success or failure of seeds to germinate in saline soil may be an indication for the success or failure of seed propagation of salt tolerant plants in saline soil.

a. Material and methods. Germination of *K. indica* seeds was experimented in petri dishes (9 cm diam.) containing filter paper moistened with distilled water or with solutions of different concentrations of NaCl [0.02 M (0.117%); 0.03 M (0.175%); 0.04 M (0.234%); 0.1 M (0.585%); 0.2 M (1.17%); 0.3 M (1.755%); 0.4 M (2.34%) and 0.5 M (2.925%)].

Ten ml of distilled water (or NaCl solution) was

added to each petri dish. Duplicate dishes were used for each treatment and fifty seeds of *K. indica* were sown in each petri dish. The experiment started on December 12, 1975 and continued for 3 weeks. Observations were made every 24 hours. Emergence of radicle and plumule was taken as the criterion of successful germination.

b. Results. Effect of different salinity levels on the germination of *K. indica* seeds show that the germination was highest when *Kochia* seeds were watered with distilled water (control dishes) or diluted solutions of NaCl. All seeds germinated, but increased NaCl solution caused lengthening of the periods of 100% germination from 7 days in the control dishes to 13, 17 and 17 in the dishes treated with 0.02 M, 0.03 M and 0.04 M NaCl solutions, respectively.

After 3 weeks, 90, 80, 68, 34 and 12% of seeds

germinated when they were watered with 0.1, 0.2 0.3, 0.4 and 0.5 M NaCl solutions, respectively. This proves that *Kochia* seeds can germinate under salinity as high as about 3% NaCl.

(ii) Mineral nutrition experiment
a. Material and methods. On March 28, 1977, the mineral nutrition experiment started using 480 *Kochia* seedlings which were transplanted in 48 pots (10 seedlings in each pot), half of which were filled with sand and the remainder with *Kochia* field silty soil. All of the pots were irrigated with fresh water once every 72 hours for about one month. First addition of nutrient solution (Arnon solution diluted to 1/5) was on April 27, 1977.

Pots of sandy soil were divided into four sets (each set of 6 pots). The same pattern was repeated for the pots of silty soil. The first set of each type of pots were irrigated with fresh water (control pots), the second with complete nutrient solution, the third with nutrient solution lacking nitrogen element ($-N$) and the fourth with nutrient solution lacking phosphorus element ($-P$). The experiment continued for three months during which irrigation of plants in each pot was carried out with 1 l of fresh water every 72 hours and with 1 l of the specific nutrient solution every 15 days.

Harvest was carried out twice. Half of the plants were harvested on June 25, 1977 and the second half on July 25, 1977.

b. Results. Table 13 shows the following:
1. Generally, *Kochia* plants of control pots showed the lowest vegetative yields when compared with the plants treated with nutrient solutions.
2. *Kochia* plants showed better vegetative yields in pots filled with silty soil under all treatments than those of sandy soil.
3. Phosphorus seems to be the most important element for the vegetative yield of *K. indica* plants. This is obvious when we compare the dry weights of plants growing in pots treated with nutrient solution containing phosphorus only with the other pots treated with complete and $-P$ nutrients. The presence of nitrogen seems to hinder the uptake of phosphorus by the plant from soil.

(iii) Salinity tolerance experiment
a. Material and methods. On April 15, 1977, eighty seedlings of *K. indica* were transplanted in 16 small pots (5 seedlings in each pot), 8 pots were filled with sand and the remainder with *Kochia* field soil. The plants were irrigated with fresh water once every 72 hours for 6 weeks. On May 26, 1977, the sandy pots were categorised into 4 sets (each set of 2 pots) irrigated with equal amounts 1 l of tap water (control set) and 0.5, 2 and 3% NaCl solutions for the second, third and fourth sets, respectively. The same sequence was done for the pots containing *K. indica* soil. Application of NaCl solution was carried out once, thenceforth the plants were irrigated

Table 13. Vegetative growth of *Kochia indica* at the beginning (A), at first harvest (B), and at second harvest (C) after the supply of mineral nutrients.

Parameters		Control		Complete nutrient		Nutrient without nitrogen		Nutrient without phosphorus	
		Silt	Sand	Silt	Sand	Silt	Sand	Silt	Sand
Height (cm)	A	14.60	7.50	18.80	8.8	19.00	9.40	18.2	8.2
	B	36.54	20.60	43.00	26.00	42.00	27.90	39.50	24.90
	C	44.37	24.80	50.30	29.74	46.50	27.70	45.70	28.40
Fresh weight (g)	A	1.17	0.60	1.57	0.64	2.12	0.79	1.80	0.61
	B	5.04	2.50	7.57	3.80	8.68	4.44	6.70	3.12
	C	10.28	3.70	12.40	5.20	12.20	8.17	11.80	4.15
Dry weight (g)	A	0.31	0.16	0.51	0.25	0.56	0.27	0.51	0.16
	B	1.60	0.72	2.33	0.95	2.76	1.22	2.25	0.88
	C	2.94	0.91	3.43	1.46	3.94	1.70	3.30	1.49

with fresh water at regular intervals in a way to keep the soil at its field capacity. The experiment continued for about 2 months. *K. indica* aerial parts were harvested on July 26, 1977.

b. Results. Table 14 includes the results of salinity tolerance experiment on *K. indica*. These demonstrate that seedlings succeeded to grow when irrigated with saline solution as high as 3% NaCl. The highest fresh and dry weights were determined in pots filled with silty (mean = 3.67 and 0.99 kg/pot, respectively) and with sandy (mean = 2.42 and 0.43 kg/pot, respectively), soils watered with 3% and 2% NaCl solutions, respectively. Under different salinity treatments, plants gave relatively higher vegetative yields in silty pots than in sandy ones.

The heights of plants were not identical with their fresh and dry weights. Dwarf individuals gave higher fresh and dry weights than relatively longer ones. This was due to the extensive side branching of stems of the shorter individuals.

(iv) Chemical investigations
Available literature showed that many workers did chemical investigations on the different species of genus *Kochia. K. prostrata* was studied by Shishkina and Tegisbaev (1970), Rasulov (1971), and Golyadkin et al. (1974). They determined that it contains triterpenes, glycosides, flavinoides, lipids, crude proteins, etc. Its seeds are rich in alanine, histidine, glycine, glutamic acid, isoleucine, lycine, proline, serine and threonine. Investigation of *K. scoparia* proved that its green parts contain alkaloids, saponine, amino-acids, etc. (Borkowski and Drost, 1965; Kernan et al., 1973). In *K. arenaria*, 4 alkaloides were detected while 3 in *K. trichophylla*.

The concentrated alcoholic extract of the latter species yielded a hygroscopic amorphous glycoside which on hydrolysis with 5% H_2SO_4 yielded oleanolic acid. Its sugar contents consisted of glucose, arabinose and rhamnose types (Tandon and Agarwal, 1966).

In Egypt, the root, shoot, pericarp and seeds of *K. indica* were analysed chemically (Abu Ziada, 1978). The mean values of moisture, ash, water-soluble ash, acid-insoluble ash, soluble sugars, total carbohydrates, total nitrogen, crude protein, total lipids and crude fibers are shown in Table 15.

The moisture content of seeds was relatively high (10.86%) and low in the roots (8.79%). The moisture content of shoots and pericarps were 9.57 and 10.15%, respectively.

The shoot system of *K. indica* contains a relatively high percentage of ash and water-soluble ash being 17.260 and 13.840%; whereas the seeds give low values reaching 5.70 and 1.50%, respectively.

The root system contained a relatively high percentage of acid-insoluble ash (2.410%), total carbohydrate (6.295%) and crude fibers (37.72%). The seeds contained high concentration of soluble sugars, total nitrogen, proteins and lipids reaching 1.387, 3.36, 21.00 and 11.40%, respectively.

The shoot system contains 4.738% carbohydrates, 1.56% total nitrogen, 9.75% protein, 3.39% lipid and 20.32% crude fiber (Table 15).

Alkaloid contents of *K. indica* were investigated qualitatively and quantitatively. The maximum amount was determined in the seeds being 512 mg%, followed by that of pericarp (387 mg%). The shoot system contains upto 294 mg% alkaloides while root system contains the minimum amount (85 mg%).

Table 14. Salinity tolerance in *Kochia indica*.

Parameters	Distilled water		0.5% NaCl		2% NaCl		3% NaCl	
	Silt	Sand	Silt	Sand	Silt	Sand	Silt	Sand
Height (cm)	19.7	8.5	16.4	15.4	17.1	11.2	16.2	8.4
Fresh weight (kg)	2.93	0.69	2.98	1.42	2.84	2.42	3.67	1.14
Dry weight (kg)	0.83	0.20	0.88	0.38	0.66	0.43	0.99	0.24

Table 15. Mean values of the different constants and constituents of *Kochia indica*.

Contents and constituents	g%			
	roots	shoots	pericarps	seeds
Moisture content	8.790	9.570	10.150	10.860
Ash content	10.480	17.260	14.420	5.700
Water-soluble ash	4.560	13.840	6.460	1.500
Acid-soluble ash	2.410	1.240	1.780	0.470
Soluble-sugar content	0.373	0.706	1.300	1.387
Total carbohydrate	6.295	4.738	1.778	5.167
Total nitrogen content	1.420	1.560	1.560	3.360
Crude protein nitrogen	8.870	9.750	9.750	21.000
Total lipid content	1.190	3.390	3.120	11.400
Crude fiber content	37.720	20.320	20.850	9.800

Sixteen amino acids and five organic acids were detected in the seeds, pericarp, root and shoot of *K. indica* namely: lysine, histidine, arginine, aspartic acid, serine, glycine, glutamic acid, threonine, alanine, proline, tyrosine, methionine, valine, phenylalanine, isoleucine, and leucine, (amino-acids) and maleic, oxalic, tartaric, citric and succinic (organic acids). Also, free (glucose and sucrose) and combined (glucose, galactose, fructose, arabinose and rhamnose) sugars were determined in *K. indica* organs. Flavonoides were detected only in the shoots, seeds and pericarp.

III. Conclusions

The results of studies carried out on *Kochia* made it clear about its economic potentialities. The species belonging to this genus are either drought tolerant, e.g. *K. scoparia* or salt tolerant, e.g. *K. indica*. Durham and Durham (1979) state '*K. scoparia* grows profusely and is very drought tolerant. Although rainfall at Durham Farms (Texas State, U.S.A.) was less than 175 mm during the past twelve months, large number of cattle are still grazing on a small acreage. One hundred and thirty cattle grazing 4.86 hectares for twenty nine days, gained 265 kg per hectare. Native grass has made no growth. Past research showed that *Kochia* was high in protein, lower in fiber than alfalfa, high in seed production, drought tolerant, winter germinating, and having stand establishment and seed matura-

tion problems. Our solution to the problems is stirring in soil in December and grazing heavily late August. All forage from *Kochia* can be consumed, leaving only enough seeds to germinate in the following year. It exhibits excellent regrowth, lending itself to strip grazing. One hundred and thirty cattles were grazed on successive four acre strips for two days period during the severe drought. Regrowth during the drought occurred in fourteen to twenty days. Many leaf and seed shoots regrow after a single stem is cut off by grazing. Since germination is early, grazing begins in April and continues until December. *K. scoparia* contains saponin and oxalates; however, no losses have occurred in four years of grazing experience.' Similarities are rather greater than differences between *K. scoparia* and *K. indica*. 'The alkaloid and protein components of *K. scoparia* seemed to be very similar to that of *K. indica*' (Abu Ziada, 1978).

Both species can be considered as potentially range plants. Germination and pot experiments showed that *K. indica* is a salt tolerant plant. Its propagation on saline soil is likely to be successful. Accordingly, our conclusion is that *K. indica* is another halophyte that may play an important role for the welfare of the people of arid and semi-arid countries. Its cultivation on saline non-productive soil (or on sandy soil using saline or brackish water for irrigation) will increase the areas of green forage and consequently, will increase the meat production of these countries.

Literature cited

Abu Ziada, M.E.A. 1978. Autecological and phytochemical studies on *Kochia indica* Wight. M.Sc. Thesis, Botany Department, Faculty of Science, Mansoura University, Mansoura, Egypt.

Bamber, C.J. 1961. *Plant of the Punjab. A descriptive Key to the Flora of Punjab, N.W. Frontier Province of Kashmir*. The Superintendent Government Printing, Punjab.

Borkowshi, B. and Drost, K. 1965. Alkaloid occurrence in species of *Kochia. Acta Polon. Pharm.* 22: 181–184.

Boulos, L. 1960. *Flora of Gebel El-Moghara, North Sinai.* General Organisation for Government Printing Office, Ministry of Agriculture, Egypt.

Boyko, H. 1966. Basic ecological principles of plant growing by irrigation with highly saline or seawater. In *Salinity and Aridity*, ed. H. Boyko, pp. 131–200. Dr. W. Junk Publ., The Hague.

Chapman, V.J. 1966. Vegetation and salinity. In *Salinity and Aridity*. ed. H. Boyko, pp. 23–42. Dr. W. Junk Publ., The Hague.

Coxworth, E.C.M. and Salmon, R.E. 1972. *Kochia* seeds as a component of the diet of Turkey poults. Effect of different methods of saponin removal or inactivation. *Can. J. Anim. Sci.* 52: 721–729.

Dahadiez, E. and Maire, D. 1934. Catalogue des plants due Moroc. En. vente a Paris Chez P. Lechevaller, Labraire, 12 Rue, de Tournon.

Drar, M. 1952. A report on *Kochia indica* Wight in Egypt. *Bull. Inst. Deserte d' Egypte* 2: 45–58.

Draz, O. 1954. Some desert plants and their uses in animal feeding: *Kochia indica* and *Prosopis juliflora. Pub. de l'Inst. du Deserte d'Egypte* 2: 1–95.

Durham, R.M. and Durham, J.W. 1979. *Kochia:* Its potential for forage production. In *Arid Land Plant Resources*, eds. J.R. Goodin and D.K. Northington, pp. 444–450. Texas Tech University, Lubbock, Texas.

El-Bagouri, I., Zahran, M.A. and Abdel Wahid, A.A. 1976. Transplantation of *Juncus* spp. in calcareous soils, Egypt. *Bull. Fac. Sci. University of Mansoura* 4: 59–61.

El-Batanouny, K.H. and Zaki, M.E.F. 1969. Root development of two common species in different habitats in the Mediterranean sub-region in Egypt. *Acta Bot. Aca. Sci. Hungariacea* 15: 217–226.

El-Batanouny, K.H. and Abul Soud. 1972. Ecological and phytosociological study of a sector in the Libyan desert. *Vegetatio* 25: 335–356.

El-Demerdash, M.A. 1978. Comparative studies on the transplantation and economic potentialities of *Juncus rigidus* C.A. Mey and *J. acutus.* M.Sc. Thesis, Mansoura University, Egypt.

El-Shishiny, E.D.H. 1953. Effect of temperature and desiccation during storage on germination and keeping quality of *Kochia indica* seeds. *J. Expt. Bot.* 4: 403–406.

El-Shishiny, E.D.H. and Thoday, D. 1953. Inhibition of germination on *Kochia indica. J. Expt. Bot.* 4: 10–22.

El-Shishiny, E.D.H., Noseir, M.R. and Barakat, S.D.Y. 1958. Viability of *Kochia indica* seeds under different stage conditions. *Bull. Inst. Deserte d'Egypte* 8: 23–32.

Fink, A. 1977. Soil salinity and plant nutritional salts. *Int. Salinity Conf. Texas Tech University*. pp. 199–210.

Girgis, W. Zahran, M.A., Reda, Kamelia A. and Shams, H. 1971. Ecological notes on Moghra Oasis. *UAR J. Bot.* 14: 147–155.

Golovchenko, S.G. and Makhamadzhanov, I. 1972. Effect of freezin on vitamin R activity of *Kochia prostrata* seeds with fast loss of germination capacity during storage. *Dorl. Akad. Nauk Uzb. SSR* 29: 51–52.

Golyadkin, A.I., Ismailov, A.K., Klyshev, L.K., Frantsev, A.P. and Gilmenov, M.K. 1974. Elec. field effect on amino acids of germinating seeds of *Kochia prostrata. Vestn. Skh. Nauk. Kaz.* 17: 31–34.

Hasanain, S.Z. and Rahman, O. 1957. Plants of Karachi and Sind. Dept. of Botany, University of Karachi, *Monograph No.* 1.

Jackson, M.L. 1962. *Soil Chemical Analysis.* Constable and Co. Ltd., London.

Khamdamov, I.Kh. 1976. Amino acid composition of some fodder species of wormwood *Artemisia* and *Kochia. Dokl. Akad. Nauk. Uz. SSR* 1: 40–42.

Kashyap, S.R. and Joshi, A.C. 1936. *Lahore District Flora.* University of Punjab, Lahore.

Kassas, M. 1966. Plant life in deserts. In *Arid Lands A Geographical Appraisal*, ed. E.S. Hills, pp. 145–178, UNESCO, Paris.

Kassas, M. and Zahran, M.A. 1967. On the ecology of the Red Sea littoral salt marsh, Egypt. *Ecol. Monog.* 17: 297–316.

Kernan, J.A., Coxworth, E. and Felming, S. 1973. Microdetermination of triterpene sapongenin content of *Kochia scoparia* seeds using gas chromatography. *Agr. food Chem.* 21: 232–234.

Massoud, F.I. 1977. Basic principles for prognosis and monitoring of salinity and sodicity. *Proc. Int. Salinity Conf. Texas Tech University*, Lubbock, Texas, pp. 432–454.

Migahid, A.M., El-Batanouny, K.H. and Zaki, M.A.F. 1971. Phytosociological and ecological study of a sector in the Mediterranean coastal region in Egypt. *Vegetatio* 23: 113–134.

Migahid, A.M. 1978. *Flora of Saudi Arabia*, Vol. I. Riyadh University Publ.

Montasir, A.H. 1937. Ecology of Lake Manzala. *Bull. Fac. Sci., Egyptian University* 15: 205–236.

Osman, F., Zahran, M.A. and Fayad, S. 1975. Potentialities of the seeds of the flora of Egypt for oil production. *Bull. Fac. Sci. University of Mansoura* 3: 85–95.

Quezel, P. and Santa, S. 1963. Nouvelle Flora De l'Algerie. Fac. de. Sci. de Marseille, Centre National de la Recherche, Scientifique.

Rasulov, I.R. 1971. Food values of different ecological forms of *Kochia prostrata* to karakul sheep. Nauch. Tr. Samarakant. *Sel'Skokhoz Inst.* 23: 159–162.

Rydholm, S.A. 1965. *Pulping Processes.* Inter-science Publishers, New York.

Sadek, L.A. 1974. Autecological studies in *Kochia indica* Wight. M.Sc. Thesis, Univ. of Alexandria, Egypt.

Salem, A.G. 1960. Agricultural residue pulps- *Kochia* pulps. *J. Chem. U.A.R.* (Egypt) 3: 85–93.

Shishkina, V.Ya. and Tegisbaev, E.T. 1970. Phytochemical studies on *Kochia prostrata* growing in Kazakhstan. *Tr. Alma. Med. Inst.* 26: 449–500.

Snedecor, G.W. 1956. *Statistical Methods Applied to Experiments in Agriculture and Biology.* Iowa State College Press.

Steward, R.R. 1972. *Flora of West Pakistan – An Annotated Catalogue of the Vascular Plants of West Pakistan and Kashmir.* Principal Emeritus, Gordon College, Rawalpindi, Pakistan.

Sutcliffe, J.F. and Baker, D.A. 1974. *Plants and Mineral Salts.* The Camelot Press Ltd., Southampton.

Szabolcs, I. 1976. Present and potential salt affected soils – An introduction. *Soils Bull.* 31, FAO, Rome.

Täckholm, Vivi and Drar, M. 1950. Flora of Egypt. *Bull. Fac. Sci. University of Cairo.* 2: 453–469.

Täckholm, Vivi and Drar, M. 1954. Flora of Egypt. *Ibid.* 3: 644–670.

Täckholm, Vivi. 1974. *Student's Flora of Egypt.* Anglo-Egyptian Bookshop, Cairo.

Tadros, T.M. 1956. An ecological survey of the semi-arid coastal strip of the Western Desert of Egypt. *Bull. Inst. Deserte d'Egypte* 6: 38–56.

Tandon, J.S. and Agarwal, K.P. 1966. Chemical examination of *Kochia trichophylla. Indian J. Chem.* 4: 545 p.

TAPPI 1960. *Technical Association of the Pulp and Paper Industry*, New York.

Thoday, D. 1956. *Kochia indica* in Egypt. *Kew Bull.* 1: 161–163.

Thoday, D., Tadros, T.M. and El-Shishiny, E.D.H. 1956. *Kochia indica* Wight and its dispersal in Egypt. *Bull. Inst. Deserte d'Egypte* 6: 57–66.

Walter, H. 1961. The adaptation of plants to saline soils. In *Salinity Problems in the Arid Zones. Proc. Teheran Symp. UNESCO*, Paris, Arid Zone- Research 14: 129–134.

Zahran, M.A. 1967. On the ecology of the east coast of the Gulf of Suez. I- Littoral salt marsh. *Bull. Inst. Deserte d'Egypte* 17: 127–252.

Zahran, M.A. and Girgis, W.A. 1970. On the ecology of Wadi El-Natrun. *Bull. Inst. Deserte d'Egypte* 18: 229–267.

Zahran, M.A., Kamal El-Din, H. and Boulos, S. 1972. Potentialities of fiber plants of Egyptian flora in national economy. *Bull. Inst. Deserte d'Egypte* 22: 193–202.

Zahran, M.A. 1972. On the ecology of Siwa Oasis. *Egypt J. Bot.* 15: 223–242.

Zahran, M.A. and Negm, S.A. 1973. Ecological and pharmacological studies on *Salsola tetrandra* Forsk. *Bull. Fac. Sci. Mansoura University, Egypt.* 1: 67–75.

Zahran, M.A. 1975. On the germination of seeds of *Juncus rigidus* C.A. Mey and *J. acutus.* L. *Bull. Fac. Sci. Mansoura University, Egypt* 3: 75–84.

Zahran, M.A., El-Bagouri, I.H., Abdel Wahid, Amal A. and El-Demerdash, M.A. 1977. Transplantation of *Juncus* spp. on saline soil in Egypt. *Proc. Int. Salinity Conf. Texas Tech University*, Lubbock, Texas. pp. 142–154.

Zahran, M.A. 1982. Ecology of the halophytic vegetation of Egypt. In *Contributions to the Ecology of Halophytes*, eds. D.N. Sen and K.S. Rajpurohit, pp. 3–20. Dr W. Junk Publ., The Hague, Netherlands.

Zahran, M.A. and El-Habibi, A.M. 1979. A phytochemical investigation of *Juncus* spp. *Bull. Fac. Sci. Mansoura University, Egypt* 5: 1–14.

Zahran, M.A., Abdel Wahid, Amal A. and El-Demerdash, M.A. 1979. Economic potentialities of *Juncus* plants. In *Arid Lands Plant Resources*, eds. J.R. Goodin and D.K. Northington, pp. 244–260. Texas Tech University, Lubbock, Texas.

Zohary, M. 1962. *Plant Life in Palestine, Israel and Jordan.* The Ronald Press. Co., N.Y.

257

Subject index

Plant index

H 165.00